The Malthusian Moment

**Studies in Modern Science, Technology, and the Environment,
edited by Mark A. Largent**

The increasing importance of science and over the past 150 years—and with it the increasing social, political, and economic authority vested in scientists and engineers—established both scientific research and technological innovations as vital components of modern culture. Studies in Modern Science, Technology, and the Environment is a collection of books that focuses on humanistic and social science inquiries into the social and political implications of science and technology and their impacts on communities, environments, and cultural movements worldwide.

Matthew N. Eisler, *Overpotential: Fuel Cells, Futurism, and the Making of a Power Panacea*

Mark R. Finlay, *Growing American Rubber: Strategic Plants and the Politics of National Security*

Jill A. Fisher, *Gender and the Science of Difference: Cultural Politics of Contemporary Science and Medicine*

Finn Arne Jørgensen, *More than a Hole in the Wall: The Story of What We Do With Our Bottles and Cans*

Gordon Patterson, *The Mosquito Crusades: A History of the American Anti-Mosquito Movement from the Reed Commission to the First Earth Day*

Jeremy Vetter, *Knowing Global Environments: New Historical Perspectives on the Field Sciences*

The Malthusian Moment

Global Population Growth
and the Birth of American
Environmentalism

THOMAS ROBERTSON

RUTGERS UNIVERSITY PRESS
NEW BRUNSWICK, NEW JERSEY, AND LONDON

LIBRARY OF CONGRESS CATALOGING-IN-PUBLICATION DATA

Robertson, Thomas, 1967–
  The Malthusian moment : global population growth and the birth of American environmentalism / Thomas Robertson.
    p. cm. — (Studies in modern science, technology, and the environment)
  Includes bibliographical references and index.
  ISBN 978–0–8135–5271–2 (hardcover : alk. paper) — ISBN 978–0–8135–5272–9 (pbk. : alk. paper) — ISBN 978–0–8135–5335–1 (e-book)
  1. Overpopulation—History. 2. Overpopulation—United States—History. 3. Environmentalism—United States—History. I. Title.
  HB871.R57 2012
  363.9′1—dc23   2011028832

A British Cataloging-in-Publication record for this book is available from the British Library.

Copyright © 2012 by Thomas Robertson

All rights reserved

No part of this book may be reproduced or utilized in any form or by any means, electronic or mechanical, or by any information storage and retrieval system, without written permission from the publisher. Please contact Rutgers University Press, 100 Joyce Kilmer Avenue, Piscataway, NJ 08854–8099. The only exception to this prohibition is "fair use" as defined by U.S. copyright law.

Visit our Web site: http://rutgerspress.rutgers.edu

Manufactured in the United States of America

To Mom, who loves nature,
and to Dad, who loves politics

CONTENTS

List of Illustrations   ix
Preface   xi
Acknowledgments   xvii

	Introduction. From Rubbish to Riots	1
1	Malthusianism, Eugenics, and Carrying Capacity in the Interwar Period	13
2	War and Nature: Fairfield Osborn, William Vogt, and the Birth of Global Ecology	36
3	Abundance in a Sea of Poverty: Quality and Quantity of Life	61
4	"Feed 'Em or Fight 'Em": Population and Resources on the Global Frontier during the Cold War	85
5	The "Chinification" of American Cities, Suburbs, and Wilderness	104
6	Paul Ehrlich, the 1960s, and the Population Bomb	126
7	Strange Bedfellows: Population Politics, 1968–1970	152
8	We're All in the Same Boat?!: The Disuniting of Spaceship Earth	176
9	Ronald Reagan, the New Right, and Population Growth	201

Conclusion. The Power and Pitfalls of Biology   221

Epilogue   230

Notes   233
Index   285

ILLUSTRATIONS

1. Cover, *Environmental Action: April 22*, 1970 — 3
2. Jay N. "Ding" Darling, "The Most Critical Race in History," 1917 — 14
3. "It Depends How You Run the Machine," CIO, *Economic Outlook*, 1946 — 30
4. Jay N. "Ding" Darling, "Speaking of Labor Day," 1939 — 44
5. Jay N. "Ding" Darling, "The Only Kettle She's Got," 1947 — 58
6. "The U.S. Outgrows Its Resource Base," *Resources for Freedom*, 1952 — 63
7. "U.S. Imports of Strategic & Critical Materials," *Partners in Progress*, 1951 — 64
8. Herblock, "Race," 1950 — 87
9. Georg Borgstrom, "World Population," *Too Many*, 1971 — 145
10. Art Rickerby, A student for ZPG trying to save the local marshland, *Life*, 1970 — 161
11. Beverley Hall, "Have You Had Your Pill Today," *New York Times*, 1970 — 192
12. Taylor Jones, "Population Dud," *Hoover Digest*, 2001 — 202

PREFACE

In the months before the first Earth Day in 1970, as an explosion of environmental activism was reconfiguring the American political landscape, *Time* magazine, looking for a way to explain the concept of ecological interconnection, turned for an analogy to the realm of international relations. The ecological process by which chemicals like DDT worked their way through—and up—the food chain, it wrote, mirrored the Cold War political-economic system in which an outbreak of communism in one niche of the world could spread quickly through and up the global food chain. "The 'domino theory,'" the magazine explained, "is clearly applicable to the environment."[1]

Later that year, at a rally on Earth Day, the massive series of demonstrations that signaled the arrival of the American environmental movement, a woman held up a sign based on a well-known *Pogo* cartoon: "I have met the enemy and he is US." The famous line crystallized a common sentiment of the postwar environmental movement, that humans were to blame for recklessly interfering in nature. But by turning the lower-case word "us" into the uppercase initials "US" during the height of protests about the U.S. war in Vietnam, the Earth Day participant added an extra layer of meaning to her message about nature: the capital letters suggested that the enemy of the environment was not just human beings but, more particularly, the United States. Like Americans in Vietnam, human beings were arrogantly making up their own rules and deploying technological tools of horrible power to inflict massive destruction upon a defenseless enemy. Human beings were to nature as the United States was to the rest of the world.

The early 1970s was not the first time that war, international relations, and ideas about America's connections to the rest of the world had shaped environmental thought and politics in the United States. The Romantic writer and philosopher Henry David Thoreau retreated to Walden Pond and put to paper some of the most influential words about nature an American has ever produced as the United States was headed to war with Mexico, something we know he thought deeply about because of his essay on civil disobedience. The naturalist John Muir experienced one of his most important epiphanies about nature while in Canada after fleeing the draft during the Civil War.

George Perkins Marsh, whose 1864 book *Man and Nature* laid the intellectual foundation for the late nineteenth-century conservation movement, derived his ideas about protecting American nature by thinking about environmental problems over the previous two millennia in Europe, where he had lived for many years as a diplomat. As part of that conservation movement, Americans began to establish national parks in part to create symbols of national greatness to counter nationalistic European claims of superiority. Theodore Roosevelt saw conserving and protecting nature as crucial for both the material and moral strength of the nation in the international arena.

American foreign relations and environmental politics overlapped after World War II, as well. During the Cold War, presidents and biologists alike invoked the same grand theme: interconnection. In his 1953 inaugural address, President Dwight Eisenhower spoke of the "basic law of interdependence." Echoing him twelve years later, Lyndon Johnson pronounced: "The unity we seek cannot realize its full promise in isolation for today the state of the Union depends, in large measure, upon the state of the world." The idea of interconnection was also a core tenet of ecology, the subset of biology that informed much of the postwar environmental movement. Ecology held that all of nature—living and nonliving, human and nonhuman—was linked together in cycles of energy and nutrient flows.[2]

Usually, the interconnected realm of international relations and the interconnected realm of nature seemed far removed from each other, but sometimes they explicitly overlapped. "We have learned in politics," famed conservationist and ecological pioneer Aldo Leopold proclaimed in 1946, "that preoccupation with the nation, as distinguished from mankind, defeats its own end. We label this fallacy isolationism. Perhaps we have now to learn that preoccupation with mankind, as distinguished from the [natural] community of which man is a member, defeats its own ends." The overlap was not just rhetorical: two years earlier, as policymakers and diplomats were busily making blueprints for a new postwar political and economic order, Leopold had warned about the environmental impact of plans to spread American industrial methods and consumption patterns around the world. Nature and international relations would overlap again twenty-five years later during the environmental activism surrounding Earth Day, after Leopold's warnings had gone mostly unheeded.[3]

Historians of the environmental movement of the 1960s and 1970s—the movement that gave us the slogan "think globally, act locally"—have themselves not always thought globally. The American environmental movement developed in the years after World War II and exploded in the middle of the Cold War. These "total" wars reshaped nearly all aspects of American life, including relations with nature. "War and nature coevolved," historian Edmund Russell has pointed out. "The control of nature expanded the scale of war, and war expanded the scale on which people controlled nature." War and environmental policy also interconnected. "As war has concentrated power in the hands of the military and the state," the historian Richard Tucker writes, "it has also enhanced the power of governments to regulate the use of natural resources." And yet, aside from mentions of nuclear weapons, historians have rarely explored the international context of the movement.[4]

The potential to do so seems great. On one hand, historians have made great strides in showing how World War II and the Cold War touched all aspects of life on the "homefront." Many examples jump to mind, but the works of Mary Dudziak, Thomas Borstelmann, and Carol Anderson stand out. These historians have illustrated how the civil rights movement of the 1950s and 1960s—a political phenomenon normally analyzed within the borders of the United States—cannot be fully understood without considering Cold War politics, especially American hopes to win allies in Africa and the rest of the decolonizing world. This global context, they have shown, was a double-edged sword: it opened new political possibilities but also narrowed options. On the other hand, historians of environmental politics in other times and places have found important overlaps with international relations. In separate books, Richard Grove and Peder Anker have shown how relations between imperial European powers and the rest of the world in the nineteenth and early twentieth centuries had also, paradoxically, spurred environmental reform. Although both scholars deplored the ecological devastation wrought by empires, they stressed how, in certain instances, the special imperatives of maintaining long-term economic and political dominance on a global scale required a degree of planning that yielded new ways to conserve and protect nature. The work of Dudziak, Borstelmann, Anderson, Grove, and Anker spurs a provocative question: how did the global context of World War II and the Cold War both advance and constrain the American environmental movement of the 1960s and 1970s?[5]

Few subjects can reveal more about the overlap of American foreign relations, here conceived of more broadly than just intergovernmental relations, and postwar environmental politics than does concern about population growth—"Malthusian" worries most clearly articulated by the English political economist Thomas Robert Malthus in the very late eighteenth century. Malthusian concerns had cropped up occasionally in the United States during the nineteenth century, especially as the frontier appeared to be closing in the 1890s, but gained much wider currency after two global wars pulled the United States into international affairs as never before. In the late 1940s, bestselling books appeared that blamed the recent planetary conflagrations on overpopulation and warned of renewed war if environmental problems went unremedied. Environmental management became a national security issue. Malthusian concerns gained momentum during the 1950s and took a central place on the public agenda in the late 1960s. During these formative years for the postwar environmental movement—between Rachel Carson's *Silent Spring* (1962) and the first Earth Day in 1970—scores of books, articles, and even movies appeared warning of population growth. Perhaps the most famous, Paul Ehrlich's *The Population Bomb* (1968), is packed with stories about India and reads as a primer on U.S. relations with the "third world."

Diplomatic historians will discover that *The Malthusian Moment* offers an unconventional look at the big stories of twentieth-century foreign relations—World War II and the Cold War—and a new angle on a relatively untold story, that of U.S. relations with the developing world. Population debates brought together a number of different elements—family, race, consumption, natural resources, and environment—that played a crucial role in mid-twentieth-century foreign relations, yet have not received enough scholarly attention. For environmental historians, this book offers a study of one of the most controversial elements of postwar environmental politics, as well as a new interpretation of one of the most significant environmental stories of the twentieth century—the shift from the early twentieth-century "conservation" movement to the post–World War II "environmental" movement. Situating environmental thinking in its larger national and international context not only helps explain the timing and urgency of the environmental movement of the 1960s and 1970s but also helps us see the race, class, and gender blind spots of a movement that often claims to represent universal concerns and be above the political fray.

But this book is about more than just the overlap of international affairs with environmental affairs. As I dug into these questions, what originally seemed mostly a story about overseas resources quickly became a study of a whole series of issues central to postwar American life: the rise of a Keynesian growth economy and a "consumer republic"; the growth of metropolitan areas and the eruption of urban violence; new ideas of sex, sexuality, and women's roles; the place of science and expertise in democratic society; Catholicism, the Religious Right, and debates about evolution; and the expansion of governmental authority. As I began to see these multifarious connections, I began to grasp why population growth has ignited so much controversy over the decades. In arguing about increasing numbers, Americans have struggled with some of the thorniest questions any society can face: what kind of families to create, what separates and unites different "races," how to distribute limited resources, who gets to decide questions of moral and scientific uncertainty, what powers to give the central government, and what kind of global neighbors to be. Through debates about population growth, all of these issues interlinked with environmental questions.

I must admit that I enter into these highly contested subjects with some trepidation. Discussions about population are often deeply polarized. What I try to do here is not so much choose sides or make judgments (although I do try to call a spade a spade) as shed light on the origins of various arguments, especially environmental arguments about population. I come neither to bury nor to praise Paul Ehrlich, the most outspoken of all environmental Malthusians, but to understand him. Curiously, that is not how I first imagined this project. Besides wanting to show the international context of postwar environmental thought, I wanted to reveal the ways American environmentalism often reinscribed imperial relations overseas and reinforced racial, class, and gendered hierarchies at home. I wanted to expose the dark underbelly of environmentalism. I found, though, that this story had been told repeatedly and that its other aspects also deserved emphasis. Thus readers of this book will find much evidence of the racism and social myopia of environmental Malthusians, but also evidence that their main concern was not racial or class composition but slowing a headlong rush for economic growth. Responding to the massive economic expansion that, especially since the end of World War II, was devastating ecosystems around the world, they pioneered thinking about "limits to growth" and sustainability. Of course, more awareness of their blind spots would have made them more effective in their pursuit of their central goals.[6]

Although this story is deeply rooted in the past, it carries significant meaning for how we think about the future. Increasingly, American daily actions have global environmental consequences—our materials come from elsewhere, our products are made elsewhere, our waste is disposed of elsewhere, and we often chose to vacation elsewhere. And, increasingly, our international policy is shaped by environmental concerns such as climate change. Better understanding the debates and controversies flowing from the overlap of environmental and international affairs in the second half of the twentieth century can help us better understand the challenges of the future.

ACKNOWLEDGMENTS

While I was finishing this book, every time I went out running I reopened a wound on my foot. That's a little like what working on this project felt like. Of course, it was not always like this: there were also those gleeful moments I thought I could run forever, as well as the later satisfaction of having gotten exercise.

I owe a lot of people a lot of gratitude for helping me with this long-distance run. My advisor at the University of Wisconsin–Madison, Bill Cronon, has been a model for scholarship and teaching. I can only hope that my research, writing, and teaching live up to his high standards. Thanks, too, to the other members of my committee for rigorous and wide-ranging suggestions: Judy Houck, Nan Enstad, Jeremi Suri, and Gregg Mitman. This work also benefited indirectly from my master's thesis committee: Paul Boyer, John Fiske, and Bill Cronon. I'd also like to thank Jim Schlender, Teri Tobias, and Linda Johnson at UW–Madison for humanizing the academic bureaucracy.

I've been lucky to work with several other great advisors. Les Olinger, a history teacher at Bethesda–Chevy Chase High School in Maryland, first introduced me to the study of the past. His dedication to his students taught me one of the essential components of good teaching. Two historians at Williams College also deserve my gratitude. Thanks to Tom Kohut for showing me how rich and complicated history can be, and thanks to Robert Dalzell for introducing me to the rewards of studying American culture during the 1960s. I also owe thanks to Richard Tucker for responding so enthusiastically during a field trip to the Denver Aquarium when I asked him advice about internationalizing American environmental history. In addition, Michael Adas helped make my year at the Rutgers Center for Historical Analysis a rich experience, and he has been a wonderful mentor since.

Many archivists helped my research in ways they will never realize. Thanks to Dave Tambo, Dave Russell, and the staff at UC–Santa Barbara Archives; Maggie Kimball, Pat White, and the excellent people at Stanford University's archives; Bob Battaly and Tom Rosenbaum at the Rockefeller Family Archives; Steve Johnson and Dale Boles at New York Zoological Society; Zuali Malsawma at Population Reference Bureau; Dawn McCleskey at the World Wildlife Fund;

Lisa Marine at the State Historical Society of Wisconsin; David Null at the University of Wisconsin–Madison Archives; David Hays at the University of Colorado–Boulder; and the friendly people at Princeton's Mudd Library, the Denver Public Library, Population Connection, and the Library of Congress.

Financial support has been hard to come by, and I would like to thank all who helped: the National Science Foundation's Science and Technology Studies dissertation support program, the UW–Madison History Department, the UW–Madison Graduate Council, the Friends of the Princeton Library, the Rockefeller Family Archive, the Friends of the UC–Santa Barbara Library, the American Society of Environmental Historians, the Organization of American Historians, and Worcester Polytechnic Institute (WPI).

I owe thanks to a number of scholars and institutions for providing me special venues for sharing my ideas: Fred Logevall and the UC–Santa Barbara Cold War conference; Ravi Rajan at UC–Santa Cruz; Karl Brooks and the Truman Library; the Massachusetts Historical Society's environmental history series; the German Historical Institute; and Mark Lawrence, Frank Gavin, and the Lyndon Johnson Library.

A number of friends and colleagues took time out of their busy schedules to read parts of the work in progress, including David Kinkela, David Stradling, Dawn Biehler, Adam Rome, Richard Tucker, Curt Meine, Nathan Sayre, Ellen More, Jim Reed, Thomas Moore, Michael Egan, Jeff Filiak, Finis Dunaway, Pat Sharma, Matt Klingle, Andrew Preston, Paul Sutter, Mark Harvey, David M. Wrobel, Matthew J. Garcia, Edward Linenthal, and the anonymous reviewers at the *Journal of the History of Biology*. During the homestretch, Kurk Dorsey and Derek Hoff read the entire manuscript and gave fantastic feedback, for which I'm very grateful.

My colleagues at Worcester Polytechnic Institute have been a constant source of good ideas and support: Constance Clark, Steve Bullock, Peter Hansen, Bland Addison, Jim Hanlan, Kris Boudreau, Mike Sokal, David Spanagel, Jennifer Rudolph, Dean Oates, Rob Krueger, Seth Tuler, Dominic and Claire Golding, Christine Drew, and Laura Hanlan. I owe special thanks to Margaret Brodmerkle, Karen Hassett, and Mary Cotnoir for always coming through.

I'd like to thank the anonymous reviewers at Rutgers University Press, and especially my editors Doreen Valentine, Peter Mickulas, and Margaret Case. Kyrill Schabert provided very helpful independent editing of the manuscript.

All along, my friends have been just tremendous. In Madison, I owe special thanks to my dissertation group—Mike Rawson, Kendra Smith, Will Barnett, and Jim Feldman—and Chris Wells, Scott Burkhardt, and Keith Woodhouse for being such great intellectual companions and friends. Many other people helped this project, as well: Jeff Miller, Eric Schranz, Blake Harrison, Rebekah Irwin, Steve Volz, Paige Shipman, Thomas Andrews, Peter Youngs, Kevin Kumashiro, Greg Bond, Honor Sachs, Paul Erickson, Jessica Tollner, Zoltan Grossman, Ken Croes,

Mary Makaruska, Maureen Mahoney, Michel Hogue, Sarah Menkedick, Lisa Cline, Eric Carter, Eric Olmanson, Juanita Trujillo, and patrons of "the corner room." I would also like to thank the members of my extended Madison family: Jamuna Shrestha, Bill and Rita Lloyd, Jessica Becker, Jeff and Annie Potter, and my Frisbee buddies Anthony, Nathan, and Issac.

Friends scattered around the country and the world have also been a great source of support: Jessica Greene, Matt Kelley, Bill O'Brien, Dave and Rafina Reiss, Elizabeth Baez, Meg Shriver, Jay Stanley, Libby Houle, Anne Rademacher, Andrew and Lisa Gifford, Martina Anderson, Andrew Seligsohn, and Carolyn Finney. In addition, a number of people have generously opened their homes to me in my travels: Julio and Carole Baez, Simon Donner, Dexter Miller, Morgan and Christie Lloyd, Tom and Sandra Moore, Jim and Betsey Williamson, Ali and Mishtu Rangwalla, and Joe Krakouskas. I also would like to thank my friends in Aiselukharka and Jalpa, Nepal, for their interest and encouragement in this project. In particular, I need to thank Hemanta of Neha Cyber Shop for opening his internet café in the middle of the night so that I could submit my NSF application on time.

Finally, my family has also been a source of enormous strength. Wherever I go—even if I don't always call or e-mail—I know that they are there with me. I owe my love of history and politics to my father, and my love of the outdoors to my mom and my grandmother. My brother, Steve, sister, Anne, and sister-in-law, Deborah, and their families have always been a source of good cheer. My brother Ted read every draft and helped me through every difficult moment. My great uncle Lou, aunt Judy, and cousin Ken have also helped this project in important ways. Thanks to you all.

The Malthusian Moment

Introduction

From Rubbish to Riots

"Every generation . . . writes its own description of the natural order, which generally reveals as much about human society and its changing concerns as it does about nature."
—Donald Worster, *Nature's Economy*, 1994

During her college commencement, normally a moment of optimism, Stephanie Mills delivered an address so grim that it made headlines. In her short speech from the spring of 1969, "The Future Is a Cruel Hoax," Mills declared that she was "terribly saddened that the most humane thing for me to do is to have no children at all." Paradoxically, Mills was born in 1948, at the beginning of one of the most prosperous periods in American history. Growing up in Phoenix, Arizona, she belonged to a generation of Americans more familiar with Cheerios and Schwinn bicycles than with breadlines and wartime rationing. But having recently read Paul Ehrlich's *The Population Bomb*, published in 1968, Mills had grown concerned about the threats that a growing population posed to both the United States and the world. "Our days as a race on this planet are, at this moment, numbered, and the reason for our finite, unrosy future is that we are breeding ourselves out of existence. Within the next ten years, we will witness widespread famines and possible global plagues."[1]

Mills was part of a wave of concern about population growth that, after gathering for some time, swept over the country in the late 1960s and 1970s. President Lyndon Johnson signaled its arrival in his State of the Union address in 1965, warning of an "explosion in world population" and a "growing scarcity in world resources." Concern escalated as the decade unfolded. In July 1968, a *New York Times* editorial spoke of "a population explosion" that threatened "to plunge the world into hopeless poverty and chaos." Population growth, President Richard Nixon told Congress in 1969, is "one of the most serious challenges to human destiny in the last third of this century." Few nations have been more aware of—and anxious about—population growth than Americans in the late 1960s and 1970s.[2]

Mills worried about population growth for environmental reasons. Although many who warned of increasing numbers of people at the time often mentioned resource shortages, the most strident had a strong environmental logic, often drawn from modern biology, that emphasized carrying capacity, ecological interconnection, overconsumption, degradation, and hard limits to growth. A young, intense, and articulate Stanford University biologist named Paul Ehrlich led the charge. "Our problems would be much simpler," Ehrlich wrote in *The Population Bomb*, "if we needed only to consider the balance between food and population. But in the long view the progressive deterioration of our environment may cause more death and misery than any conceivable food-population gap." This logic led Ehrlich to a legendary pessimism. "It is highly unlikely," he stated in a *Newsweek* interview in the early 1970s, "that we will get through the next two decades without a major disaster resulting in the deaths of hundreds of millions of human beings." Other biologists such as Garrett Hardin of the University of California–Santa Barbara voiced even more extreme positions.[3]

Mills, Ehrlich, and Hardin were part of a quickly growing force in American politics in the late 1960s: the environmental movement. In 1962, biologist and nature writer Rachel Carson had published *Silent Spring*, her attack on DDT and chemical pesticides. In April 1970, as many as twenty million Americans participated in the first Earth Day "teach in," one of the largest rallies in American history. This movement was unlike anything Americans had seen before. It differed in substantial ways from the conservation movement led by Theodore Roosevelt and U.S. Forest Service chief Gifford Pinchot in the 1890s and early 1900s: it was more comprehensive, more broad based, and more pessimistic.

Led by Adam Rome, historians have begun to reexamine where the movement came from. Writing in the *Journal of American History* recently, Rome pointed out that the explanations of environmentalism that historians normally point to—postwar abundance, new technologies, and new ecological thinking—do not say much about why the movement exploded when it did. Why did Earth Day happen in 1970 and not 1962 or 1975? For Rome, this was not a pedantic question about dates, but a mystery that, if solved, could shed new light on the movement. He concluded that environmentalism cannot be understood without connecting it to broader patterns of American postwar life, such as the liberal policies of John F. Kennedy and Lyndon Johnson, as well as the specific currents of 1960s life, such as the counterculture.[4]

The story of concern about population growth among environmentalists can also shed light on both the timing and the larger cultural context of postwar environmentalism. Earth Day came just two years after the biologist Paul Ehrlich's *The Population Bomb*, which sold over two million copies and developed a passionate following, including Senator Gaylord Nelson of Wisconsin, Earth Day's initiator and prime mover. Indeed, the very month Nelson made his famous call for what would become Earth Day, he placed an article by Ehrlich called "Eco-Catastrophe!" within the *Congressional Record*. Like Stephanie Mills, many of the young people who made Nelson's Earth Day a mass movement worried deeply about population growth and the "limits to growth." Not coincidentally, the environmental movement crested at precisely the same time as concerns about overpopulation.[5]

FIGURE 1 The cover design of the magazine *Environmental Action: April 22*, from March 3, 1970. Bill Garner's image was used by a number of organizations around the country.

Courtesy of Wisconsin Historical Society, Image ID 80854.

Examining environmental concern about population growth can help reveal the movement's roots in the broader currents of twentieth-century American life, especially the early postwar decades. During these years, in ways that historians have mostly forgotten, concerns about population growth both reflected and remade the intimate contours of households and neighborhoods as well as the grand landscapes of international relations. In the 1950s and 1960s, concerns about overpopulation touched everything from family structure and race relations to ideas about poverty and Cold War strategy. Indeed, many Americans, especially environmentalists, used population to make sense of such crucial postwar spaces as the third world, the inner city, the suburb, and the modern household. "From rubbish to riots to starvation," Paul Ehrlich wrote in 1969, "we are faced with an array of problems, all of which can be traced, at least in part, to too many people." Stephanie Mills's concern about population growth, expressed so passionately at her college graduation, must be seen within this larger cultural context.[6]

Malthus and Environmental Malthusianism

In the grand sweep of American history, Mills's stark pessimism about population growth was unusual. It embodied a gloomy strand of thought that had spread widely through Europe in the century and a half since Thomas Robert Malthus, an English pastor and professor of political economy, had first warned about the dangers of overpopulation at the beginning of the nineteenth century. Noticing that births in his parish far outnumbered deaths, Malthus took issue with the optimistic ideas flowing out of the French Revolution suggesting that human reason could bring order, peace, and prosperity to modern society. Because he believed population grew faster than food supply, Malthus saw little on the horizon but decline and even disaster.[7]

Pastor Malthus's arguments consisted of a complicated brew of new ideas about technology, poverty and poor people, trade and international borders, birth control, and environment. Malthus held little faith in technology, which helps explain why he believed that food production would always lag behind population. He had little faith, as well, in human beings, especially poor people, who he believed lacked the discipline to restrain their sexual urges. The poor had no one to blame for their poverty but themselves. Indeed, Malthus saw poor people as almost a different kind of human being. According to biographer Patricia James, Malthus thought of the poor boys of his parish as "a different race from the lads who played cricket at Cambridge," where he studied. Not surprisingly, he disapproved of charity, believing it would only encourage irresponsibility. Malthus also thought trade between regions and countries could not be depended upon to make up for the overproduction of people. He favored "moral self-restraint" instead of birth control. Thus disaster seemed imminent: he predicted that uncontrolled population growth would lead to war, starvation, and disease.[8]

Malthus transformed modern thinking not only about poverty and society but also about land and limits. "Man is necessarily confined in room," he wrote in 1803. "When acre has been added to acre till all the fertile land is occupied, the yearly increase in food must depend upon the amelioration of the land already in possession. This is a stream which, from the nature of all soils, instead of increasing, must be gradually diminishing." Elsewhere he added, "The power of population is infinitely greater than the power in the earth to produce subsistence for man." When later environmentalists cited Malthus, they emphasized this logic. Malthus's ideas also played a crucial role in the discovery of evolutionary processes, perhaps the most important biological breakthrough of the modern era and crucial for understanding environmental politics. His idea of growing population pressing against scarce resources led both of the mid-nineteenth-century founders of evolution, Charles Darwin and Alfred Russel Wallace, to their ideas of speciation and natural selection. In fact, in later years, environmentalists would sometimes derive their Malthusianism not directly from Malthus, but rather by applying Darwin's ideas of nature back onto human society.[9]

In the nineteenth century, Malthusian pessimism, especially of the extreme kind, appeared only occasionally in the United States, as Americans tended to assume their birthright to include an abundance of natural resources. Thomas Jefferson's vision of an expanding republic—a vision that pervaded much of nineteenth-century America—grew from the promise of almost limitless resources, especially to the west. Indeed, the historian David Potter titled a 1954 book about American culture *People of Plenty*. More so than anything else, Potter wrote, a history of abundance defined Americanness.[10]

Yet, because Americans depended so heavily, both materially and culturally, on abundant resources, the threat of shortages sometimes loomed large in their imaginations. Americans relied on abundance so much that the fear of impending shortages could take on great power. In the late nineteenth century, for instance, especially after the American frontier was perceived to close in the 1890s, many Americans displayed concern about population growth. Following the historian Frederick Jackson Turner's idea that the frontier had given birth to American democracy and prosperity, they worried that the land was filling up. "We have practically reached the limit of our available free land supply," Minnesota populist Ignatius Donnelly wrote in 1893. "That free-land has been the safety-valve of Europe and America. When the valve is closed, swarming mankind every day will increase the danger of explosion. Nothing can save the world but the greatest wisdom, justice and fair play."[11]

Curiously, however, although the turn of the century saw the emergence of the conservation movement led by Theodore Roosevelt and Gifford Pinchot, conservationists generally did not display deep concern about population growth, certainly not in its gloomiest form. Early conservationists, historian Samuel Hays has written, "expressed some fear that diminishing resources would create critical shortages in the future," but "they were not Malthusian prophets of despair and gloom." Instead, conservationists tended to follow Gifford Pinchot's dictum to deliver "the greatest good to the greatest number for the longest time" by way of cutting waste and increasing technological efficiency. Above all, they held tremendous faith in scientific expertise and planning. Nothing exemplified this better than the dream of "making the desert bloom" by using irrigation to reclaim arid lands. "The reclamation dream," historian William Cronon has noted, extended a "vision of frontier plenty even into the drylands of the arid West, making it possible to discover abundance even in the face of seemingly irrefutable scarcity." Fearing a shortage of resources but displaying none of Malthus's concern about reliance on overseas sources, Americans also increasingly looked abroad for raw materials. In the 1890s, the United States took control of the Philippines, Cuba, and Puerto Rico and expanded trade in Latin America and the Far East.[12]

Malthusian prophets of despair and doom became more common in American life after World War I. "A whole group of careful and reputable scholars," Frederick Jackson Turner noted in 1924, "have attempted to demonstrate quantitatively that before the year 2000, so great is the increase of population and so rapid the exhaustion of resources and such the diminishing production of food relative to population, our present standards of life must be abandoned or the birth rate decreased if we are not to feel the pressure of want and even of universal famine and war." During these years, Malthusians took on many different shapes and sizes. Some emphasized economic well-being, others eugenic concerns and differential fertility among races, yet others family welfare. The scale could be familial, local, national, or global.[13]

Of these, only a few made resource limits and environmental degradation their priority. While most warned of overrapid population growth and even the imbalance of people and resources, this group stressed aggregate population limits and the end of resources. This subset of Malthusians—led by Raymond Pearl and Edward Murray East writing about human beings and Aldo Leopold writing about wildlife populations—pioneered a robust kind of environmental Malthusianism that would grow in importance in the wake of World War II.

In the late 1940s, this environmental Malthusianism informed two bestsellers: Fairfield Osborn's *Our Plundered Planet* and William Vogt's *Road to Survival*, both from 1948. In these books, Osborn and Vogt popularized proto-environmental ecological ideas long before Rachel Carson's *Silent Spring*, the book historians usually cite as the beginning of environmentalism. Ultimately, however, despite striking a cultural nerve, Vogt and Osborn's environmental Malthusianism was rejected by the U.S. government, which, like the turn-of-the-century conservationists, embraced the logic of planning and technology, this time on a global scale. Indeed, even those concerned about population growth, such as the founders of the Population Council in 1952, eschewed the hard limits of which Vogt and Osborn wrote.

All this changed by the mid-1960s. By this time, a litany of international and domestic problems prompted the U.S. government to get into the business of family planning and population limitation. Population growth appeared to threaten peace and prosperity abroad and "quality of life" at home. At the same time, many environmental activists also saw an urgent need to rethink inherited approaches to conservation because of population growth. "Much of what is called conservation today," naturalist Joseph Wood Krutch wrote in 1962, "is no more than a useful holding action." Conservationists, he explained, "must face the fact that behind almost every problem of today lies the problem of population." "In the absence of a reasonable and successful human population policy," one Sierra Club member wrote in the club's magazine in 1965, "conservation is doomed before it begins." Not long afterward Ehrlich published his book, Mills delivered her startling speech, and the environmental movement emerged full bore.[14]

A Cultural and Political Bomb

Concern about population developed in the 1960s in response to an unprecedented climb in human numbers, both overseas and at home. The world population first reached one billion sometime between 1800 and 1830, hit two billion 100 years later in 1930, three billion thirty years later in 1960, and four billion sixteen years later in 1976. Never before had the world's population grown as fast or as high—at some points adding as much percentage *each year* as it did *each century* before 1600. Never before the twentieth century had a person ever lived through even a doubling of the population. Stephanie Mills and others born near 1950 would see the world's population double before their fortieth birthdays.[15]

The United States also broke records in the postwar decades for total numbers and growth rates. For decades, the per decade growth rate had dropped steadily until reaching a low of 10 percent in the 1930s, then climbed to 18 percent in the 1950s during the "baby boom." The absolute numbers were startling: Americans numbered 100 million in 1917, 200 million fifty years later in 1967, and 300 million forty years after that.[16]

Rising numbers alone, however, cannot explain the wave of Malthusian concern that swept over the United States in the postwar years. Not everyone reacted to the demographic charts and graphs with the same amount of concern, or recommended the same remedies. Some were more sober and restrained, others more fearful and urgent. Indeed, even as the population continued to reach new heights in the 1970s, a strong counter reaction to the

Malthusians grew on both the right and the left, and even among environmentalists. This resistance culminated in 1984, when President Ronald Reagan famously declared that population growth had only a "neutral" effect on human society.

New babies simply meant something different in the late 1960s and 1970s than they did before or afterward.

In order to understand the cresting of concern about population during these years, we need to understand what Paul Ehrlich once called the "feel" of overpopulation. In *The Population Bomb,* Ehrlich wrote that he had understood the population problem intellectually for years, but had come to grasp it emotionally only through personal experience, in his case on a taxi ride with his family through a slum in New Delhi. This was the "feel" of overpopulation—the varied meanings that Malthusians gave to babies and population growth. The postwar population explosion was not just a demographic bomb but also a cultural, scientific, and political bomb. The mushrooming of concern grew as much from the alignment of potent international and domestic ingredients—ideas about poverty, war, racial difference, technology, sex, motherhood, and the role of the government—as from numbers on a chart.

This is a book, then, about the larger cultural, social, and political contexts of postwar Malthusianism, especially as those contexts have shaped environmental thinking. The history of postwar Malthusianism shows that environmental ideas inevitably reflected the moment in which they were born, including events seemingly unrelated to environmental thought, such as international affairs, race relations, and the women's movement. To talk about the environment in the 1950s and 1960s was to talk about social relations—between nations, classes, generations, and genders. "If we are to formulate an appropriate land conscience," Secretary of the Interior Stewart Udall wrote in *The Quiet Crisis*, an early environmental call to arms, in 1963, "we must redefine the meaning of 'neighbor.'" Conversely, to talk about neighborhoods and nations in these years often meant talking of population growth and related questions of consumption, natural resources, and environment. The environment did in fact link people together, as environmentalists in the 1960s and 1970s liked to emphasize, but that did not mean that everyone always saw eye to eye.[17]

The history of environmental Malthusianism shows that the environmental movement of the 1960s grew not just from concern for "nature" but also from concern about international affairs, especially poverty and war. In the wake of World War II, concern about overpopulation-induced poverty and war combined with new ecological models to bring about path-breaking "environmental" ways of thinking. In the 1950s and 1960s, as colonial empires collapsed and as the Cold War turned into a struggle for the resource-rich but racially non-white newly independent nations of the "third world," many Americans came to see reining in high overseas population growth rates as crucial to U.S. national security. For others, though, overpopulation was a planetary problem that required a bridging of the great East-West divide.[18]

During these same years, concerns about population growth also emerged in regards to the United States, especially the poverty of the areas increasingly called the "inner city." In the mid-1960s, scores of American cities exploded in violence. As with the third world, many Americans, including environmentalists, turned to Malthusian explanations. Paradoxically, overpopulation also provided a way to think about the unprecedented prosperity that characterized postwar America, the "affluent society." As historians such as Hays have pointed out, this new prosperity often created new environmental values, such as intolerance of pollution and a fondness for outdoor recreation. Many saw overpopulation as a threat to this new "quality of life." Implicitly and often explicitly contrasting their well-being with the poverty of war-torn Europe and later the third world, which they often blamed on overpopulation, environmental Malthusians worried about "too many Americans." One pair of environmentalists even warned of the "Chinification" of the United States.[19]

Linking World War II, the Cold War, inner cities, and middle-class ideas of "quality of life," concerns about overpopulation highlighted one of the most fascinating aspects of the postwar years: how the global and the local intertwined. In its broader sense, one member of Planned Parenthood wrote in 1950, the effort to fight overpopulation "has significance not only for the family next door but for the entire human family as well." During these years, foreign relations often came to be understood through the lens of family and resources, and at the same time, local environments—from households to farm fields, cities to national parks—were often seen in the light of global dynamics.[20]

Assessing Environmental Malthusians

Strikingly, concern about population growth could be either extremely liberating or shockingly repressive. Especially when tied to Cold War strategy, Malthusian concerns added great momentum to the nation's birth control and abortion rights campaigns, and helped prompt new thinking about sex and family roles, especially for women. At least in part, concerns about overpopulation provided urgency for the search for the birth control pill, added energy to the fight for legal rights to contraception, and gave support to the early women's movement of the late 1960s. These breakthroughs paved the way for the vast changes in federal birth control and family planning programs during the 1960s and 1970s, programs that dramatically improved the lives of many people, especially poor women, both around the world and at home.

And yet, as these programs were implemented, a terrible contradiction emerged: the same technology that gave women and families more control over reproduction also created an opportunity for governments and physicians to commit horrible abuses. In the name of fighting overpopulation, thousands of people, often poor and minority women, were unwillingly sterilized by doctors and other health practitioners, both within the United States and overseas. In the United States, forced sterilization has a long history dating to the eugenics movement of the early twentieth century, and it gained new justification as race- and class-based concerns about overpopulation grew after World War II. The problems were even graver overseas. The push for addressing overpopulation in India in the late 1960s, often championed by Americans, led to the overuse and careless implementation of the IUD, causing injuries and even deaths. It also led, at least in part, to widespread coercion, such as in the forced sterilization campaigns run by Indira Gandhi's government during India's "emergency" in the mid 1970s.[21]

In sum, the family planning and population limitation movement fostered both overdue advances and heinous abuses. In her 2005 book *Choice & Coercion*, the historian Johanna Schoen brilliantly sums up the ambivalence of birth-control technology with a chapter entitled "A Great Thing for Poor Folks." Many poor women, Schoen argues, saw federally provided birth control as the key to attaining reproductive freedom and power over their lives. But all too often, she notes, family planning was used as a way for others to control their lives.[22]

Where did the environmental Malthusians of the 1960s and 1970s stand? Many believed that by applying their models both at home and abroad they were advancing a progressive agenda that included third-world development, racial equality, and women's liberation. At a time when birth control was still controversial because of traditional puritanism and the sway of the Roman Catholic Church, they were among those who pushed strongest and most effectively for government-sponsored efforts to research and spread birth-control technologies. Indeed, it was the Malthusian environmental activist Garrett Hardin, not a reproductive rights feminist, who gave us the slogan "abortion on demand."

But environmental Malthusians also often displayed two major flaws, one of diagnosis, one of remedy. Ignoring or downplaying causes such as colonialism, capitalism, poor government, local exploitation, and individual failings, many environmental Malthusians reduced poverty and other complicated socioeconomic dynamics to a simple factor of population, in effect blaming the poor for their own poverty. "All poverty," biologist Paul Colinvaux wrote, "is caused by the continued growth of population." These flaws became far more obvious in the early 1970s, when the collapse of the Cold War consensus opened up more space on the left and right of the political spectrum, and Americans gained a far more complicated understanding of poverty and the way it overlapped with race and gender hierarchies. Not coincidentally, this was just as environmental concerns about population were peaking; reaction against environmental Malthusian extremism helped spur new understandings about race, class, and gender power dynamics. In response, many environmental Malthusians adjusted their views, but not all.[23]

Second, some environmentalists concerned about population growth called for coercive measures that, in effect, would have restricted the reproductive autonomy of some of the world's most disadvantaged people. Hoping to forestall catastrophe, they pushed for the regulation of human reproduction, including government-run programs to "control" population growth. Some believed control could be accomplished without resort to coercion, such as through education and propaganda programs celebrating small families. This middle-ground position is often forgotten. But some, like Ehrlich and compatriot Garrett Hardin, called for measures that seem drastic, even draconian, by today's standards: licensing childbirth, implementing "stop at two" laws, placing sterilents in the public water supply, and cutting off food aid to famine-threatened nations. Unprecedented times, they believed, called for unprecedented measures. Such supercharged rhetoric sold books and spurred people to action, but also opened environmentalists to claims of being "chicken littles" and misanthropic, even racist, authoritarians.

And yet, in focusing on the social blind spots of some environmental Malthusians, it is easy to miss their contributions. Most important, despite their misjudgments, the Malthusian environmentalists identified and called attention to environmentally destructive patterns of modern American society—especially overconsumption—far sooner and with more clarity than others. They may have overreacted in many ways, but they were right about two important imperatives: to assess human economic activities using a broad ecological framework, and to hold individuals responsible for the environmental consequences of their actions, especially their prodigious consumption habits. "When I write 'we' I do not mean the other fellow," William Vogt wrote in 1948 about environmental responsibility, "I mean every person who reads a newspaper printed on pulp from vanishing forests, I mean every man and woman who eats a meal drawn from steadily shrinking lands." In the postwar decades, few were more vocal than environmental Malthusians in challenging Americans and their leaders to rethink their ideas of economic growth, which had undergone a dramatic change during the twentieth century, especially during the 1930s, 1940s, and 1950s. New economic ideas centered upon growth and mass consumption reshaped not only American material and cultural life but also American foreign relations, and indeed, global history. Now, after many decades of almost unbridled buying, using, and disposing of goods—some of it necessary, much of it not—we can only imagine the environmental and social problems that could have been prevented if environmentalists had been able to get this message across more effectively in either the late 1940s, or the 1960s and 1970s. Alas, exactly that which helped open their eyes to these problems, the Malthusian focus on people and their consumption, also often led to alarmist and coercive remedies that did a great deal to undermine their overall position.[24]

Population growth and the extremely complicated questions it spurs about human relations with the earth and with each other are issues not just of the past but also of the future. When Stephanie Mills gave her gloomy commencement talk in 1969, the world's population numbered roughly 3.5 billion. It is now twice as large, and projected to peak midway through the century somewhere between eight and eleven billion people. This growth, in combination with a multitude of other factors, will remake environments around the world, including the world's climate. It will also shape how nations and communities interact with each other. In the twenty-first century, we continue to live in an interdependent world, connected to our neighbors around the block and around the world through tricky environmental problems with complicated social histories.[25]

1

Malthusianism, Eugenics, and Carrying Capacity in the Interwar Period

"Forty-knot vessels, 60-mile trains, aeroplanes, wireless, and other devourers of distance, have made neighbors of us all, whether we wish to be neighborly or not."

—Edward Murray East, *Mankind at the Crossroads*, 1923

After World War I, Malthusian prophets of despair and gloom appeared in large numbers. "There is at the present moment," the biologist Raymond Pearl wrote in 1925, "a great recrudescence of public interest in the problem of population. Books and articles on population growth have been appearing in the last five years with an abandon which could fairly be called reckless if the protagonists were not so obviously in deadly earnest."[1]

Several factors drove this "great recrudescence." First was a sense that the world had grown smaller because of global integration. "The world has changed," fellow biologist Edward Murray East wrote at the opening of *Mankind at the Crossroads* (1923), one of the prominent Malthusian books of the decade. "The present age is totally unlike any previous age. There is no longer isolation; space has been annihilated.... Thanks to steam and electricity the world as a whole is more of a single entity than were some of the smaller kingdoms of Europe in the fifteenth century." More than ever before, East pointed out, people and resources moved around the globe.[2]

East pointed to a second factor: a new abundance of population statistics. "We are just coming to be able to judge population matters," he wrote, "by knowledge rather than by guess, owing to a more plentiful supply of facts." When Malthus wrote his classic essays over a century earlier, such data did not exist. These data enabled people, especially state actors, to make sense of the accelerating changes in the world. In the late nineteenth century, states began registering citizens, issuing passports, and in some cases, restricting immigration. "Bookkeeping on a national scale," East pointed out, "is rather a new thing to the world."[3]

FIGURE 2 "The Most Critical Race in History," 1917, by Jay N. "Ding" Darling, shows the connection between World War I–related famine and American agriculture. Darling, an avid conservationist, later expressed worry about the toll of American foreign policy programs on U.S. soil and streams.

Courtesy of the Jay N. 'Ding' Darling Wildlife Society.

World War I also drove the surge of interest in population growth. Nothing showed that the world was interconnected, or the importance of statistics and planning, more than the war. And for Malthusians, nothing so dramatized the consequences of overpopulation. Fear of renewed war loomed over this generation of Malthusians much as it would for Malthusians of the late 1940s and early 1950s.

In retrospect, the Malthusian arguments of the interwar period can be clustered into at least three groups. One group stressed the political-economic aspects of population growth. In his influential *The Economic Consequences of the Peace* (1919), British political economist John Maynard Keynes saw the origins of World War I in the population explosions of turn-of-the-century Germany, Austria-Hungary, and Russia. A second group emphasized the ethnic and racial consequences of population growth. In *The Rising Tide of Color against White Supremacy* (1920), Lothrop Stoddard warned that nonwhite peoples were winning demographic superiority around the world. A third strand, best represented by American birth-control activist Margaret Sanger, stressed the consequences for women and family. These three strands often overlapped. Sanger, for instance, wrapped together her concern for war, growing numbers of the unfit, and women's health.[4]

To varied extents, the environmental Malthusians of the late 1940s and 1950s—the generation including William Vogt and Fairfield Osborn, who would lay the groundwork for 1960s environmental Malthusians such as Paul Ehrlich and Garrett Hardin—drew from these political, racial, and family concerns. But even more they borrowed from an often overlapping but ultimately distinguishable set of concerns that a group of biologists, ecologists, and scholars developed in the interwar years: concerns about consumption, soil fertility, and environmental degradation. It was this emphasis on degradation and limited carrying capacity that most separated environmental Malthusians from other sorts of Malthusians.[5]

Some of those who pioneered these ideas focused on human beings, others on animals. The best known of the first group included biologists Pearl and East. Most Malthusian arguments, East noted at the beginning of *Mankind*, had overlooked the most central issue: they "concern only birth-rates and death-rates, and . . . neglect the importance of agriculture." Studying not just food supply but also agriculture was crucial, East insisted, because the planet's capacity was declining. Historians of population have noted the prominence of Pearl and East in debates about population, mostly to point out their race and class fears, but have rarely stressed their emphasis on environmental limits.[6]

A second group of theorizers about carrying capacity and environmental limits were ecologists who studied animals. Best known was Aldo Leopold, whose influence on the postwar environmental movement ranks with that of Rachel Carson. Investigating population dynamics—especially of deer—led Leopold to some of his most influential breakthroughs and one of his most memorable essays, "Thinking Like a Mountain." Even more than the ideas of Pearl and East, Leopold's ideas about population, consumption, limits, and environmental quality influenced postwar environmental Malthusians such as Vogt and Osborn and their 1960s successors. Leopold's ideas created a framework that later environmentalists used to critique the economic order that dominated the second half of the twentieth century—a Keynesian emphasis on economic growth that, not coincidentally, emerged in the 1930s and early 1940s, exactly when Leopold was unraveling the mysteries of deer population dynamics. Like Pearl and East, Leopold helps show that concerns about population growth grew less from eugenics concerns than from new ideas about consumption, resource degradation, and carrying capacity in a time of war.

Raymond Pearl: Eugenics, Carrying Capacity, and Population Planning

Raymond Pearl (1879–1940) was one of the best known biologists of the early twentieth century. Conducting research at the Maine Agricultural Experiment Station from 1907 to 1918, and at Johns Hopkins University during the 1920s and 1930s, he pioneered breakthroughs that have shaped modern biology. He was also, according to one scholar, "unquestionably the most vocal exponent of population control ideology within the scientific community at the time." During the 1920s, Pearl wrote several articles about population problems as well as *The Biology of Population Growth* (1925). In 1928, he helped found the International Union for the Scientific Investigation of Population Problems. His thinking helps show the social politics wrapped up in early concern about population growth but also the emergence of the idea of carrying capacity.[7]

A key influence on Pearl's career was the British scholar Karl Pearson, with whom Pearl studied in London in 1905, not long after he earned a Ph.D. in zoology from the University of Michigan. From Pearson, Pearl learned two key things. First was biometrics, the statistical analysis of biological phenomena. As Pearl later wrote, Pearson made scientists "for the first time truly conscious that some sort of a logically coherent statistical calculus was indispensable." Pearl also learned about eugenics from Pearson, who was, according to one historian, a "chief disciple" of its founder, Francis Galton. Eugenics was a Progressive-era movement that hoped to spread genes seen as desirable and contain those thought to be destructive by encouraging those of the "right stock" to have lots of children and discouraging the "unfit" from doing so. Combining new discoveries about genetics with older ideas of racial and ethnic hierarchy, eugenicists believed that human reproduction could be monitored and shaped in order to prevent social problems like poverty, crime, and prostitution.[8]

After returning from Europe, Pearl made his reputation investigating the genetics of chicken populations. As a researcher in Maine, he studied how to increase egg output by hens. Drawing from the work of Danish botanist Wilhelm Johannsen, he helped discover the difference between the phenotype of an individual, its outward appearance, and its genotype, its internal genetic composition. Because of this difference, offspring did not always appear exactly like their parents. This realization was important in its own right, but also pushed Pearl toward the study of populations. He realized that studying the fertility patterns across an entire population could reveal as much as studying individual pairs. He began to focus on gathering population-wide data—a focus on groups, not individuals, that was a prerequisite for adopting a Malthusian view.[9]

While conducting this research, Pearl also started publishing articles about eugenics, quickly becoming a leading voice in the movement. "It is of prime importance," he wrote in 1908, "for the welfare of state or nation that those stocks which are on the whole endowed with the best traits should contribute more, many more, individuals to the next generation than should those stocks whose characteristics are on the whole bad." Eugenics historian Daniel Kevles writes that Pearl was part of the "eugenic priesthood," along with Charles Davenport, Herbert S. Jennings, Clarence C. Little, William E. Castle, and Edward Murray East.[10]

World War I helped Pearl see the importance of using population analysis as a tool for understanding human society. Due to his background in biometrics, he was selected to head the statistical division of the U.S. Food Administration under Herbert Hoover from 1917 to 1919. He drew two major conclusions from the war: overpopulation played a key role in causing international conflict, and good population-wide statistics were needed for postwar assistance. Numbers helped him see how close Europe had come to famine.[11]

These insights—combined with a 1919 fire that destroyed his lab and almost twenty years of records—pushed Pearl to begin to study the population dynamics of not just animals but also human beings. In what became a famous article, Pearl and fellow Hopkins researcher Lowell J. Reed claimed in 1920 to have found a "law" of population growth that applied equally to human beings and animals. All populations, they argued, followed a similar pattern of slow growth at first, followed by rapid growth, and then a plateauing. "The half-pint universe of the laboratory Drosophila," Pearl wrote in *The Biology of Population Growth*, "is without doubt a simpler world than that in which we carry on, but in both the great realities of birth and death are much the same. And it is these that count." On a graph, this traced a flattened, elongated "S." This "law" was criticized roundly, but has become a useful starting point for many population biologists.[12]

Over the next decade, in addition to promoting his idea of the logistical curve, Pearl warned frequently of unsustainable human population growth. In particular, two issues worried Pearl, the first being that the unfit of the world, both individuals and races, would take over the planet. "What kind of people are they to be," Pearl asked in 1923, "who will then inherit the earth?" Eugenics was the answer: "Man, in theory at least, has it now completely in his power to determine what kind of people will make up the earth's population." Most historians have focused on this aspect of Pearl's thinking. "It was not population growth of all people which worried Pearl," Garland Allen writes, "it was the growth of that segment of the population that he thought was defective."[13]

This may be so, but Pearl was no simple eugenicist. As he was developing his ideas about population in the mid-1920s, he was also rethinking—in fact narrowing—whom he thought of as defective, and in the process reforming the eugenics movement. His genetic research over the previous decade had indicated that offspring would not necessarily be identical to their parents, and he moved away from some of his more crudely formulated eugenics positions, criticizing eugenicists such as Madison Grant for using flawed science to attack poor people. "It has yet to be demonstrated," Pearl wrote in a 1927 article, "that either poverty or lack of membership in a social aristocracy are biologically inherited traits, though the inference is too often drawn that they are." In *American Mercury* the same year, he attacked the "biology of superiority," arguing that eugenics had "largely become a mingled mess of ill-grounded and uncritical sociology, economics, anthropology, and politics, full of emotional appeals to class and race prejudices, solemnly put forth as science."[14]

Nonetheless, Pearl did not completely renounce eugenics. He still claimed to adhere to "Galton's view that heredity plays the principal role." His point was to expose the flawed genetic thinking of mainstream eugenicists, especially the idea that an individual's intelligence and character could be predicted on the basis of his parents' traits. Genetics worked in more complicated ways. "The almost infinite manifoldness of germ-plasmic combinations can be relied on to produce in the future, as it has in the past, Shakespeares, Lincolns, and Pasteurs, from socially and economically humble origins," he wrote. Pearl himself had come from humble origins: his father worked as a grocery clerk and foreman in a shoe factory.[15]

Pearl's concern about global resource limits also drove his approach to population growth. Whereas some Malthusians spoke of resource shortages and imbalances, Pearl stressed limits. In a 1922 essay examining population growth side by side with rates of resource consumption, he asked, "Where Are We Going?" With this question, Pearl was noting and criticizing the revolution in American patterns of consumption that that had developed in the early decades of the twentieth century and intensified in the 1920s. As many historians have described since, during these years more and more Americans became consumers, purchasing the goods they relied upon for daily needs, especially food and clothing. The 1920s also saw the spread of urbanization, cars, and electricity. The increased consumption that came with increased population focused Pearl's mind on the limits of the earth. "The volume and the surface of the planet on which we live are strictly fixed quantities," Pearl stressed. "This fact sets a limit." He often spoke of "closed universes." Pearl was using the concept of carrying capacity, if not the actual term, applying it to the world as a whole and its human inhabitants. Carrying capacity refers to the number of animals that a given landscape can support without deterioration. In the late nineteenth century, ranchers and rangeland managers began applying this concept to cattle and land, but only in crude fashion. Pearl was now giving the concept wider application.[16]

Scholars often ignore Pearl's emphasis on carrying capacity, or suggest that it was merely a more politically acceptable package for his main concern, eugenics. But the two concerns were not identical and deserve to be distinguished. "It is not only desirable in the eugenic interest of the race to cut down, indeed completely extinguish, the high birth rate of the unfit and defective portions of mankind," Pearl wrote in *The Biology of Population Growth* (1925), his major work on the subject. "But it is also equally desirable because of the menacing pressure of world population, to reduce the birth rate of the poor, even though that unfortunate moiety of humanity be in every way biological sound and fit." The first point addressed proliferation of the unfit and the population composition. The second regarded not the "quality" of the population but rather its quantity, which Pearl worried about despite the lower classes being biologically sound and fit. He wanted to reduce their birth rate because he wanted to decrease the overall population, not because the poor were genetically inferior.[17]

In a 1927 essay, Pearl again distinguished his views of population composition from those about carrying capacity. After critiquing the flawed science of eugenicists in favor of a reformed vision of eugenics based less on group and more on individual deficiencies, he moved on to discuss carrying capacity: "It should be emphasized that what has been said in this paper related entirely and solely to the *relative* or *differential* aspect of fertility . . . and *not* to the absolute fertility of the population as a whole." For Pearl, the sustainability of continued aggregate population growth differed from the relative population size of different groups. "That the population of the United States," he wrote, "as a whole cannot go on increasing at its present rate per unit of time, and its component elements continue to enjoy the standards of living which they have in the past and do now, would seem to be obvious." Pearl's eugenic concerns and his concern about limits to growth, which he connected to increased consumption rates, may have overlapped, but they were not the same.[18]

In 1928, the population biologist Royal Chapman, one of Pearl's chief competitors, worked this concept of carrying capacity into an influential paper. "We have in nature a system," he wrote, "in which the potential rate of reproduction of the animal is pitted against the resistance of the environment, and that the quantity of organisms which may be found is a result of the balance between the biotic potential, or the potential rate of reproduction, and the environmental resistance." Chapman did not apply the concept to human populations, but later ecologists who did, including William Vogt and ecologist Eugene Odum, cited his works.[19]

In the 1920s, in the context of a recent global war, resource competition, growing populations, growing consumption, and the spread of new technologies of agriculture and new agricultural sciences, ideas of carrying capacity seemed a useful tool for understanding the changing world.

Edward East: Malthus and Soil Conservation

When Edward East (1879–1938) sat down to write *Mankind at the Crossroads* (1923), he originally had in mind a study of immigration, population composition, and eugenics. Far more than population growth, these questions dominated demographic discourse in the first half of the twentieth century. Like Pearl, East was a eugenicist; *Mankind* includes many racist, elitist remarks. But it is not primarily a book about eugenics. East came to believe the problem of population growth to be separate and far more pressing. "The broader aspects" of the problem, he wrote, "continually asserted themselves." The population of the world "is advancing in a tidal wave the like of which has never before been seen."[20]

Like Pearl, East saw humanity filling all available spaces and reaching limits. But East went beyond Pearl. He warned not just of limits being reached but also of environmental degradation that would severely reduce carrying capacity—the argument that most defined environmental Malthusians. Moreover, even more clearly than Pearl, East showed the gap between eugenics and Malthusian population concerns.

East made his name as an expert in corn genetics, just after the rediscovery of Gregor Mendel's pea experiments had kicked off a genetic revolution in biology. He grew up and studied in Illinois, and worked at agricultural extension stations there and in Connecticut before joining Harvard's faculty in 1909. He made two discoveries that sealed his place in the history of modern biology. First, while researching inbreeding and outbreeding in corn, he showed that blended inheritance could be explained by genetics, thus removing one of the key doubts about Mendelian theories. He also discovered that sexual reproduction leads to a recombination of genes and thus to more variations—what has become a bedrock idea in genetic theories of evolution. Ironically, given his pessimism about population growth and food supply, East's experiments and writings helped lay the foundation for hybrid corn and a revolution in food production in the United States and eventually around the world.[21]

As with many of the Malthusians in the 1920s, East's concern about population grew out of World War I. Like Pearl, East held administrative posts during the war that offered him a bird's-eye view of the world food situation: chief of statistics for Food Administration and chairman of the Botanical Raw Products Administration. His work in the food administration showed him, according to one colleague, "how narrow is the margin between the world's food supply and its ever increasing needs." During the war, East also read Thomas Malthus's essays on population.[22]

In *Mankind*, like Malthus and other Malthusians, East offered stern warnings: "If something is not done ... the day of reckoning may arrive with astonishing suddenness in the shape of a breakdown ... of the system." He prophesized that "If world saturation of population, which approaches speedily, is not prevented, in its train will come more wars, more famine, more disease." Too many people, he stressed, failed to realize "the gravity of this situation."[23]

Like Pearl, he saw the planet as a finite place: "The vessel is somewhat elastic, it is true, but it is a closed vessel nevertheless." Yet East took Malthus in a different direction from other Malthusians, even Pearl. What most worried him most was not differential fertility, or even the filling the planet—but the diminishing returns of agriculture as populations grew. "Soil fertility," he wrote, "is being exploited with high speed and unnecessary wastefulness." In the 1890s, the United States reached the "era of diminishing returns in agriculture." The world, he wrote, could do so shortly.[24]

In many ways, East anticipated late twentieth-century concerns about sustainable agriculture. Devoting an entire chapter of *Mankind* to what he called "permanent agriculture," East highlighted soil fertility and called for putting agriculture on "a permanent scientific basis for all time." In particular, he wanted to remind people of agriculture's chemical foundation. Crops, he said, citing German chemist Justus von Liebig, drained the soil of various elements, especially nitrogen, phosphorus, and potassium. Without replenishing them, land would simply not produce. "In the last analysis," he noted, "the future food-supply of the world depends upon the conservation of soil fertility."[25]

Several factors led East to stress soil. He had grown up in the rural Midwest, surrounded by farms and farmers. He had trained and worked as a chemist, so he fully understood soil science, and knew that crops drained the soil of vital nutrients. And finally, as one of the inventors of hybrid corn, he could discern better than most the world's agricultural future. He was not optimistic; hybrids would increase yields but not dramatically. "There will be no revolution," he wrote. "There is no royal road to raising turnips." East doubted modern science's ability to improve agriculture, criticizing "those credulous day-dreamers who expect all future troubles to be straightened out by the genius of the test-tube shaker." Again and again, he returned to the basic issue of soil fertility: "All increase will be temporary, and even current production cannot be maintained, unless the essential elements of soil fertility are conserved by every method possible." Those who expect an agricultural revolution lacked real knowledge of agriculture. "The end of expansion," he stressed, "is in sight."[26]

East called for a new kind of conservation. Traditional conservationists focused too much on minerals: "Conservation[ists] ... usually expend their energies solely upon this type of natural wealth. Forestry is the only agricultural specialty included in their exhortations." Instead, he pushed conservationists to focus more on agriculture: "It is agriculture rather than mechanical industry that is in need of these evangelists to-day."[27]

In addition to a new emphasis on conservation, East advocated several other measures that foreshadowed post–World War II environmentalism. He called for "rational birth control," which he said was "just as much a fundamental need of the nation as conservation of resources, equitable laws, and healthy social customs." He also wanted "a severe permanent restriction on immigration." Overseas, he called for "an agricultural foreign policy" that would have as its primary aim "the stabilization of prices and the conservation of soil wealth."

He also worried about exporting health measures to poor countries. He felt that the Rockefeller Foundation, a leading philanthropy working in Asia, had "gone naively into China bringing her the blessings of Western medical art and sanitation" without thinking through how "they are going to support the people they save."[28]

Like many biologists, East stressed that human beings were essentially animals, as genetics had proven. "Man is an animal," he wrote. "His life, growth, and death are subject to natural law." East underscored in particular the "imperious instinct" of reproduction. "The fundamental nature of the sex instinct," he wrote, "is no less apparent in man than in the lower animals." Like many in the later population planning movement, he felt that humans had no choice but to acknowledge and accept the sexual aspects of humanity. "Nature," he noted, "is not to be denied."[29]

Like Pearl, East often criticized eugenicists for bad science, but remained a racist, elitist, and ardent eugenicist, arguing that intermarriage between whites and blacks could undermine social progress. He wanted birth control distributed "at the lower end of the social scale" because, as he said, social problems came "disproportionately from the least desirable elements." "Parentage," he insisted, "must not be haphazard."[30]

And yet, like Pearl, East shows that the relationship between population planning and the racism and elitism of the eugenics movement was not straightforward. Although they often overlapped, his arguments about overpopulation were at times distinct from his racial and ethnic calculations. In *Mankind*, he did not fear a decline of the white population relative to others. Indeed, he believed that the white population was outgrowing the nonwhite population. Directly rejecting Lothrop Stoddard's argument that Asians, Africans, and other nonwhites were taking over the planet, he pointed out that whites tended to double their population every fifty-eight years, while Asiatics only in eighty-seven years. Moreover, unlike Asians, whites had more space in which to grow. In *Mankind* at least, East worried about the world filling up, not so much the world filling up with the wrong races. To him, the real problem with population growth was not differential fertility, but soil fertility.[31]

In one passage of the book, East crystallized the core problem. He saw "two great opposing forces" driving history. On one side were genes encouraging reproduction, high population growth, and, as for Pearl, increasing consumption: "the natural human desires to live an individual life of comfort and to exercise the instinct of reproduction." On the other side, he saw environmental degradation and "the immutable law of diminishing returns." This law "rules in agriculture more strictly than in other industries, pulling down and simplifying the standards of living, and requiring greater and greater efforts for mere existence as time goes on."[32]

Many of East's views and values—ideas of an interconnected, closed vessel of a planet with diminishing carrying capacity; of people as animals, especially reproductive animals, yet with the capacity to plan; of hopes for birth control technologies but overall technological skepticism; of concern about agriculture as well as new health measures overseas; of racism and calls for immigration restriction; and especially of increasing consumption—would reappear in the environmental Malthusians in later decades. But East also shows that concerns about carrying capacity, limits, and environmental degradation were not identical to concerns about differential fertility and weeding out the "unfit."

Aldo Leopold: Population, Consumption, and Wildlife Management

Aldo Leopold's thinking about population dynamics and consumption also shaped his ecological vision in profound and often overlooked ways. By studying wildlife population fluctuations, Leopold extended many of the ideas about carrying capacity and environmental degradation that Pearl and East had articulated, and these ideas, in turn, shaped his path-breaking ideas of ecological interconnection. Moreover, although later readers associate Leopold with wildlife ecology, his career helps show how Malthusian ideas of human society intertwined and overlapped with ideas of nature. He was greatly influenced by ecologists who gleaned ideas from Malthusian models of human society, and he himself often thought of human events—especially the Great Depression and World War II—in terms of the models of population and consumption that he was developing for animals.[33]

Leopold (1887–1948) is remembered as a transitional figure between Theodore Roosevelt's conservation movement and the 1960s environmental movement. He studied at Yale, the main training ground for Gifford Pinchot's U.S. Forest Service, the heart of the conservation movement. In 1909 he began fifteen years of work for the forest service in Arizona and New Mexico, which at the time were territories on the edge of the nation's growing empire. After moving back to his home state of Wisconsin and becoming the nation's first professor of game management, he became involved in many 1930s New Deal conservation projects. Interpreting these experiences through the lens of his specialty, wildlife conservation and ecology, Leopold pioneered ecological and ethical ideas that charted a path for environmentalists in subsequent decades. He synthesized his ideas in a famous collection of essays called *A Sand County Almanac* (1949).

Leopold's first exploration of population and carrying capacity came in the Southwest, where he worked in the U.S. Forest Service from 1909 until 1924. In 1913, after several years of work, Leopold experienced an epiphany: he realized that foresters fundamentally misunderstood their real task. They were, he wrote in a Forest Service newsletter, confusing the means and ends of conservation. They had become so focused on the gospel of efficiency in managing land for production that they had lost sight of the true end of conservation—the long-term health of the forest and other resources under their protection. "The sole measure of our success," he wrote, should be "the *effect* . . . on *the Forest*." "My measure," he reiterated in capital letters, "is THE EFFECT ON THE FOREST."[34]

Leopold's thinking about human society, especially the tragic history of the Blue River Valley in Arizona, helped him refine his ideas. Settled by Anglo farmers and ranchers in the 1880s, the Blue River Valley had initially provided fertile bottomland soil, cottonwoods, and pines. But by the first decade of the twentieth century, when Leopold first visited, it routinely suffered flash floods and severe erosion. By the 1920s, settlements were literally washing away. The place, Leopold wrote, was "mostly boulders, with a few shelves of original bottom land left high and dry between rocky points." Only forty years after settlement, less than 8 percent of its arable land remained. In scores of other colonial settlements in the Southwest, Leopold encountered a similar pattern. He came to realize that, as one biographer put it, "the test of true civilization . . . was whether it could endure, whether its citizens could prosper for generations in a place."[35]

What he saw in these valleys—small islands of habitation whose limits had been reached—prompted Leopold to refine the idea of carrying capacity, which previously had mostly been applied in simplistic fashion. The settlers, and even the Forest Service, had overestimated the number of animals that could graze in a certain area. They had been looking at a single resource, the amount of available forage, but really needed to look at the health of the entire fabric of life in the area, especially the health of the soil. Summarizing his conclusions in 1923, Leopold stressed "the interdependent function of the elements"—what later would be called ecology. His conception of carrying capacity resembled Edward East's. Supply of food alone was not the best measure of carrying capacity; the condition of the environment had to be considered. There's no evidence that Leopold knew of the work of East and Pearl, but a decade later he would make a quick reference to Royal Chapman's 1928 article about carrying capacity in *Game Management*, his 1933 textbook on wildlife ecology.[36]

After moving back to his native Midwest in the mid-1920s, Leopold struggled for the next fifteen years with questions regarding wildlife populations and carrying capacity. During the 1920s, wildlife populations around the country but especially in the Midwest had declined alarmingly. The reigning explanation for this decline, most associated with the famous naturalist William Hornaday, was overhunting by humans. But using the ideas he had developed in the Southwest, Leopold stressed that habitat loss and degradation were just as important, if not more so. The main point here—that a population's size was dependent on the quality of its land and resource base—would eventually become crucial to understanding human populations and sustainability.³⁷

The problem was not just population declines but population explosions, as exemplified most dramatically on the Kaibab plateau in Arizona. The story of what happened there, later made famous by Leopold in his essay "Thinking Like a Mountain," would become a parable repeated by Rachel Carson and especially environmental Malthusians. The plateau, an "island" ecosystem bounded by the Grand Canyon on one side and by desert on all other sides, came under federal protection during the 1890s and became a wildlife refuge in 1906. In the 1920s, the deer population spiked then fell off dramatically. Having neither worked on the Kaibab nor visited it, Leopold relied upon others for information. At that time and later, scientists and wildlife officials disputed both the size and drop of the population and the dynamics involved. Estimates of the peak number of deer ranged from 20,000 to 100,000, and of the population crash from a few thousand to tens of thousands of animals. On the basis of the work of D. I. Rasmussen, Leopold eventually decided upon the higher range.³⁸

Leopold struggled to understand the Kaibab and other population irruptions for most of the 1930s. He found a great source of insight, especially about the understudied role of consumption, from Charles Elton's *Animal Ecology*, an instant classic from 1927. Leopold and Elton met in 1931 and developed a good friendship. In his book, Elton established several core tenets of animal ecology, including "food chains" and the "pyramid of numbers"—concepts that, according to historian Thomas Dunlap, "put flesh on the skeleton of the 'balance of nature.'" Both concepts highlighted not just ecological interconnection but the mechanism for interconnection: animals were linked to each other through cycles of consumption. In a food chain, to put it simply, one species lived off of a second, which lived off of a third, which itself lived off of the first species. A species with fewer numbers sustained itself on—and was therefore limited by—those species with far greater numbers: thus, the pyramid of numbers. In Elton's wake, wildlife managers began to focus much more on consumption; wildlife research became, according to Dunlap, "studies of who ate whom." Management practices also moved toward concern with populations in a given environment, and toward quantification.³⁹

Elton had developed many of his ideas thinking about human sociology and economics, especially in an imperial context (as had Leopold). In much the same way that Charles Darwin drew from Thomas Malthus's ideas, Elton drew from the British sociologist Alexander Carr-Saunders and his well-known 1922 book, *The Population Problem: A Study in Human Evolution*. Elton adapted ideas about food chains, cycles, and the pyramid of numbers directly from similar patterns in Carr-Saunders's analysis of human population history. In fact, Elton once described Carr-Saunders's work as the "sociology and economics" of human beings and his own work as the "sociology and economics of animals."

"Elton's achievement," historian Peder Anker has argued, "was to read by analogy Carr-Saunders's sociology into nature."[40]

During 1930s, as he continued to struggle with the question of population irruptions, Leopold began to use ideas of consumption and carrying capacity to criticize human society. "Man thinks of himself as not subject to any density limit," Leopold wrote in his pioneering 1933 textbook *Game Management*. "Industrialism, imperialism, and that whole array of population behaviors associated with the 'bigger and better' ideology are direct ramifications of the Mosaic injunction for the species to go to the limit of its potential, i.e., to go and replenish the earth. But slums, war, birth-controls, and depressions may be construed as ecological symptoms that our assumption about human density limits is unwarranted." Leopold took aim at modern civilization in general, not just capitalism. "As nearly as I can see," Leopold wrote in a 1933 essay, "all the new *isms*—Socialism, Communism, Fascism, and especially the late but not lamented Technocracy—outdo even Capitalism itself in their preoccupation with one thing: The distribution of more machine-made commodities to more people." He specifically warned about consumption. People in all these forms of society, he elaborated, "all proceed on the theory that if we can all keep warm and full, and all own a Ford and a radio, the good life will follow." The stakes, he stressed, couldn't be higher: "There are now wars and rumors of wars which foretell the impending saturation of the earth's best soils and climates."[41]

By the late 1930s, Leopold felt he had finally figured out the mysteries of population explosions. In an important 1939 essay called "A Biotic View of Land," he outlined an ecologically informed model of population explosion and carrying capacity. Nature, he wrote, should be understood as energy circulating in the form of nutrients through Elton's food chains and pyramids. Species should be defined by their consumption—not "where they came from, nor . . . what they look like, but rather . . . what they eat." Leopold connected these insights about consumption to his earlier insights about soil erosion, which had by this time become one of the most glaring conservation problems of 1930s because of the devastating Dust Bowl storms on the Great Plains. The health of land, he stressed, could not be separated from animal dynamics, especially animal consumption patterns. When the circulation of energy and nutrients from animals was healthy, plants and soils were healthy. But if these circulations were disturbed, soils suffered. "Land, then, is not merely soil; it is a fountain of energy flowing through a circuit of soils, plants, and animals. Food chains are the living channels which conduct energy upward; death and decay return it to the soil."[42]

"A Biotic View" displays a shift from the conservation focus on increasing productivity to the environmental focus on ecological interconnection and environmental quality. Conservationists, Leopold believed, had focused too much on cultivating species deemed useful and extirpating those deemed injurious, instead of on the health of the land and the overall system. In particular, he stressed, removing predators was counterproductive because they helped maintain the circulation of nutrients within animal and plant communities, which in turn maintained the health of the soil. This was the problem on the Kaibab Plateau. All throughout the Southwest, federal officials, hoping to increase the yield of a desired species—in this case deer—had disrupted the ecological food chain and thereby degraded the entire ecology. He himself had killed many a wolf in such federal programs. "A Biotic View" warned of an "epidemic of new Kaibabs." The real goal of conservation, Leopold said, should be to achieve land health by maintaining ecological relationships. "Only wolves and lions," he concluded, "can insure the forest against destruction by deer and insure the deer against self-destruction."[43]

Here, too, the influence of larger events was visible. In the late 1930s, as Europe was falling into the chasm of war, a dark tone crept into Leopold's writings, a gloominess that appeared in "A Biotic View" as a deep concern with violence, especially human violence toward nature, a theme that drives the essay. "Man's invention of tools," Leopold wrote, "has enabled him to make changes of unprecedented violence, rapidity, and scope." The work of conservation should lead us "toward a nonviolent land use." Combining this idea with his understanding of population dynamics, he noted that human violence corresponded to population density. "Violence," he wrote, "would seem to vary with human population density: a dense population requires a more violent conversion of land." He criticized the "pioneering philosophy" in the United States, "which assumes that because a small increase in density enriched human life, that an indefinite increase will enrich it indefinitely."[44]

In subsequent years, Leopold spoke more explicitly about ecology, overpopulation, unsustainable consumption, and war. In a lecture to his University of Wisconsin students in 1941 called "Ecology and Politics," he stressed that when ultimate limits are reached, populations fall, sometimes dramatically. "One of the most emphatic lessons of ecology is that animal populations are usually self-limiting." For an example he pointed to "overpopulated" Europe: "Perhaps the present world-revolution is the sign that we have exceeded that limit, or that we have approached it too rapidly. If so ... why not call a moratorium on human increase? ... Why not bend science more toward new understandings, less toward new machines? ... The technologists' cure for war is more technology." He then attacked ideas of unlimited production, consumption, and population growth: "We assume, I think naively, that increasing 'take' (i.e. more extraction, conversion, and consumption of resources) always raises standards of living. Sometimes it merely raises population levels. Perhaps this is a bear chasing his own tail."[45]

The mass patterns of human society, Leopold wrote in a 1943 essay, needed to be uncovered. Animals, he wrote, have behavior patterns of which an individual animal is unaware but which he nevertheless helps to execute. These patterns were not visible by looking at just one animal, but only by "scrutiny of the mass through decades of history." This idea raised for Leopold a "disquieting" question: "Do human populations have behavior patterns of which we are unaware, but which we help to execute? Are mobs and wars, unrests and revolutions, cut of such cloth?" Historians and philosophers, he pointed out, persist in interpreting the "mass behaviors" of humans as the collective result of individual acts of volition. He recognized that humans have higher volitional content than animals, but thought that "it's important to look for analogues in higher animals." In future years, searching for such lessons would be one of the chief missions of ecologists. Doing so, he emphasized, might be "of potential importance to the whole human enterprise." With these questions, Leopold was broaching an old question with tremendous new consequences: In what ways were human beings similar to nonhuman animals and in what ways were they different?[46]

In April 1944, at the urging of one of his former students, Leopold sat down to summarize his intellectual journey from his early days in Pinchot's forest service to an ecological framework at odds with most 1930s conservation. The result was an essay about animal carrying capacity and overpopulation that became one of the most famous parts of *A Sand County Almanac*: "Thinking Like a Mountain." Leopold opened the essay by recounting shooting several wolves while crossing a river in the Southwest because, as he explained, "I thought that because fewer wolves meant more deer, that no wolves would mean hunters' paradise." He then painted a vivid picture of how early federal wolf extirpation programs in the Southwest had backfired, creating a massive deer population explosion and die-off. "In the end," he wrote, "the starved bones of the hoped-for deer herd, dead of its own too-much, bleach with the bones of the dead sage, or molder under the high-lined junipers." Instead of focusing on one or two desirable species, Leopold stressed, land managers needed to think more "like a mountain"—to manage from the long-term perspective of the land and the entire system. In particular, Leopold indicted environmental manipulations that backfire because of a lack of ecological knowledge, such as predator controls that increase a population to where it ends up dying of "its own too much."[47]

By stressing interconnection, limits, and land health, Leopold's "Thinking Like a Mountain" departed from the single-species and utilitarian conservation values that he had implemented in the Southwest three decades earlier as one of Pinchot's foot soldiers and that, importantly, had come to dominate New Deal conservation, only now on a much larger scale. It was this new ecological model, with consumption patterns at its center, that became a textbook example of wildlife management—and for some, a way to think about human populations—for decades to come. "'The Kaibab,'" Thomas Dunlap has written, "became a classic conservation horror story, repeated in sportsmen's magazines and game management and ecology texts for the next forty years." Later environmentalists would draw from Leopold's analysis of the Kaibab in textbooks, articles, and museum exhibits. Even activists on Earth Day in 1970 would borrow from it.[48]

That in other venues Leopold had blamed World War II on overpopulation adds an unusual layer of meaning to "Thinking Like a Mountain." In the context of the war, a story about misguided human actions and a population that dies of "its own too much" carried a darker warning about humanity's future. It is perhaps because of this subtext that the essay has moved so many people over the years. Indeed, *A Sand County Almanac* became popular during historical moments of war or threatening war that made it easy for readers to infer this meaning. The book appeared in 1949, just a year after Osborn's *Our Plundered Planet* and Vogt's *Road to Survival* blamed the recently ended global war on overpopulation and resource depletion. And it gained its greatest popularity in the 1960s and 1970s, the decades of another disturbing American war, and a moment in which overpopulation ranked as one of the environmentalism's top concerns.

Keynesian Growth Models and the Environment

During the Depression and World War II, American policy both at home and internationally was moving in exactly the direction that Pearl and East, and especially Leopold, were growing alarmed about: toward an economic system based on consumption-based growth that paid little regard to environmental limits. These programs grew out of the ideas of economist John Maynard Keynes, who, by the 1930s, had flipped his earlier Malthusian assumptions. Leopold and Keynes are rarely juxtaposed, and yet in the 1930s and early 1940s, they laid the foundation for rival philosophies about consumption and economic growth that would compete with each other the rest of the century both within the United States and around the world. They form the necessary context for understanding postwar environmental Malthusianism.

Few people have shaped the economic, political—and the environmental—history of the United States and the larger world, especially since the 1930s, as much as Keynes and his intellectual descendents. Hoping to revive economies around the world and improve the lot of the working class through government action, Keynes (1883–1946) more than anyone else invented the tools of an economic order based on planned consumption and economic growth, and advertised its advantages. Along with his allies, he was the main architect of not just the American economy that emerged after World War II but also the postwar international economy. Expanded by others into advocacy of a permanent consumption-driven growth machine, Keynes's insights would remake the policies of nations around the world, and also their forests, soils, rivers, and atmospheres. Many environmental Malthusians of the next half century—including

FIGURE 3 This image from the March 1946 issue of the Congress of Industrial Organization's *Economic Outlook* depicts a Keynesian economy. While most Americans and organizations like the CIO focused on the good wages and high purchasing power that greater consumption would bring, some conservationists began to focus on resources required and pollution emitted.

Courtesy of The George Meany Memorial Archives/SN70.

William Vogt, Fairfield Osborn, and Paul Ehrlich—reacted as much against this new obsession with growth as from concern about population per se.

Keynes's thinking, which he outlined in his best-remembered work, *The General Theory of Employment, Interest, and Money* (1936), emerged from his diagnosis of the Great Depression, which had stumped traditional economists and policymakers. By the early 1930s, unemployment rates had reached 25 percent for several consecutive years. To many observers, social turmoil threatened to turn to upheaval and even revolution. The tools that economists of the period relied upon—the microeconomic analysis of how supply and demand played out—enabled them to suggest nothing better than to wait for the business cycle to finish its rotation. Keynes thought otherwise. Looking at the economy as a whole, he noticed that the supply of goods far exceeded the aggregate demand. While others saw this as a problem of *overproduction*, he instead emphasized *underconsumption*. The problem was not overstocked warehouses but shopperless stores. Eliminating overproduction, he acknowledged, could in fact realign supply with demand, stabilize prices, and end the rash of bankruptcies plaguing western economies. But doing so would do nothing for the millions of jobless Americans who relied on high production for employment. If, on the other hand, aggregate level of demand could be increased to meet the level of production, prices would stabilize and businesses would hire millions of workers to replace depleted inventories. Increased consumption could end the economy's freefall *and* create jobs.[49]

But how to increase consumption? Keynes's answer transformed the nation's economy and the world's environment for decades to come: government-sponsored consumption. Through his study of savings and the flow of money through the web of relations that constitute an economic system, Keynes realized that, in a stagnant economy, each new purchase had a "multiplier effect" on the economy—that is, that in an interconnected economic system, because each new purchase led to a series of additional purchases, each original purchase ultimately yielded an economic impact several times the original purchase amount. Thus, Keynes predicted that governments could jumpstart stagnant economies through carefully planned deficit spending. Because of the multiplier effect, each new governmental purchase could influence the economy far more than it might seem. Here was born the "growth fetish"—the obsession with government-fueled economic growth, which, according to environmental historian J. R. McNeill, has influenced the environment more than any other twentieth-century idea besides national security, to which it would soon become attached.[50]

Ironically, given the opposition that Keynes's growth economy would eventually receive from conservationists worried about population growth, Keynes owed many of his key insights to Thomas Malthus. Malthus, too, had studied economic "gluts" similar to the Great Depression, where supply seemed to vastly outpace demand, and, like Keynes, favored holistic approaches to understanding them. He even developed a notion akin to Keynes's idea of aggregate demand and recommended increased government spending in order to escape imbalances. Between October 1932 and January 1933, just before he started work on *The General Theory*, Keynes rewrote an essay on Malthus.[51]

Indeed, aggregate population analysis formed a key part of Keynes's diagnosis of the Depression. In a 1937 article, "Some Economic Consequences of a

Declining Population," Keynes argued that, by decreasing aggregate demand, the precipitous decline in birth rates during the 1930s had contributed to the Depression. "When [the] devil . . . of Population is chained up, we are free of one menace," Keynes wrote, "but we are more exposed to the . . . devil . . . of Unemployed Resources than we were before." Keynes eventually argued for pro-population-growth policies as a way to increase consumption and promote full employment and economic prosperity.[52]

A "gross national product war," World War II convinced policymakers of the efficacy of Keynes's prescriptions, and his ideas dominated early postwar economic thinking. As the end of the war drew near, planners worried about how to replace the government's wartime spending began to call for a dramatic expansion of consumption. "The volume of demand and production . . . will have to increase steadily," Harvard economist Alvin Hansen and Gerhard Colm of the Bureau of the Budget's Fiscal Division wrote. In *Mobilizing for Abundance* (1944), liberal economist Robert R. Nathan echoed this sentiment, "Only if we have large demands can we expect large production. Therefore . . . *ever-increasing consumption* on the part of our people [is] . . . one of the prime requisites for prosperity. *Mass consumption* is essential to the success of a system of mass production." Many Roosevelt advisors had bought into the Keynesian approach by the late 1930s, and in 1943, President Roosevelt himself spoke of the virtues of "an expanded economy." No doubt Roosevelt noticed how well Keynesian economics played politically, since it appealed to both big business and big labor.[53]

After the war, the U.S. government promoted Keynesian consumerism through several mechanisms: the G.I. Bill of Rights, which subsidized veterans' house purchasing; the Federal Housing Authority, which underwrote a massive expansion of homebuilding; and the Employment Act of 1946, which was designed "to promote maximum employment, production, and purchasing power." During the late 1940s, some powerful economic planners even went beyond Keynes's ideas of consumption-driven full employment to champion what historian Robert Collins has called "growthmanship." Whereas Keynes had pushed for increasing demand to help the economy reach full productive capacity and full employment, President Harry S. Truman's newly formed Council of Economic Advisors (CEA) hoped to use Keynesian tools to significantly *expand* productive capacity. By the end of 1947, the CEA had begun to call for not just full employment but "maximum production." In time, according to Collins, this "self-conscious emphasis on economic growth" became "the centerpiece of the postwar political economy."[54]

It's hard to exaggerate the environmental impact of growthmanship. Although other factors were involved, much of the material history of the 1950s—and indeed, the entire postwar period—owed no small debt to a consumption-based growth economy that saw few material limits. Government-sponsored highway building and home buying, for instance, ate up prodigious amounts of farmland for new subdivisions, forests for balloon-frame houses, steel for appliances and cars, and oil for heating and transportation. Moreover, these economic policies did not fade after a year or two, but persisted over a half century. It was not for nothing that McNeill emphasized that next to national security, no idea reconfigured the environment more than the growth fetish.

Keynesian-based calls for planned consumption also played a crucial, yet often overlooked, role in the new international order at the end of the war. Remembering the Versailles Treaty and the Depression, and especially Keynes's prescient warnings in *The Economic Consequences of the Peace* (1920), many politicians and diplomats saw conscious development of an interconnected and growing international economy as a prerequisite for global prosperity and peace. Presenting this vision himself at the Bretton Woods conference in New Hampshire in 1944, Keynes articulated a vision of a rationally planned economy that connected societies instead of separating them. With the right tools, he believed, world leaders could do on a global level what the U.S. government was doing nationally: through calculated macroeconomic interventions, they could generate a ripple effect of economic activity around the world. The resulting growth would prevent the vast inequities between "haves" and "have-nots" that often fueled instability and war.[55]

Keynes found a great deal of support for his international program from Franklin Roosevelt, among others. In 1941, calling for "a wider and constantly rising standard of living," Roosevelt had declared "Freedom from Want" as a top war aim. After Bretton Woods, Roosevelt supported the creation of two organizations to oversee the world's economy: the International Monetary Fund (IMF) to coordinate currencies and market integration, and the World Bank to provide low-interest loans and grants around the world. Two American programs from the late 1940s—the massive Marshall Plan, initiated in 1948, and the Point Four foreign aid programs for the underdeveloped world, launched in 1949—also became foundations of international Keynesianism. "For roughly a quarter of a century after the Second World war," historian Robert Skidelsky notes, "Keynesian economics ruled triumphantly."[56]

By the late 1940s, most American policy makers believed that a mass consumption-oriented economy provided the best route for forestalling a return to economic depression and global war. Domestically, the United States was well on its way to becoming what historian Lizabeth Cohen has called a "consumers' republic." By this, Cohen meant not just a society of consumers but a society in which consumption had become something of a civic religion, or where, as she puts it, "the consumer satisfying personal material wants actually served the national interest, since economic recovery ... depended on a dynamic mass consumption economy." It might also be claimed that because of the Keynesian influence on American foreign policy, the United States was well on its way to developing a "consumer world system" as well. By this time, Americans not only consumed more than ever before but they also tried to convince others around the world to consume more than ever before. They also did so in a remarkably self-conscious fashion. In the postwar years, more than ever before, Americans talked and thought about the societal benefits of mass consumption.[57]

Consumption and Interconnection

Curiously, although Keynes and Leopold reached opposite views about consumption and limits, their worldviews emerged out of the same conditions and deployed many of the same analytical tools. Both took aim at the tendency to study things in isolation and emphasized how consumption created systems. Within economics, Keynes went beyond the traditional focus on individual transactions to focus on the "multiplier effect" and aggregate demand. In doing so, he helped invent the field of "macro" economics. Similarly, Leopold focused on aggregate analysis and the cycles of production and consumption within interconnected systems, only in his case to understand nature, not economic life. Previous conceptualizations of nature, which had focused on individual plants and animals in isolation from each other, were too descriptive. "The early attempts to apply biology," Leopold wrote in 1933, "soon disclosed the fact that science had accumulated more knowledge of how to distinguish one species from another than of the habits, requirements, and inter-relationships of living populations." Instead, in books like *Game Management* (1933), Leopold began stressing what he called "the fundamental behavior of all *aggregations* of living things."[58]

Ultimately, however, Keynes and Leopold understood interconnected systems differently. One stressed limits, the other did not. One saw consumption solving many of modernity's problems, the other saw it creating yet more problems. Like East and Pearl, Leopold worried about population growth and increased consumption undermining carrying capacity. All this came at a time when, because of Pearl Harbor and many new technologies, the United States was itself becoming less isolated from the rest of the world. In the age of interconnection that followed, the differences between Keynes and Leopold took on global importance.

For the most part, the two schools of thought that Keynes and Leopold developed did not clash directly in the 1930s and early 1940s. Occasionally, however, certain moments foreshadowed the disputes to come, to be carried out by their disciples. In 1944, writing in *Audubon* magazine, Leopold sketched what he saw as the chief conservation issue of the postwar years: the global spread of American-style industrialization. "The impending industrialization of the world, now foreseen by everyone," he wrote, "means that many conservation problems heretofore local will shortly become global." He warned of the consequences for rivers, forests, and air: "No one has yet asked whether the industrial communities which we intend to plant in the new and naked lands are more valuable, or less valuable, than the indigenous fauna and flora which they, to a large extent, displace and disrupt." In the foreword to *A Sand County Almanac* (1949), Leopold attacked blind faith in a "higher standard of living": "our bigger-and-better society is now like a hypochondriac, so obsessed with its own economic health as to have lost the capacity to remain healthy. The whole world is so greedy for more bathtubs that it has lost the stability necessary to build them."[59]

Later decades would see more clashes between the viewpoints pioneered by Keynes and Leopold. Critics such as William Vogt and Fairfield Osborn would combine the insights about overconsumption and environmental degradation coming from Pearl, East, and Leopold into a distinct kind of environmental Malthusianism. The first of these clashes would come in the late 1940s, at a difficult moment of transition when the United States was beginning to reconstruct the domestic and global economy along consumption-based models of industrial growth.

2

War and Nature

Fairfield Osborn, William Vogt, and the Birth of Global Ecology

"Greater production is the key to prosperity and peace."
—President Harry Truman, Inaugural Address, 1949

"[I] wonder if we shouldn't properly call production destruction and try to do something about it before we all starve to death."
—Angus McDonald, *New Republic*, 1948

The Furnace of War

Malthusian worries about overpopulation-driven scarcities exploded in the United States after World War II, even more strongly than after World War I. "The ghost of a gloomy British clergyman, Thomas Robert Malthus, was on the rampage last week," *Time* magazine announced in November 1948. "Cresting a wave of postwar pessimism, it flashed through the air on the radio [and] rode through the mails." Publishers, *Time* went on, "opened their arms and presses to 'Neo-Malthusian' manuscripts prophesying worldwide overpopulation and hunger." *Time* could also have pointed to the hundreds of newspaper articles on Malthus and overpopulation, the references to conservation in the 1948 presidential election, or the theme that editorial writers adopted for their annual conference in 1948: the "national problem of natural resources." In 1948, the *Economist* noted, American Malthusianism had taken on the "virulence and high excitement of a fever."[1]

No one exemplifies this Malthusian rampage better than Fairfield Osborn and William Vogt, two conservationists who, drawing from Aldo Leopold and Raymond Pearl, among others, published separate bestsellers in 1948 warning about overpopulation and resource scarcity. "The tide of the earth's population is rising," Osborn, the quirky director of the Bronx Zoo and New York Zoological Society, argued in *Our Plundered Planet,* and "the reservoir of the earth's living resources is falling." "Where human populations are so large that available land cannot decently feed, clothe, and shelter them," Vogt, an outspoken ornithologist, wrote in *Road to Survival*, "man's destructive methods of exploitation mushroom like the atomic cloud over Hiroshima." Appearing within months of each other, these two books created a stir. "Glowingly reviewed and selling like hot cakes," *Time* remarked, "Their influence has already reached around the world."[2]

Strikingly, although Osborn and Vogt's books focused on a traditional conservation concern—resources—they exhibited many features that would define the environmental movement of two decades later: a distrust of progress and human technology, a sense of apocalyptic urgency, and a focus on overconsumption, sustainability, and limits to growth. Indeed, at the heart of their thinking lay a set of ideas that many historians believe separated postwar "environmentalism" from prewar "conservation": ecology. Because of *Our Plundered Planet* and *Road to Survival*, one American editor noted in 1948, "citizens everywhere have been consulting dictionaries to learn the meaning of such words as ecology." Ecology is the study of how organisms—including human beings—relate to and depend on the organisms and physical materials around them. "Man," Osborn wrote, "has been, is now and will continue to be a part of nature's general scheme." Too many Americans, Vogt stressed, think "in compartments," and therefore miss the "dynamic, ever-changing relationships between the actions of man and his total environment." Ecology helped Osborn and Vogt highlight the imbalance of people with resources that Malthus had outlined, yet emphasize a part of Malthus's thought that others overlooked: an understanding of environmental degradation. Man is not just running short of resources, Osborn noted, he "is destroying his own life sources." Vogt and Osborn articulated these proto-environmental views—and found a broad audience for them—fifteen years before the 1962 publication of Rachel Carson's *Silent Spring*, the book historians normally credit with launching the "age of ecology."[3]

At the time, however, many of the problems most associated with the environmental movement of the 1960s and 1970s—pesticides, radioactive fallout, suburban sprawl, roadside litter, and polluted streams—were almost unknown. Instead, Osborn and Vogt's ecological focus on human society must be seen in the larger political and social context of the Great Depression and especially World War II. Concern about war was crucial, as it had been for East, Pearl, and Leopold. Extending the lessons of wildlife ecology developed in the previous two decades by scientists such as Leopold and Charles Elton to the human world of poverty and war, Osborn and Vogt blamed World War II on overpopulation and overexploitation of nature, and offered a strong warning about the new world order emphasizing consumption and economic growth that was emerging under America's growing dominance. They stressed that many of the new international policies proposed between 1946 and 1950—especially new Cold War foreign assistance programs such as aid to Greece and Turkey, the Marshall

Plan, and Point Four programs—would devastate ecosystems and undermine national security. Osborn and Vogt pleaded with policymakers to think through the environmental consequences of exporting industrialization and consumption-based economic growth models worldwide. Ignoring sustainability issues, they warned, would bring more war. "When will the truth come out into the light in international affairs?" Osborn asked on the last page of his book, "When will it be openly recognized that one of the principal causes of the aggressive attitudes of individual nations and of much of the present discord among groups of nations is traceable to diminishing productive lands and to increasing population pressures?" Vogt was even more blunt: "If we continue to ignore [ecological] relationships, there is little probability that mankind can long escape the searing downpour of war's death from the skies."[4]

Vogt and Osborn's concern about resource scarcity and overpopulation invites investigation of two areas not usually brought together: environmental history and U.S. foreign relations history. The big stories of American foreign relations in the 1940s—World War II and the Cold War—are usually told without an environmental dimension. But nature and foreign relations came together in many important and surprising ways during the decade of Pearl Harbor, D-Day, Hiroshima, and the Berlin airlift. They joined directly, as when environmental conditions influenced actual battles and fighting devastated ecosystems, and indirectly, through an American military-industrial complex dependent upon ecosystems around the world. "Industrialization between the two world wars," the historian Richard Tucker has written, "enabled militarized states from 1939 to 1945 to mobilize far greater resources from around the world than a quarter-century before and to inflict new levels of destruction." Osborn and Vogt brought new attention to this little-recognized dimension of the war. "Obviously the areas where war is actually being fought are violently injured," Osborn wrote. "Yet the injury is not local but leaves its mark even in continents far removed from the conflict because of the compelling demand that war creates for forest and agricultural products. These are in truth poured into the furnace of war." Total war not only remade human society, it also rearranged nature. Understanding the Malthusian views of Fairfield Osborn and William Vogt requires understanding how they saw war and foreign affairs through the lens of nature.[5]

Thinking about Wildlife: The Zoo-Keeper and the Ornithologist

Osborn and Vogt developed their ecological critique of World War II and the postwar order not just because they were conservationists but also because they had a background in the new ideas of consumption and carrying capacity developed by wildlife ecologists during the 1930s.

Osborn (1887–1969) gained his experience running a zoo, inheriting the reins of the New York Zoological Society (NYZS) during the late 1930s and holding them until the mid-1960s. The Bronx Zoo, however, was not just any zoo. It was one of the world's most important centers for zoological research and wildlife conservation. Not only had it pioneered the display of animals within urban areas in the United States, but since the turn of the century it had played a leading role in wildlife conservation. Its early leaders—William Hornaday, Madison Grant, and Osborn's father, Henry Fairfield Osborn—helped save the American bison. By the end of the 1930s, however, these giant personalities were all gone, and the institution found itself looking for new directions. Fairfield Osborn was in many ways the perfect fit: he shared the organization's elite social and organizational roots, but was also a leader with novel ideas.[6]

Osborn's father, Henry Fairfield Osborn, was the scion of a wealthy New York family. By 1908, he had become the nation's foremost paleontologist and the president of the American Museum of Natural History (AMNH) in New York. He developed the museum into one of the top scientific organizations of the day, yet also excelled at presenting information to the larger public, a skill his son would share. He was reputedly the man who made "dinosaur" a household word. From his post at the AMNH, Osborn also promoted eugenics. As president of the American Eugenics Society, he was a great friend and colleague of Madison Grant, a fellow founder of the New York Zoological Society and author of *The Passing of the Great Race*, a 1916 book that Hitler often lauded. In the late 1930s, several years after his father's death, this family history may have helped Fairfield Osborn land the job of president of the Zoological Society. In later years, however, it would push some critics to link his population ideas with the scientific racism of eugenics.[7]

The younger Osborn honed his zoo-keeping skills at an early age. As a child, he kept a small menagerie of animals, including a flying squirrel and an alligator, at the family's brownstone on the corner of Madison Avenue and Seventieth Street in Manhattan. As a schoolboy he often accompanied his father on research expeditions. And while at Princeton, he kept a horse for recreational riding. After college, however, Osborn did not pursue a career as a scientist. After a stint as captain of a field artillery unit in France during World War I, he spent most of the next two decades working for an investment banking firm with global interests. These early experiences were important: Osborn's memories of war and his knowledge of international commodity trails would later undergird his Malthusian warnings about international resource scarcity. Later at the zoo, Osborn rubbed shoulders with well-known scientists and researchers, including the naturalist and explorer William Beebe, who tutored Osborn in ecology.

Two exhibits at the Bronx Zoo from the early 1940s showed how Osborn absorbed—and advanced—the lessons from 1930s wildlife ecology about the dependence of animals on their surroundings. Doing away with barred cages, the African Plains exhibit, which opened in 1941, was a Bronx version of an African savannah, complete with zebras, springboks, mountain reedbucks, lechwe waterbucks, impalas, nyalas, hartebeests, warthogs, blue cranes, crowned cranes, demoiselle cranes, guinea fowls, marabou storks, ostriches, secretary birds—and lions, which were separated from potential prey by a system of moats. The exhibit was novel because, instead of displaying animals by *type*—sheep with sheep, goats with goats—it arranged them by *habitat*, thereby emphasizing the connections animals had to other animals and their proper physical surroundings. The exhibit, which Osborn referred to as an "environmental" approach, was among the first to highlight ecological relationships so directly.[8]

A second exhibit shows Osborn's pioneering belief that wildlife ecology had crucial lessons for human society. Having long noticed the detachment of urban residents from the resources they depended on—for too many city residents, Osborn decried, "a Jersey cow is just about as strange and rare as a Malay seladang"—he designed an exhibit to bridge the gap, "a gap which should never exist." With the "Farm in the Zoo," a full working farm complete with Jersey cows, Dorset sheep, horses, pigs, turkeys, ducks, tractors, and farmers, Osborn aimed to highlight "what the country means to us city people." He wanted to use the zoo to show not just how wild animals depended on their surroundings but how human beings depended upon nature, in this case how New Yorkers depended on their city's hinterlands. He would later extend this idea globally.[9]

William Vogt (1902–1968) owed his expertise in foreign relations to an unusual source: bird-watching. Born in Mineola, New York, Vogt grew to love birds as a young polio victim with a fondness for Ernest Thompson Seton stories. Years later, lodging a complaint about a city park, he so impressed Parks Commissioner Robert Moses that the famous road and park builder took him on as curator of the Tobay Wildlife sanctuary in Jones Beach State Park. From 1935 to 1939, Vogt worked as a field naturalist and lecturer for the National Association of Audubon Societies and edited *Bird Lore*, its national newsletter. Through his birding, Vogt came into contact with leaders in ecology and conservation, such as Robert Cushman Murphy, Paul Sears, and Aldo Leopold—and, at the suggestion of Ernst Mayr, started a novel system of monitoring bird populations.[10] This eventually led to experience managing the bird populations of the Guano Islands in Peru, and to responsibility for broader conservation programs in Latin America.

Vogt, too, came to stress how animals depended upon their natural surroundings. One incident during the late 1930s—a threat posed to birds through a federal marsh drainage program—hammered home to him the importance of habitat. The drainage programs, which Vogt campaigned to change, were not only ineffective at malaria control, their intended mission, but were also decimating bird populations. The threat to birds came, however, not from direct killing, but from the destruction of their habitat and food supply. Vogt faulted the engineers who designed the program for lacking "knowledge of the biological problems and values involved."[11]

The lessons Osborn and Vogt learned during the 1930s about how animals depended on their resources gave them a new way to think about human society—a new ecological vision that shaped their understanding of World War II and altered their goals for conservation.

A "Carnival of Misery": Osborn, Vogt, and World War II

Osborn, a veteran of the Great War, often wrote in messages to the supporters of the Bronx Zoo of the "world catastrophe," the "critical days this country is passing through," and the "carnival of misery" the world was enduring. Even before the advent of nuclear weapons, the war fostered within Osborn two values often associated with the 1960s environmental movement—an antimodern distrust of progress, and a sense of alarm about human survival.[12]

Whereas prewar museums, zoos, and films had contrasted the violence of nature—nature "red in tooth and claw"—with the basic benevolence of human beings, during the war Osborn began to see humans as essentially brutal and nature as essentially peaceful. During "an hour of recreation, snatched from these troubled days," he noted at the opening ceremony of the African Plains exhibit in 1941, visitors to the zoo would be "refreshed" for a while from "the spectacle of Man's cruel and needless destruction of himself." "We should have no patience," he elaborated, "with those unthinking persons who rant that Man, in his present cruelties, is reverting to primitive nature—to the so-called law of the jungle. No greater falsehood could be spoken. Nature knows no such horrors."[13]

Over the next few years, Osborn often reiterated these views. "Nature knows nothing comparable to war's destruction," he wrote in 1942. "Her ways are more balanced. They provide no injustices so sudden or so bitter. Combat among other living things is, with rare exceptions, neither so general nor so ruthless as that in which man engages." In *Our Plundered Planet*, Osborn devoted an entire chapter to man's evolution as a violent predator and destroyer. "The uncomfortable truth," he summarized, "is that man during innumerable past ages has been a predator—a hunter, a meat eater and a killer." Indeed, he began to conceive of conservation itself in terms of violence and peacefulness. "Conservation," Osborn wrote in 1944, "may be thought of as a symbol—*a symbol of kindliness* which, denied for the moment as between man and man, can be extended to other living things, mute and without anger."[14]

The war also made Osborn believe that humanity threatened its very survival by destroying its resource base. Discovering that American soldiers on small atolls in the Pacific were using birds and other wildlife for target practice, he started compiling a natural history handbook for servicemen. Through this project, eventually called *The Pacific World*, Osborn discovered widespread devastation—and not just near battlefields. Extending the logic of the Farm-in-the-Zoo exhibit globally, this former businessman aware of the flow of resources around the planet realized that society was destroying the resources it depended upon. In 1944, he started speaking of another worldwide war: humanity's "destruction of the 'living resources' of nature." Not long afterward, Osborn ventured that humanity's "own existence" was at risk, and that the damage to resources was "infinitely greater" than generally believed. He started writing *Our Plundered Planet* at this time.[15]

For Osborn, the war changed the very meaning of conservation. Humans, he stressed, must protect resources for the preservation "not only of wild animal life but of man himself." In earlier decades, conservation groups such as the Bronx Zoo had aimed "solely at the protection of wildlife itself." But recent events had shown that the protection of wildlife was "inter-related with the protection of forests, water sources and other natural elements." Consequently, conservation organizations had to take "a broader aspect, with purposes more far-reaching than those previously envisioned." Wildlife protection, he stressed in 1948, must include protecting "the other related and interdependent resources of nature . . . the living environment as a whole." Moreover, Osborn stressed that those who love nature must reject the conservationist faith that humans could lend order to nature. They must never forget that the main problem was "man . . . the destroyer."[16]

During the war, William Vogt also gave a lot of thought to resource overuse. In 1939, after spending most of his life in the New York area, Vogt was hired by the government of Peru to study the bird populations on the Guano Islands, rocky outcroppings off the coast of Peru famous for fertilizer. His job was to investigate ways to increase the birds' manure production—or as he often put it, to "augment the increment of the excrement."[17]

Vogt's South American laboratory surely ranks as among the most fascinating ecosystems in the world. A collection of thirty-nine islands, the Guano Islands were home to some 11 million birds, mostly white-breasted cormorants (known as the *guanay* in Peru) and Peruvian brown pelicans. These birds lived in fantastically dense flocks. On average, each square yard on the islands held three separate nests. When leaving the island each morning to look for fish, the birds flew so close together that, Vogt once claimed, "if they could be frozen in flight it would be possible for a boy to jump from back to back." This flight might last for an hour or more. Not even the bison on the Great Plains, Vogt pointed out, existed in such numbers. Because the reproductive habits of these birds determined their numbers, Vogt once described his work as a study "of the love life . . . of 11,000,000 guano birds."[18]

Eleven million birds produce a lot of guano. A century before, prior to the heaviest harvesting, some of the islands lay under guano deposits nearly 200 feet thick. Vogt once said, "I'm doubtless one of the few men who ever spent three years on a manure pile in the interest of science." In good years, the birds produced 350,000 tons. Harvesting the guano was inexpensive and brought in lots of money (by the middle decades of the twentieth century guano sold for $100 a ton). The Peruvian cormorants were aptly known as the "billion-dollar birds." After massive harvesting during the nineteenth century, however, the guano mountains had been reduced to profitless valleys. Boom turned to bust. Harvests dropped from twenty-three million tons to thirty thousand tons. In response, the Peruvian government in 1909 formed the Peruvian Guano Administration Company, which protected the birds and managed the harvests. It proved so successful that, by the time Vogt arrived in the late 1930s, the business produced large and steady profits. This history reaffirmed a key tenet of Vogt's New Deal progressive faith: state planning could prevent a capitalist system from destroying the resources on which it depended.[19]

Not long after arriving, Vogt witnessed one of the most traumatic and mysterious aspects of the Guano Island ecosystem: a "famine" in which millions of birds perished. Such massive die-offs periodically decimated the Guano flocks, yet perplexed the Peruvian administrators, who disliked the reduced profits. To decipher this mystery, Vogt turned to tools of animal ecology put forth by Elton and Leopold. Because of their similar interests, Vogt and Leopold developed a warm friendship. The two engaged in a lively correspondence from Vogt's time in Peru through the late 1940s, when Vogt was preparing *Road to Survival* and Leopold was working on *A Sand County Almanac*. Vogt even visited Leopold in Wisconsin in the mid 1940s and considered earning a doctorate with him.[20]

Wildlife ecology taught Vogt that he had to see the Guano flocks as part of a much larger cycling of nutrients, which included other species and the abiotic environment. Vogt determined that the die-offs were due to a change in the Humboldt Current, the massive ocean current that flowed by the islands. In what would later be called an El Niño event, a shift in the current depleted local plankton and anchovy populations, which in turn starved the birds. Vogt concluded that little could be done: because the bird population had outstripped its resource base, a certain amount of death was inevitable.[21]

Studying the boom and busts of bird populations just as war was breaking out in Europe, Vogt realized that animal ecology held crucial lessons for human society, as well. In a 1940 grant application, he proposed writing a book on the ecological laws that govern "the living together of plants and animals on the earth" because, "in a day when nations war for 'lebensraum,' such understanding is more than ever important."[22]

FIGURE 4 Along with Fairfield Osborn and William Vogt, the political cartoonist and conservationist Jay N. "Ding" Darling also noted the toll that the war exacted on natural resources. In this 1939 cartoon, "Speaking of Labor Day," he suggested that nature would be carrying the burden for the next hundred years.

Courtesy of the Jay N. 'Ding' Darling Wildlife Society.

Vogt's Latin American expertise soon brought him into American diplomatic circles. In 1942, he became an advisor for the U.S. War Department, and shortly thereafter, for Nelson Rockefeller's Office of Inter-American Affairs. In 1943, he became the chief of the conservation section of the Pan American Union, the forerunner of the Organization of American States. In this post, which he retained until 1949, Vogt researched population and resource balances among both animals and humans in Latin America and sounded the alarm about resource depletion in the region. Latin America, he pointed out in the *New York World-Telegram* in 1946, was "far from being the rich storehouse of untapped resources that is generally supposed."[23]

Vogt and Osborn were not the only conservationists who lamented the war's impact on nature. Jay N. "Ding" Darling, a political cartoonist for the *Des Moines Register* and *New York Herald Tribune* (and friend of Vogt), also noted with alarm how American overseas obligations, especially wars, depleted American landscapes. During World War I, Darling created cartoons showing American crops enlisting in the army and American farmers in a race with European famine. A cartoon from 1936 showed two feckless kids—"civilization" and "politics"—pigging out on pantry supplies that Mother Nature had spent millions of years creating, with the result a devastated landscape and a war for scarce resources. Another from 1939 showed hundreds of people—with "natural resources" and "the next 100 years" written on their backs—struggling to pull a massive cannon at the behest of a whip-wielding giant named War. Darling noted in a short book published in the early 1940s that World War II "put upon our natural resources the greatest burden that had ever been known." He also predicted that Roosevelt's "freedom from want" war aim and postwar rebuilding efforts would become a drain on U.S. natural resources.[24]

Guy Irving Burch and Elmer Pendell also mixed conservation into their interpretation of World War II in *Population Roads to Peace or War* (1945). Citing Raymond Pearl several times, they argued that overpopulation fueled the war. They, too, emphasized soil erosion. "Over and over again in many parts of the world," they said, "men have overcrowded and abused their most precious natural resource until the top soil became exhausted." Growing population pressures, they warned, threatened another war. Vogt would later thank Burch in *Road to Survival* for being "extraordinarily helpful with advice, bibliographical suggestions, and critical discussion."[25]

As the war was ending, Osborn and Vogt grew increasingly concerned about planning for the postwar world. "The third of the Four Freedoms, 'Freedom from Want,' Dumbarton Oaks, the San Francisco Conference, the U.N.O. meetings—all of these reachings of the human mind and spirit for a better world can well prove futile unless the conservation of renewable resources becomes a cornerstone of cooperative effort, of governments and people alike," Osborn told a group of biologists. In May 1945, just after the first-ever United Nations meetings, a frustrated Vogt complained in a *Saturday Evening Post* article that world diplomats were overlooking the real causes of the war: overpopulation and resource scarcity. "Many explanations have been offered for Japanese aggression," he explained, "but I have never seen Louis Pasteur mentioned, nor American foundations, nor our own schools of public health, which have contributed so much to building up the Japanese population. . . . Can anyone deny that population pressures set off the explosion?" Even after the war, he said, population and resources were like two trains heading for collision: "Is it possible to doubt that, as in an old-time melodrama, Peace lies bound to the tracks where the two locomotives will come together?" The "best insurance countries could have against future disaster," Vogt morosely concluded, was "a high death rate."[26]

The role of nature in America's foreign relations, both as actor and victim, did not stop when the war came to an end. As long as total war—now thought of as total preparedness—required a strong military-industrial base, demands on the natural environment were tremendous. Moreover, as part of the reconstruction of the postwar order, American officials began to export a vision of economic relations that also had dramatic consequences for nature: an economic order based on government-sponsored growth through increased consumption. They did so through the Bretton Woods agreements that set up the International Monetary Fund (IMF) and the World Bank in 1944, and through the Marshall Plan and Point Four programs of 1948 and 1949. Even with international health programs, historian Matthew Connelly has written, "the ultimate goal was to change the lethargic and fatalistic sharecropper or peasant into a modern worker and consumer who could participate in global markets." Planned economic growth, especially through increased consumption, became the mantra that would last, with slight variations, for decades. "The preferred policy solution after 1950," historian John McNeill has written, "was yet faster economic growth and rising living standards: if we can all consume more than we used to, and expect to consume still more in the years to come, it is far easier to accept the anxieties of constant change and the inequalities of the moment. Indeed, we erected new politics, new ideologies, and new institutions predicated on continuous growth." As with war and preparations for war, these policies reshaped landscapes and communities around the world.[27]

Concerns about conservation sprang up in response. In June 1947, two prominent Republicans—the chairman of the Senate Foreign Relations Committee, Arthur Vandenberg, and former president Herbert Hoover—expressed concern that the Marshall Plan might mean resource scarcities within the United States. "Trying to feed the world," the *Hempstead Newsday* noted in 1948, "has intensified America's interest in conservation of natural resources." President Truman set up a committee to investigate, but ultimately disagreed. Indeed, he redoubled efforts, calling for "a strong productive effort." Testifying before Congress in January 1948, Secretary of the Interior Julius Krug reiterated this position: "An all out production drive here and in the rest of the world is needed at this time."[28]

This was exactly what Aldo Leopold had warned against in his 1944 article in *Audubon*, and what Osborn and Vogt would criticize in their 1948 books *Our Plundered Planet* and *Road to Survival*. That they were taking issue with Secretary Krug, a well-known conservationist, suggests the divisions that were emerging among conservationists.

The Nature of Modern Society: *Road to Survival* and *Our Plundered Planet*

Both *Our Plundered Planet* and *Road to Survival* devoted half their space to explaining nature's dynamics and half to a continent-by-continent survey of the world. They made similar arguments. Drawing upon their ecological understandings of the Depression and World War II, both Osborn and Vogt concluded that unless nature's patterns were respected more, the new world order based on the spread of industrialization and growth-based economies would yield not peace, but greater instability and conflict. Both books articulated a shift in conservation thinking toward interconnection and ecology, overpopulation, overconsumption, limits and sustainability—and potential apocalypse.

Like the Keynesian planners so influential after World War II, Osborn and Vogt started from the premise that the world had grown, in Osborn's words, "constantly smaller." According to Osborn, Americans lived in "a world society" where "the boundaries or barriers between localities, nations, even continental populations, are dissolving." Atomic weapons had helped usher in this consciousness, Vogt added, but other forces were more important: "Columbus, more than the atomic scientists, made this one geographic world. Woodrow Wilson saw that we all live in one world in a political sense, and [Republican presidential nominee] Wendell Willkie popularized the concept for the man in the street." In this smaller world, the happenings of one country could touch the lives of people in far-off lands. Conditions "which exist among peoples in one section of the earth," Osborn wrote, "now have a bearing on the lives of peoples of far distant nations." Interconnection created interdependence. Because of "world-wide systems of commerce," Osborn explained, all nations are now "dependent upon others in varying degrees for products, materials or goods." A smaller, interdependent world meant that the United States could not return to prewar isolation. "No longer," Osborn wrote, "is an American unaffected by the trends of living conditions of other peoples."[29]

Economists and foreign policy specialists had also stressed interconnection, interdependence, and the impossibility of American isolation. But Osborn and Vogt added an understanding of environmental limits. The "once apparently inexhaustible natural assets" of the United States were, for Osborn, "now little more than sufficient." He criticized those who saw the world as "illimitable." Osborn employed the concept of carrying capacity without using the term. Vogt, on the other hand, introduced the term *carrying capacity* in his first chapter and deployed it repeatedly through the book. He defined it as the ratio of *biotic potential* (the ability of land to produce shelter, clothing, and food) to *environmental resistance* (the limitations that any environment, including the part of it contrived and complicated by man, places on productive ability). To make these arguments, Osborn and Vogt drew from the population biologists and wildlife ecologists of the 1920s and 1930s. Osborn cited Raymond Pearl, and Vogt borrowed the idea of carrying capacity, biotic resistance, and environmental resistance from a 1928 article by Royal Chapman, a population biologist and one of Raymond Pearl's main colleagues and interlocutors.[30]

Their sense of interconnection emphasized environmental degradation. "No longer," Osborn wrote, "can 3,000,000 people in India die of starvation, as they did in 1943, without a specific and cumulative effect on an Englishman in Sussex. The spoiling of the land and the ensuing destruction by floods in the great Yellow River Valley of China sooner or later, in one manner or another, impinge on the well-being of peoples one thousand horizons away." Vogt seconded this idea: "Few of our leaders have begun to understand that we live in one world in an ecological—an environmental—sense." In a complex modern world "where no part lives unto itself," he continued, environmental degradation anywhere on the planet could touch the lives of Americans. Choosing an example with potent meaning to a nation that had endured devastating storms a decade earlier, Vogt pointed out that "dust storms in Australia have an inescapable effect on the American people."[31]

This emphasis on degradation distinguished Osborn and Vogt from other Malthusians. In *The World's Hunger* (1945), Vogt wrote, Frank Pearson and Floyd Harper relate the pressure of population to land resources with enviable clarity and succinctness, but they "neglect destruction of resources." "We must realize," Vogt wrote, "that not only does every area have a limited carrying capacity—but also that this carrying capacity is shrinking and the demand growing."[32]

To illustrate how humans were degrading the planet's carrying capacity, both Osborn and Vogt devoted large portions of key chapters to the importance of soil and soil erosion. Describing soil as the most vital part of nature's economy, Osborn called erosion "sinister." Vogt pointed out that modern agriculture had "*enormously* increased" environmental problems, to the point of destroying hundreds of millions of productive acres. Soil erosion had been the key conservation challenge of the 1920s and 1930s. As their bibliographies make clear, Vogt and Osborn built upon the interwar soil conservation writings of a great number of scholars, especially Paul Sears, Hugh Hammond Bennett, Charles Kellogg, and R. O. Whyte, and G. V. Jacks.[33]

In particular, Osborn and Vogt drew on Aldo Leopold's work about soil health and population irruptions. Osborn cited several of Leopold's works, including "A Biotic View of Land," the 1939 essay Leopold wrote as war was emerging in Europe. Soil biology, Osborn wrote, provides "a clue to another relationship, that between the soil and the health of human beings as well as of all other animals that feed upon the earth's products." For an example, Osborn pointed to the dramatic increase of rabbits in Australia after human introduction, from 24 in 1859 to roughly 30,000 just six years later. This situation "caused injury to the land resources of Australia, much of it even of a permanent nature, that is beyond computation." Better understanding of nature was crucial. "Nothing short of a thorough advance knowledge of the intricacies of wildlife ecology could have prevented the avalanche of trouble that continues even to this day."[34]

Vogt used Leopold's story about the deer irruption on the Kaibab to illustrate the checks and balances that hold a natural system together. More conversant with the ecological literature than Osborn, he had even visited the Kaibab plateau in 1943. He described how, after the government's predator control program, the population of mule deer climbed from four thousand to one hundred thousand in fourteen years. The result was devastating: "Consumption of all browse within reach was followed, in two years, by a 60 per cent reduction in the herd through starvation." Population irruptions were a phenomenon that "the human animal should ponder well."[35]

Osborn and Vogt drew a straight line from resource scarcity and degradation to war. Both believed that environmental problems had ignited the two world wars. Ecological imbalances, Osborn wrote in *Our Plundered Planet*, threatened "man's very survival." The "spawn [of such imbalances] are armed conflicts such as World Wars I and II." Vogt explained why: Industrialization in England, Japan, and Germany had formed large urban classes whose voracious consumption had outstripped their resource bases, creating not only ravaged landscapes but also expansionism. "Great Britain, Japan, and Germany were three of the most heavily industrialized nations in the world," Vogt wrote. "None of them was able to maintain a high living standard through industrialization without access to adequate areas of productive land." Vogt even likened Europe in the 1940s to the Kaibab plateau during the 1920s: "Had the deer of the Kaibab plateau been provided with guns and munitions, and a cerebral cortex to free them from the restraint of instinctive behavior and allow them to develop a master-race psychology, they might well have started a campaign of world conquest."[36]

Ironically, Osborn and Vogt were taking wildlife ecology back full circle to its origins in ideas about human society. They had drawn many of their ideas about nature from Leopold and others who, themselves, extended the work of Charles Elton. Elton, in turn, had modeled parts of his book *Animal Ecology* on Alexander Carr-Saunders's sociological study of human societies, *The Population Problem: A Study in Human Evolution* (1922), in which he had interpreted World War I as a result of overpopulation. Now Vogt and Osborn were reapplying these ideas to human society.[37]

Osborn and Vogt saw the environmental ingredients for war almost everywhere, including Mexico, the rest of Latin America, China, India, Italy, and Greece. A "violent explosive upsurge" in the nineteenth century, Osborn explained, had left most of Asia, Russia, Europe, and the Americas overpopulated. Increasing populations presented "perhaps the greatest problem facing humanity today."[38]

Foreseeing catastrophe, Osborn and Vogt implored the architects of the postwar world to consider nature much more seriously. "If we continue to disregard nature and its principles," Osborn argued, "the days of our civilization are numbered." The war on nature "threaten[s], at the end, even man's very survival." Vogt complained that too often international organizations overlooked "such indispensables as geography, climate, psychology, carrying capacities, folkways, and population developments, all of which are, themselves, dynamically interrelated." Above all, he concluded in *Road to Survival*, Americans must add ecology to their international thinking: "We must weigh our place in the society of nations and our future through the decades to come in the scale of our total environment."[39]

Both authors repeatedly stressed the importance of not just nature but also an understanding of nature as an ecologically interconnected system. Vogt spoke of the "living web," the "total environment," and the "environment-as-a-whole." He devoted an entire chapter to tracing energy flows through closed systems. In *Our Plundered Planet*'s epigraph, Osborn wrote that nature was a "co-ordinated machine" in which "each part is dependent upon another, all are related to the movement of the whole." The "interdependence of elements," he stressed elsewhere in the book, was "the basic law of nature."[40]

With these books, Osborn and Vogt reinforced and extended the reevaluation of conservation Leopold and others had begun in the 1930s. In particular, they attacked urban alienation, overconsumption, and overpopulation far more than conservationists had previously. Drawing on a core tenet of recent wildlife ecology—that cycles of production and consumption must be studied together—they singled out the consumer demands that drove high production and fueled war. "The fact that more than 55 per cent of our population live in cities and towns," Osborn noted in *Our Plundered Planet*, "results inevitably in our detachment from the land and apathy as to how our living resources are treated." In other venues, he attacked consumption more specifically. The "struggle to keep up with the Joneses," he noted in a 1948 speech, had "hasten[ed] the drain upon soil, timber and other resources." Vogt agreed, but went further. Urban residents, he wrote in *Road to Survival*, display a "waster's psychology." They tended to see little "beyond the butchershop, the can of spinach, and the milk bottle" and had "little understanding of their dependence on the soil, grasslands, forests, wildlife, and underground waters." The American landscape once rested in balance, but then came "the rise of cities with millions of inhabitants needing to be fed" and "the concept of the American standard of living." Vogt cautioned against spreading American consumerism around the planet. To anyone who "thinks in terms of the carrying capacities of the world's lands," the American wartime promise of a higher standard of living embodied in

President Roosevelt's idea of "freedom from want" was "a monstrous deception." This promise did nothing more than send peoples around the world "awhoring after strange gods."[41]

Historically, conservationists had mostly focused on the *productive* side of human interactions with nature. In 1917 Richard Ely, an economist specializing in natural resources, observed that conservation "has to do first of all with that division of economics which we call production." But whereas early twentieth-century conservationists had approached scarcity through managing forests or water supplies or other components of production, Osborn and Vogt stressed overconsumption. Whereas earlier conservationists such as Theodore Roosevelt had mostly hoped to decrease waste by making productive systems more efficient, thereby reducing scarcity and protecting nature, Osborn and Vogt hoped to reach much the same ends by reducing consumption. "I found that I could not think about conservation," Osborn noted in 1952, "without thinking about the factor of consumption." When he did, he said, he arrived naturally at "the question of the consumer and the population problem."[42]

Thinking about consumption, Osborn and Vogt warned of overpopulation, even in the United States. The "once apparently inexhaustible natural assets of this continent," Osborn explained, "are now little more than sufficient to support its own increasing population." In a section of *Road to Survival* called "Too Many Americans," Vogt made a similar point: Like rodents and insects, human beings could devour a region's bounty through their appetites alone. "In terms of what we like to think of as American standards . . . we are now probably overpopulated."[43]

Race, Eugenics, and Regulating Reproduction

Although Osborn and Vogt agreed that policymakers abused nature at their peril, they differed on numerous points. Osborn, who came from a family of capitalists and whose work at the Bronx Zoo depended on donations from wealthy patrons, ultimately believed that businessmen and conservationists could work closely together. Vogt, in contrast, lashed out repeatedly at America's "headlong, Paul Bunyan" free enterprise system. In the United States, he wrote, land is managed on the basis of economic concerns in "general disregard of the physical and biological laws." America's system of free enterprise "must bear a large share of the responsibility for devastated forests, vanishing wildlife, crippled ranges, a gullied continent, and roaring flood crests." Vogt hammered at this point in the book's concluding chapter: "The freebooting, rugged individualist, whose vigor, imagination, and courage contributed so much of good to the building of our country (along with the bad), we must now recognize, where his activities destroy resources, as the Enemy of the people." Indeed, Vogt made such a strong criticism of capitalism that he made some observers uncomfortable during the

McCarthyist days of the late 1940s and early 1950s. Vogt believed he had to leave his job at the Pan American Union in 1949 in part because of his strong views about capitalism.[44]

Our Plundered Planet and *Road to Survival* also differed slightly about the role of governmental planning. According to Osborn, "The interdependence of all the elements in the creative machinery of nature" called for "a supreme co-ordinated" effort. He imagined TVA-style comprehensive planning applied on a "world-wide" basis. Because "renewable resources are the property of all the people," he argued, "land use must be co-ordinated into an over-all plan." Vogt was more skeptical about the TVA, although overall he did not disagree with Osborn about the importance of planning. Ecological health required using renewable resources "on a sustained-yield basis." But Vogt also emphasized the need to adjust "our demand to the supply, either by accepting less per capita (lowering our living standards) or by maintaining less people." Successful nations, he pointed out, usually maintained a relatively favorable relationship "between carrying capacity and population" by "checking the increase of the latter."[45]

In other words, Vogt was much more vocal about population limitation programs, especially birth control programs. Whereas Osborn said nothing about contraception in *Our Plundered Planet*, Vogt repeatedly called for more and better contraception. A cheaper, more dependable, easier-to-use contraceptive was "indispensable," as was the enactment of more liberal contraceptive laws—what he called the "freedom of contraception." Contraception, he said, "should, of course, be voluntary." Vogt made the case for birth control by tying it to national security. "If the United States had spent two billion dollars developing . . . a [powerful] contraceptive, instead of the atom bomb, it would have contributed far more to our national security while, at the same time, it promoted a rising living standard for the entire world."[46]

Such views led Vogt into fraught terrain. He called for greater access to birth control at a time when birth control was hard to get and sometimes illegal in the United States, and even harder to find in Latin America, where Vogt had worked. But, as a number of critics have pointed out, he often seemed to be advocating a kind of eugenics on a global scale. It is worth looking at his ideas in these areas in some detail.[47]

Vogt questioned foreign aid, especially health programs. He criticized Western doctors for spreading public health measures such as DDT for malaria control without regard to the problems of population growth. He worried about the unintended consequences of the World Health Organization, which was established in 1948 to fight infectious diseases like malaria. To Vogt, modern medicines represented "a power that was perhaps as dangerous as that of the atomic bomb." He pointed to American health programs in Puerto Rico—"our cherished slum"—where United States aid had caused "an appalling increase of misery for more people every year." He worried that keeping people free from disease would only add to the problems of overpopulation and resource shortages. "Was there any kindness in keeping people from dying of malaria so that they would die more slowly of starvation?" He believed that biologists had a much broader viewpoint than doctors. "The biologist sees his role more clearly.... He hesitates as to whether he should loose this new, uncontrolled force [modern disease prevention] upon the world." Vogt saw these dynamics through the prism of wildlife ecology. Just as removing predators had caused the population irruptions of deer on the Kaibab Plateau, now human managers were removing one of the most powerful checks on human population through public health measures such as malaria control. To Vogt, it seemed almost a law of nature that a population irruption—with potentially dire consequences—would result.[48]

Vogt wanted to make foreign aid contingent on population limitation programs. Consider his views on U.S. aid to Greece: It would be stupid not to give aid to Greece, he wrote, "but we shall be even more stupid if we do not recognize that the overpopulation that has contributed so much to past European disorders is a continuing and growing threat." There was a difference between coercing governments and coercing people: "Any aid we give should be made contingent on national programs leading toward population stabilization through *voluntary* action of the people. We should insist on freedom of contraception as we insist on freedom of the press; it is just as important. And as we pour in hundreds of millions of the American taxpayers' dollars we should make certain that substantial proportions make available education and functional contraceptive material." More generally, he believed that the United States should lead "in making available to all the peoples of the world the most modern information on contraception, and the services of its health and educational experts in organizing birth-control campaigns." Wherever the Food and Agriculture Organization (FAO) finds overpopulation, its conservation and food-production schemes should include contraception.[49]

Vogt often used dramatic and condescending arguments. The FAO "should not ship food to keep alive ten million Indians and Chinese this year, so that fifty million may die five years hence." Vogt thought ill-advised public health measures had increased the population in Japan and created war with the United States. Asia was now overpopulated with Chinese—"ignorant, backward peoples"—and Indians—people who bred "with the irresponsibility of codfish." The region's appallingly primitive "Asiatic" standard of living threatened to bring down the rest of the world. American plans to increase foreign aid in the region would merely "extend to the rest of Asia the formula that made Japan an explosive mixture." In Vogt's view, overpopulation and the resulting ecological overreach made Asia into "the greatest threat to the civilization of the modern West."[50]

Vogt advocated for voluntary contraception, but also showed a fondness for a kind of incentivized eugenics. Reviving an idea proposed by H. L. Mencken before World War II, Vogt suggested that societies pay indigent individuals $50 or $100 to be sterilized, rather than support "their hordes of offspring that, by both genetic and social inheritance, would tend to perpetuate the fecklessness." Because such bonuses "would appeal primarily to the world's shiftless," it would probably have "a favorable selective influence."[51]

As disturbing as these views are, they should be distinguished from race-based calls for forced population control. Vogt's views on Latin America, for instance, were paternalistic but not racist. Unquestionably, he saw the continent as overpopulated: Puerto Rico was an American colony "poor in resources and almost without power—except the power to reproduce recklessly and irresponsibly." But he never lumped all Latin Americans into a homogeneous mass. To Vogt, Latin America was "limitlessly complex" and defied easy generalizations. While describing the very poor standards of training in biology in Latin America, he was quick to point out that it was not "for lack of intelligence or ability."[52]

Indeed, Vogt had very high regard for his field assistant in Peru, Enrique Avila, even though he was dark-skinned and part Indian. In 1940 and 1941, while he was working in Peru, Vogt tried to help Avila find advanced training in the United States. Avila "is a terrific worker, and his notebooks put mine to shame. No day is too long for him, no job too messy. . . . He's detached and critical. And he's got imagination." He "seems to have a pretty good idea of what science is." Vogt worried about racism in the United States. He approached Aldo Leopold, whose wife was Hispanic, about opportunities at the University of Wisconsin in part because he thought Madison would be an accepting place for Avila. "He's quite dark—largely Indian, I suppose—and there may be places where that would make things difficult for him," Vogt wrote Leopold. "I hope not in Madison. He's a grand youngster, and has the makings of a good scientist."[53]

Moreover, Vogt blamed environmental problems in Latin America on the legacy of colonialism and on the upper class rather than on the reckless breeding of the poor. He understood that poverty—not just ignorance or racial predispositions—lay behind most environmental problems. In *Road to Survival*, he railed against the "purely exploitative attitude" of the region's large landowners, and traced their exploitation to the legacy of colonialism. Whereas the rich held most of the land, the vast majority of people were forced onto slopes easily damaged by subsistence agriculture. "I do not want to exaggerate," he wrote. "Conservation is not going to save the world. Nor is control of populations. Economic, political, educational, and other measures also are indispensable."[54]

In *Our Plundered Planet*, Osborn displayed a seemingly much less offensive view regarding racial and social difference, bending over backward to distance himself from the parochial views of the eugenicists, including his father, to stress the underlying similarity of human beings—what he called the "unity of mankind." Again and again, he remarked on the similarities that bound one human being to another, stressing the "basic similarity, from a biological point of view, of all peoples on the face of this earth." The saying "We are all brothers under the skin," he said, had a basis in "scientific fact." These ideas anticipated the "species-wide" biological universalism that animated both the Museum of Modern Art's "Family of Man" exhibit and the famous 1950 UNESCO statement rejecting scientific racism, as well as Martin Luther King's integrationist rhetoric in the civil rights movement.[55]

Instead of highlighting genetic differences among social groups, Osborn stressed how genetics undergirded a *universal* human nature. In particular, he believed that all human beings shared a genetic disposition toward violence, which they passed on from generation to generation. He even believed that, in terms of their violent natures, little separated Germans and Americans, or even combatant nations and noncombatants. This tendency for violence lay behind humanity's "war on nature."[56]

Although more palatable to modern readers than Vogt's noxious views, Osborn's universalism also contained flaws. Emphasizing the biological unity of mankind made him lose sight of important social and economic differences, such as the history of racial and class oppression. Lulled into complacency by the rhetoric about the unity of mankind, population planners in Osborn's tradition could also easily overlook how regulating reproduction in the name of the greater good could end up being discriminatory in practice. From the perspective of the planet, each additional body may just be a lump of consuming cells, but claims of color-blind universalism rarely satisfied ethnic and other minorities who argued that, from the perspective of history and culture, we are not all the same.

Osborn's universalism and Vogt's crude elitism shaped the way that each articulated the danger posed by overpopulation. To Osborn, global interdependence implied a single common human fate: *all* human beings lived their lives within interconnected and interdependent networks of economy and ecology. "The peoples of the earth," Osborn wrote, "are bound together today by common interests and needs." Vogt emphasized the consequences such global interdependence had for Americans in particular. The "survival" in his book's title refers as much to American civilization as to humanity as a whole: "If we are to escape the crash we must abandon all thought of living unto ourselves. We must form an earth-company, and the lot of the Indiana farmer can no longer be isolated from that of the Bantu. This is true, not only in John Donne's mystical sense, in the meaning of brotherhood that makes starving babies in Hindustan the concern of Americans; but in a direct, physical sense. An eroding hillside in Mexico or Yugoslavia affects the living standard and probability of survival of the American people." While both Vogt and Osborn mixed universalism and defensive self-interestedness, Osborn stressed the former while Vogt stressed the latter.[57]

This mixture of universalism and elitism had important political considerations. It meant that Malthusians could attract people with very different political views, both elitist whites concerned about the erosion of the racial stock *and* anti-racist universalists. This ambivalence would allow the population planning movement and environmentalism to attract many types of followers, but it would also lead to divisions and controversy in later decades.

"Some of the Most Valuable Anxiety You Ever Had": The Impact of Osborn and Vogt

Our Plundered Planet and *Road to Survival* came out in March and August of 1948, a year when many Americans held deep hesitations about the country's new global role. During this "Age of Doubt," as one historian labeled the anxious transitional period of the late 1940s, the fears of breadlines and bombs had yet to give way to the high hopes of tailfins, subdivisions, and the mighty dollar. Unease continued about the Marshall Plan, and many even worried about renewed war. Vogt and Osborn's books, one reviewer wrote in 1948, were "of special interest during these days of talk of war and destruction." *Road to Survival*, a reviewer for the *New York Times* added, "will give you some of the most valuable anxiety you ever had."[58]

In this context, the books garnered a great deal of attention. Most mainstream newspapers and magazines recommended them and many endorsed their warnings. In an editorial based on *Our Plundered Planet* entitled "Coming: A Hungry 25 Years," *Life* noted, "The stark mathematics of even the rosiest predictions do not add up to enough to feed all the people in the world." The *Saturday Review of Literature* claimed that *Road to Survival* was "the most eloquent, provocative, and informative book that has been written thus far in the United States on conservation." Malthusianism, a British paper reported, had created a "furore in America."[59]

Such press coverage, of course, did not hurt sales. Osborn's book was a bestseller, excerpted in places such as the *Atlantic Monthly*. Vogt's also became a bestseller, selling 250,000 copies during its first year. A Book of the Month Club and Nonfiction Club selection, *Road to Survival* was also excerpted (quite selectively) in the *Reader's Digest* in early 1949. Both books were widely translated—*Our Plundered Planet* into thirteen languages, *Road to Survival* into eleven. Osborn's book won awards from *Look* magazine and the National Education Association, as well as a special citation from the Gutenberg organization given to books which "which most progressively influenced American thought in 1948."[60]

Osborn and Vogt became part of a dialogue about what conservation would look like in the postwar world. Foreign aid programs like the Marshall Plan were at the center of the discussion. At a public debate organized in late 1948 between Paul Hoffman, the coordinator of the Marshall Plan, and Fairfield Osborn, Hoffman argued that the program to rebuild Europe represented "the most important and positive conservation project ever." Emphasizing his "faith in human ingenuity," Hoffman argued that "total production can be increased without diminishing productivity." Offering a different view of conservation, Osborn countered that the Marshall Plan reflected too much faith in "the dazzling triumphs of materialism and industrialization." The idea of a constantly growing standard of living was an "illusion."[61]

President Truman's Point Four program to spread economic prosperity to "underdeveloped" areas sharpened these divisions. "We must embark on a bold new program," Truman announced in his 1949 inaugural address, "for making the benefits of our scientific advances and industrial progress available for the improvement and growth of underdeveloped areas." The program called for sharing America's "imponderable resources in technical knowledge" in order to develop "better use of the world's human and natural resources." It also stressed growth. "Greater production," Truman stated, "is the key to prosperity and peace."[62]

To Secretary of the Interior Krug, Truman's Point Four program was essentially a conservation program to develop resources. President Truman, Krug announced to the U.S. delegates at the U.N. conference on conservation in Lake Success, New York, in 1949, had "made our concern with resource development the fourth point in our foreign policy." Indeed, at Lake Success, Krug called for a "new era" in conservation. Until then, he pointed out, conservation had been "largely negative." Earlier conservationists such as Theodore Roosevelt had "dramatized the waste and damage which have resulted from exploitation." But they have not put forward "constructive ideas as to how we should find the wherewithal to satisfy our ever-increasing needs for the fuels, energy, materials and foods which are the basis of a sound standard of living." He wrote Secretary of State Dean Acheson shortly thereafter, "Ways must be found for basic development of land and water resources, power and transportation upon which comprehensive and balanced economic progress of an area depend." "Wealth," he emphasized, "is created only by increasing production."[63]

William Vogt saw things otherwise. "Few words in history," he wrote, "have been more foolish than President Truman's Point Four remarks." In a 1949 *Saturday Evening Post* article, Vogt warned about the "destructive exploitation" that might accompany a global development program: "If Point Four results in speeding up soil erosion, raiding forests and land fertility, destroying watersheds, forcing down water tables, filling reservoirs . . . and wiping out wildlife and other natural beauties, we shall be known not as beneficent

FIGURE 5 According to Jay N. "Ding" Darling's "The Only Kettle She's Got" from 1947, efforts to feed the world's growing population were pressing Mother Nature's capacity to produce, especially with limited soil. Elsewhere, Darling warned that Franklin Roosevelt's war aim of "freedom from want"—the basis for many postwar U.S. foreign aid programs—would be impossible without birth control programs.

Courtesy of the Jay N. 'Ding' Darling Wildlife Society.

collaborators, but as technological Vandals." Osborn also had reservations. "One of the strongest points that the President made in his recent Inaugural Address was that our nation should help other nations with . . . technical knowledge. . . . Do we really want especially 'to greatly increase the industrial activity in other nations'?"[64]

War in Korea in 1950 intensified the debate about conservation. Worried about resource scarcities, Truman convened a high-level commission, the President's Materials Policy Commission, also known as the Paley Commission. The commission's 1952 report affirmed the prospect of shortages but ultimately rejected Vogt and Osborn's concerns about limits. The nation faced a "very real and growing conservation problem," it stated, but not a "fixed inventory" of resources. Too much concern with limits, it noted, can lead to a "hairshirt concept of conservation which makes it synonymous with hoarding." Instead of Vogt and Osborn's emphasis on limits, the commission offered a type of conservation that would have looked familiar to Gifford Pinchot's National Forest Service, one based on "efficient use of resources and of manpower and materials." But the commission also added a Keynesian call for more "growth and high consumption in place of abstinence and retrenchment." "Using resources," it concluded, "is an essential part of making our economy grow." This updated view of conservation became the guiding philosophy of the government for at least two decades, if not longer.[65]

In making these points, the Paley Commission was echoing the findings of "Partners in Progress," a 1951 governmental study of the Point Four program. "Partners" also blamed population growth for political instability in the underdeveloped world, where "population is rising more rapidly than production." "We are moving into a new period of national accountancy in which," the report stressed, "a waste of resources now may have to be paid for later in lives." Yet the report directly rejected Osborn and Vogt's pessimistic idea of limits. The task of helping a billion people improve their standard of living may, to some, "seem a hopeless one," the report declared, but reason for optimism persisted. Indeed, "Partners in Progress" found hope in exactly what Vogt and other conservationists disliked: economic expansion. "Peace, free institutions, and human well-being can be assured," the report stated, "only within the frame of an expanding world economy." In a world of shortages, "a quickened and enlarged production of materials in the underdeveloped countries is of major importance."[66]

Ultimately, Truman himself was also optimistic. After a 1951 meeting with Henry Bennett, the administrator of the Point Four program, he proclaimed his faith that food production could outpace population growth, even in the developing world. "Point Four can help some 50 countries with a population of almost a billion people double their food production in five to ten years," Truman announced. "Comparable advances can be made by these countries in public health and education, as well as in other aspects of economic development." Truman was also very aware of the political consequences of birth-control programs, especially in Catholic communities. On August 19, 1952, President Truman wrote in his diary about a discussion he had with General Eisenhower about the underdeveloped world and birth control: "Ike had told me . . . that he didn't understand Point IV and that he thought *birth control* is the answer . . . I told him to make speeches on birth control in Boston, Brooklyn, Detroit, and Chicago. He did not get the point at all." Truman added, "He is not as intelligent as I thought."[67]

But if the U.S. government was not yet ready to adopt Osborn and Vogt's new form of conservation, many conservationists were. Conservation magazines like *Audubon*, the Izaak Walton League's *Outdoor America*, and the Wilderness Society's *Living Wilderness* gave Osborn's book overwhelmingly positive reviews. In other reviews, prominent conservationists Paul Sears, Harold Ickes, and Bernard DeVoto lavished praise on the book. *Our Plundered Planet*, Sears wrote, was "perhaps the most convincing account of man's material plight that has yet appeared. . . . Its importance can hardly be exaggerated." *Road to Survival*, DeVoto wrote, "is by far the best book so far in [ecology] literature. . . . It is more basic, more comprehensive, more thoroughgoing than any of its predecessors." *Road to Survival*, one professor of conservation remarked, stood as a "worthy successor to [George Perkins] Marsh's *Man and Nature*." Aldo Leopold, too, approved of the books. After seeing a draft of *Road to Survival*, he wrote to Vogt, "the only thing you have left out is whether the philosophy of industrial culture is not, in its ultimate development, irreconcilable with ecological conservation. I think it is." He liked Osborn's book, too.[68]

Vogt and Osborn also made an especially strong impression among scientists. In 1948, the American Association for the Advancement of Science (AAAS), the most influential organization of scientists in the world, invited Osborn to speak at their annual meetings in a special session called "What Hope for Man?" One headline described the tone of the conference: "Forty Thousand Frightened People." Dr. Edmund Sinnott, president of the AAAS, appeared to sum up the new concerns about the human impact on nature when he announced that "man's command over nature has grown more rapidly than his mastery of himself. Man, not nature, is the great problem today." At the time of the meeting, one of the science-minded Americans to be most influenced by Vogt and Osborn was just finishing high school and beginning his formal studies in biology: Paul Ehrlich, the future author of *The Population Bomb* (1968). Ehrlich would later credit *Our Plundered Planet* and *Road to Survival* for stimulating his thinking about population and environment issues.[69]

Thinking about war and poverty, in addition to the atomic blasts over Japan, had pushed these conservationists and scientists to embrace Osborn and Vogt's new conservation. Contemplating, like Vogt and Osborn, the new order emerging after World War II, they adopted many of the Keynesian views then gaining ascendance, but rejected others. Like the Keynesians, they believed that in order to guarantee peace and prosperity, American planners had to grasp the interconnectedness of the planet and think more holistically. But unlike the Keynesians, they believed that humans were depleting and even degrading natural resources across the globe and would quickly reach hard limits. Increased consumption was a problem, they maintained, not the solution.

3

Abundance in a Sea of Poverty

Quality and Quantity of Life

"Abundance and scarcity both have derived a part of their meaning and significance from the existence of the other."
　　　－Robert M. Collins, *Reviews in American History*, 1988

Paul Ehrlich opened *The Population Bomb* (1968), the most famous of the environmental Malthusian books of the late 1960s, with a memorable line: "The battle to feed all humanity is over." Then, drawing upon carrying capacity, the idea that Aldo Leopold had refined and William Vogt and Fairfield Osborn had globalized, he predicted that "nothing can be done to prevent a substantial increase in the world death rate." Less well remembered, but equally important was the vision of America's role in the world that Ehrlich built this view upon. "Nothing could be more misleading to our children than our present affluent society," he wrote. "As the most powerful nation in the world today, *and its largest consumer*, the United States cannot stand isolated.... We are today involved in the events leading to famine; tomorrow we may be destroyed by its consequences."[1]

With these lines, Ehrlich evoked several themes central to environmental Malthusian thinking about population growth in the two decades after Osborn and Vogt published their classic books: the gap between American prosperity and a world of poverty, the centrality of consumption to American life and the nation's foreign relations, and, because of these, the impossibility of isolation in an interconnected world. To Ehrlich, the United States was a consuming island of affluence surrounded by and connected to a sea of poverty and famine. "We, of course, cannot remain affluent and isolated," he wrote. "At the moment the United States uses well over half of all the [planet's] raw materials consumed each year." Ehrlich left no doubt that he was alarmed by the situation, but less clear was his judgment of American consumption: was he saying that the United States consumed too much, or that in order to protect America's high level of consumption, it needed to act?[2]

Unlike earlier Malthusians, who worried as much about European famine and political instability as problems in other parts of the world, the Malthusians of the 1950s and 1960s worried primarily about the poverty and politics of a region newly prominent in world affairs and American foreign policy in the 1950s and 1960s: the "third world." These were the impoverished former colonial areas that won their independence after World War II: India, Pakistan, Burma, Korea, and Indonesia in the late 1940s; Vietnam, Laos, and Cambodia in 1954; and forty more countries, mostly in Africa, in the late 1950s and 1960s. Not since the dissolution of Spain's Latin American empire in the nineteenth century (creating a group of countries also considered part of the third world because of their economic and political dependence) had the world gained so many new nations. For Americans, this wave of decolonization put a spotlight on issues such as imperialism, racial discrimination, nationalism, and especially poverty. "Roughly 500,000,000 people in the underdeveloped countries have won their national independence since the end of the last war," a Truman administration task force explained in 1951. "In all these countries the pent-up discontents of past generations are breaking through in demands for better living." American newspapers described the struggles of new nations, philanthropic organizations began to take up their cause, and new academic fields of study such as development economics emerged. "No economic subject more quickly captured the attention of so many," economist John Kenneth Galbraith noted, "as the rescue of the people of the poor countries from their poverty."[3]

As Ehrlich's opening to *The Population Bomb* suggests yet historians sometimes forget, natural resources formed an increasingly important part of U.S.–third world relations and the concern about population growth. In 1952, the Paley Commission emphasized that the United States had become a net importer of many resources, which it needed for its large and expanding consumer economy and military. Possessing a wealth of seemingly underused resources, the third world seemed a new global frontier: Southeast Asia had tin and lumber; India had manganese; the Middle East had oil; Africa had aluminum, diamonds, gold, and cobalt; Latin America had copper and oil. Many countries had productive lands and forests. The postwar years saw a global scramble for these resources, abetted by Cold War competition. "Our own ever growing needs for raw materials," geographer Carl Sauer noted in 1955, "have driven the search for metals and petroleum to the ends of the Earth." But who would control these resources, and were there enough for local use and for the international system?[4]

Anticipating Ehrlich, some experts in the 1950s predicted that a great age of resource abundance was coming to a close. In the 1952 book *The Great Frontier*, historian Walter Prescott Webb argued that the age of discovery that had opened a world frontier and prompted "a sudden, continuing, and ever-increasing flood of wealth" since the days of Columbus had now come to an end. The year before, he had published an article entitled "Ended: Four Hundred Year Boom." But

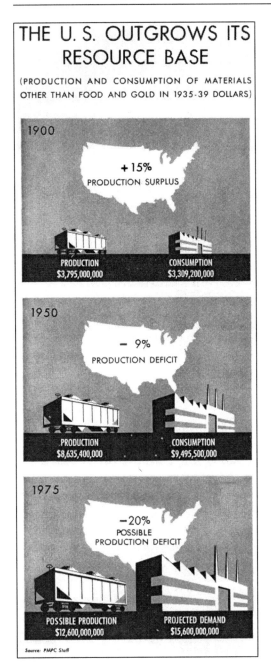

FIGURE 6 *Resources for Freedom*, the 1952 report of the President's Commission on Materials Policy, also known as the Paley Commission, warned that the United States was growing increasingly dependent upon the rest of the world for resources. From *President's Materials Policy Commission, Resources for Freedom*: vol. 1, *Foundation for Growth and Security* (Washington, D.C.: Government Printing Office, 1952).

others, such as Yale historian David Potter, professed optimism. In his unusually influential 1954 book *People of Plenty*, Potter argued that Webb, like the famous American historian Frederick Jackson Turner before him, had fundamentally misunderstood the spreading frontier, both in the United States and globally.

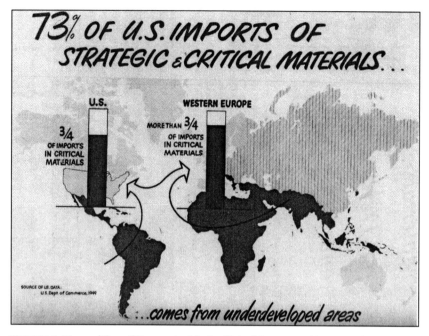

FIGURE 7 Many of the military materials the United States—and its strategic partners—needed came from underdeveloped nations.

From *Partners in Progress: A Report to President Truman by the International Development Advisory Board* (New York: Simon and Schuster, 1951), 44.

Picking up where early twentieth-century conservationists, Paul Hoffman, the Paley Commission, and Julius Krug left off, Potter argued that American abundance had not been created so much by land resources as by "an advancing technology." He wrote, "The American standard of living is resultant much less of natural resources than of the increase in capacity to produce." Other kinds of frontiers awaited, each with its own form of abundance: "Science has its frontiers, industry its frontiers, technology its frontiers." Indeed, Potter believed that America's "mission" in the world was to spread the technologies that created abundance and thereby promote democracy.[5]

During the same years these resource questions percolated, unexpectedly large population growth in the third world exceeded expectations. In *Our Plundered Planet*, Fairfield Osborn had been aghast that the globe's population might reach three billion by 2000. By the early 1960s, demographers were predicting that by that year the global population would reach five billion and possibly seven billion. (The actual population in 2000 was 6.1 billion.) Most of the increase came in the third world. During the 1950s and 1960s, in large part because of advances in personal hygiene, improved sanitation, antibiotics, and public health programs such as malaria eradication using DDT, many third-world populations skyrocketed. An American born in 1950 shared the planet with roughly 2.5 billion others. By 1990, he or she had 5.3 billion neighbors. The world grew as much during the 1960s as between 1800 and 1900.[6]

Potter was undaunted by this population growth. "No doubt, it is true that in many societies the level of living will be controlled by the scarcity of resources (e.g. by lack of soil fertility)," but ultimately, "it was the limited productivity of the worker rather than the absolute lack of resources in the environment which fixed the maximum level for the standard of living." Potter rejected the idea of limited resources: "If abundance is to be properly understood, it must not be visualized in terms of a storehouse of fixed and universally recognizable assets, reposing on shelves until humanity, by a process of removal, strips all the shelves bare."[7]

During the 1950s and 1960s, as the world shrank and its population expanded, a diverse but increasingly well-organized and well-funded population limitation movement emerged. The movement consisted of groups that have been well studied by historians—philanthropists, demographers, birth-control activists—but also some less well examined, such as naturalists, biologists, and conservationists. This movement reinvigorated old organizations such as Planned Parenthood, created new ones such as the International Federation of Planned Parenthood and the Population Council, and recast academic fields of study such as biology and conservation.[8]

These Malthusians had a complicated relationship to the debate about scarce resources and the closing of the global frontier. Compared to Potter, they placed much more emphasis on the problems posed by population growth. Indeed, what defined the movement was a belief that population growth was a much bigger problem than most policymakers acknowledged, and that better birth control needed to be developed and spread more broadly. That said, though, population groups differed among themselves over the question of resource limits. Many focused more on resource imbalances than ultimate limits, and even shared Potter's faith in spreading technology and industrialization around the world, especially if modern birth control was one of the technologies. But others, especially biologists and conservationists who drew from Edward Murray East, Raymond Pearl, Aldo Leopold, Fairfield Osborn, and William Vogt in their concern about overconsumption and their skepticism of technology, warned of environmental degradation and hard environmental "limits to growth." This fault line did not matter much as long as the movement was small and of limited influence, but it became crucial during the 1960s, especially as the environmental movement was coalescing.

In the conversation about resources and growing populations in the 1950s and 1960s, the idea of "quality of life" took on increasing importance. Suggesting that life should be about more than just subsistence, the quality-of-life idea first gained prominence in the international context in contrast to quantity-of-life concerns. Eventually, it became a central part of Lyndon Johnson's "Great Society" and a driving interest, as historians such as Samuel Hays have noted, of the environmental movement of the 1960s and 1970s. Americans pursued environmental protections as a way to improve their own quality of life. As with Ehrlich's warning about America's island of consumption in a sea of famine, however, quality-of-life concerns contained a political ambivalence: they could be framed as a self-interested defense of class and national privilege, or they could be deployed altruistically, as part of an attempt to improve all of humanity's standard of living. The larger environmental movement shared this ambivalence.

Rationalizing Reproduction

The Population Council, one of the leading population limitation organizations of the postwar decades, got its start in 1952, when—after the dismissal of population issues by the Paley Commission, as well as the Rockefeller Foundation and the World Health Organization—John D. Rockefeller III brought together a group of like-minded experts in Williamsburg, Virginia, to discuss global and domestic population growth. The grandson of the oil baron John D. Rockefeller and son of John D. Rockefeller Jr., Rockefeller (1906–1978) had come to the population issue through an interest in birth control and Asia. As a member of various governmental planning boards during World War II, he learned of the population growth expected in Asia after the war and of the worries that such growth might undermine hard-won advances in public health, for which his family's philanthropies had worked for decades. Rockefeller adopted a two-pronged attack. The first was to increase food production around the world, especially through hybrid grains, in what would became known as the "green revolution." In this effort, he worked closely with the Rockefeller Foundation, of which he was a trustee. But because of the social and political sensitivity of the second issue, population planning, he ventured out on his own. "The population field," he declared at the end of the war, "is a logical broadening of my interest in the birth control problem." During the late 1940s and 1950s, Rockefeller helped fund most of the major population organizations, including Princeton's Office for Population Research, the Conservation Foundation, the Planned Parenthood Federation of America, International Planned Parenthood, and the Population Council. During the 1950s, he gave the Population Council roughly $250,000 each year. Twenty years later, he would chair two presidential study groups on population. Rockefeller had the resources and talents to turn a personal interest into an influential social cause.[9]

Rockefeller wanted the Population Council to take a scientific, research-based approach to population issues, and to advance the relatively new academic discipline of demography. In the early 1950s, the population field barely existed. The United States had three small demographic organizations—the Office of Population Research at Princeton, which Rockefeller help fund, the Scripps Foundation, and the Milbank Memorial Fund. It also had one prominent reformed eugenics organization, the American Eugenics Society. Nationwide only seven professors researched and taught demography, almost all of whom specialized in very local interests.[10]

The attendees at the Williamsburg Conference included a number of physicians, natural scientists, social scientists, and conservationists who would become fixtures of the population movement in ensuing decades: Pascal Whelpton, Roger Revelle, Kingsley Davis, Marshall Balfour, Irene Taeuber, Fairfield Osborn, Frederick Osborn (cousin of Fairfield Osborn), Frank Notestein, and William Vogt, among others. Many had extensive experience in international public health, especially in Asia. Explaining his interest in population, Dr. R. R. Williams, a chemist, pointed to "the accident of [his] birth in India, and the accident of residence in the Philippines, as well as a number of visits to the Orient, where some of the world's worst population problems present themselves." What bound this diverse group together, according to the official convener of the meeting, Dr. Detlev Bronk, was a common concern about rapid population growth—"a problem in which most of us, at least, had to become deeply concerned because of impressions which we had gathered in Japan and in the European Theatre during and following the war."[11]

Despite its ostensibly scientific approach, many of the Williamsburg participants displayed racist and elitist views. These views cropped up in the discussion at the Williamsburg Inn, a whites-only establishment. Bronk spoke of "the potential degradation of the genetic quality of the human race." Warren Weaver worried that foreign aid to India would increase the number of poor, incompetent people, making the country "nigger rich." Frederick Osborn, the man who would oversee the daily activities of the Population Council for much of the 1950s, believed that population growth had to be curtailed because "the little groups of three or four hundred people, who produce most of the freedom of the human mind ... may be engulfed by a great mass of people to whom these conceptions are largely alien." The legacy of eugenics also informed the first draft of the Population Council's mission statement, which called for creating conditions such that "parents who are above the average in intelligence, quality of personality and affection will tend to have larger than average families." But not all those associated with the Population Council held such views. At least one trustee objected, for both political and intellectual reasons, to the language about parents of above-average intelligence. In the end, the final statement was ambiguous, calling for research "in both the quantitative and qualitative aspects of population in the United States."[12]

Indeed, a model that rejected racial explanations in favor of sociological factors came to dominate Population Council thinking. As part of the attempt to modernize their field, two demographers who became associated with the Population Council, Frank Notestein and Kingsley Davis, advanced the "demographic transition theory," a meta-narrative of population growth and decline that would dominate population planning for much of the next three decades. Echoing the resurgence of evolutionary thinking across the social sciences, the theory held that all societies—regardless of race—pass through three stages: a traditional stage of high fertility and high mortality, a transitional phase of high fertility and low mortality, and a final "advanced" phase of both low fertility and low mortality. According to the model, population levels remained stable in the first and third stages, yet expanded, sometimes threefold, during the transitional phase. By the mid-twentieth century, most Western nations had reached the stable "advanced" phase. Most underdeveloped nations, on the other hand, stood on the edge of the transitional stage of quick population growth. Such dramatic growth threatened "efficient" population replacement around the world.[13]

Although demographer Warren Thompson had articulated a very similar model in 1929, it failed to earn wide acceptance until Notestein and Davis repackaged it for postwar planners during World War II. According to demographic historian Simon Szreter, three reasons explain why. First, by the end of the war, Americans were much more willing to accept governmental economic and social planning, and those planners needed solid demographic information. Second, Keynesian concern about aggregate demand had created a new interest in the economic implications of demographic change. And finally, diplomats and strategists urgently needed to understand the social forces shaping the third world. For those interested in these areas, the demographic transition theory was, according to Szreter, "something of a philosopher's stone": it allowed planners to place a wide diversity of societies into a straightforward, ranked typology based on available data and simple concepts.[14]

On the basis of the transition model, demographers such as Notestein and Davis began to warn about demographic catastrophes. "A sober consideration of the existing situation," Notestein declared in 1948, "leads one to expect that catastrophes will in fact check rapid growth." In 1950, he predicted that high population growth could lead to "a period of disorganization" characterized by "catastrophic losses from time to time and from place to place." Kingsley Davis wrote that only a period of "strife and turmoil," which reduced the "existing demographic glut," would end the circle of poverty caused by overpopulation. During the 1950s and 1960s, the potential for such problems was never far from the minds of most demographers.[15]

As originally conceived, the demographic transition model saw little space for birth-control and population-limitation programs. Because fertility was so dependent upon the entire socioeconomic system, a single technology alone could not bring about change. But several developments helped convince policy planners otherwise: the growth of faith in technological transfer programs, the development of new birth-control technologies, and the example of Japan. During the American occupation of Japan after World War II, the Japanese were able to sharply reduce their population growth rate within just a few years through widespread abortions and condoms. Moreover, a sense emerged that the transition would simply take too long. As Frederick Osborn explained in 1958, "The process of industrialization should of itself reduce the birth rate if we are to judge by Western experience. But Asia cannot afford the time this transition took in the West."[16]

An intellectual shift in economics during the late 1950s, partially sparked by a book funded by the Population Council, encouraged this idea. Before Ansley Coale and Edgar Hoover's *Population Growth and Economic Development in Low Income Countries: A Case Study of India's Prospects* (1958), development economists and policymakers subscribed to Keynesian models holding that population growth spurred economic growth. In their study, though, Coale, a demographer at Princeton's Office of Population Research, and Hoover, a Duke University economist, concluded that population growth actually impeded economic development. Using some of the earliest simulations of future trends, Coale and Hoover predicted that average income per adult would increase about 40 percent if the rate of childbearing were to fall by 50 percent in twenty-five years. Even a small fertility drop could yield an economic windfall—an idea that development funders with an eye on cost-benefit analyses found hard to resist. They said little about hard resource limits.[17]

During the 1950s, the Population Council devoted most of its roughly one-million-dollar annual budget to conducting demographic research and training hundreds of officials from around the world. Through this global network of trainees, historian John Sharpless has noted, "the American vision of population studies became the global vision." The Population Council stressed top-down training of experts because, as Frederick Osborn put it in 1958, the intellectual authority of elites was greater "in areas of low literacy and education" than in the West, where "more of the people are able to deal with intellectual concepts on their own." This view meshed with their own elitist view of themselves as the most important movers and shakers within American and global society. But it was also good public relations, at least in the short term. By relying on scholars from around the world, rather than only Americans, the Population Council could appear less imperial, not as a Western power imposing its will but as a benevolent friend lending a helping hand. Population programs, Frederick Osborn wrote, were best conducted by insiders because the programs are so "heavily loaded with emotion."[18]

Although the Population Council did a great deal to spread concern about population growth, generally it focused on economic development and political stability, not environmental degradation. At the Williamsburg conference, Fairfield Osborn and William Vogt had voiced their concerns about resource depletion and environmental degradation. Their presence suggests some support for their positions, and indeed, a full one-third of the hefty conference workbook addressed natural resource topics, including issues such as "carrying capacity" and the "ravages of erosion." But a strong majority of the participants rejected their ideas about limiting the spread of industrial development to other parts of the world. Most believed that population limitation must be accompanied by the spread of modernization and industrialization. Reflecting these sentiments, the final resolution of the conference called for further studying "the populations of the world and the resources, cultural and material, available to give such populations a progressively improving standard of living," to consider the "relations of population to these resources and ways in which such relation can be improved in the near and more distant future," and to examine "how such balance can be secured and maintained on a short and long range basis." Kingsley Davis, a bit of an outlier in demographic circles because of his interest in ecology, complained at the time that "demographers do not know enough about resources."[19]

Indeed, many within the Population Council ultimately shared David Potter's faith that the right application of American know-how could solve third-world problems, including rapid population growth. If more people had access to modern contraception, unwanted births and poverty could be reduced or eliminated. Moreover, they held little animosity toward industrial growth, both on its own merits and also because they thought that, as the demographic transition suggested, it would slow population growth. They mostly worried about imbalances of population and resources, not aggregate numbers, overconsumption, and environmental quality.

Planned Parenthood, the other leading organization in the population movement in the 1950s, grew out of the birth-control clinic that Margaret Sanger opened in the late 1910s to provide women with greater access to birth-control information and technology. Haunted by her own mother's premature death from poverty and frequent childbirth, and outraged that the poor lacked the access to contraception that the middle class enjoyed, Sanger (1879–1966) launched a campaign in the tenements of Manhattan and Brooklyn to spread knowledge and contraception technology. Because her actions struck conservatives as immoral, she was forced into temporary exile in England. But Sanger's efforts eventually led to a string of birth-control clinics around the United States, as well as a network of supporters who pushed for more acceptance of sexuality and more liberal birth-control laws.[20]

After World War II, Planned Parenthood combined its hopes for improving women's health with other goals, especially the alleviation of overpopulation, both overseas and at home. Sanger led these Malthusian efforts. A Malthusian since her exile in Europe, Sanger grew even more concerned about overpopulation after World War II, which she blamed on overpopulation. "There is just about enough time before total chaos sets in for the world to do something about its overpopulation," she declared in 1948. "Instead of trying to feed the world without restraint, we should be devising ways of checking the growth of population." As demographic historians Dennis Hodgson and Susan Watkins have put it, Sanger's "Malthusianism spanned many decades and is difficult to doubt."[21]

In 1952, realizing the global dimension of the population problem, Sanger and international allies such as Sweden's Elise Ottesen-Jensen organized the International Planned Parenthood Federation (IPPF), an umbrella organization for raising money, sharing knowledge, and establishing clinics around the globe. Sanger first gathered supporters to organize such a group in Cheltenham, England, in 1948, but the IPPF didn't get off the ground until a 1952 meeting in India. Sanger directed the organization until 1959, when she was almost eighty. Like Planned Parenthood, IPPF was driven by a variety of concerns, including worry about overpopulation. The organization's primary goal, according to its 1953 bylaws, was "to advance through education and scientific research the universal acceptance of family planning and responsible parenthood in the interests of family welfare, community well being, and international good will." Such broad language could entail both early feminist inclinations but also, as the phrase "responsible parenthood" suggests, eugenics-style paternalism.[22]

Raising money for itself and IPPF within the United States during the 1950s and 1960s, Planned Parenthood spread a Malthusian vision of the postwar world. The organization and its affiliates organized public talks, distributed literature and films, and gave interviews to local TV, radio, and newspapers. Stories of overpopulated countries around the world laced their local, national, and international publications. "Think of it," one Planned Parenthood brochure implored in 1953, "over 85 million babies born each year on this troubled planet—over one half of them doomed to starvation, disease, miserable lives and early graves." The local newsletter of Planned Parenthood's Chicago chapter had a section during the 1950s called "Vox Population: Expressions of Opinion from Those Who Link Planned Parenthood to World Problems." "The International 'Population Explosion' is of course in all of our minds always," the newsletter explained in 1960. "Board and Staff are constantly being asked to speak to groups on the many phases of the 'population explosion' and we fill these requests gladly."[23]

More than the Population Council, Planned Parenthood and International Planned Parenthood stressed natural resource depletion and conservation. The theme of IPPF's 1953 international conference was "Population and World Resources in Relation to the Family." Although the title was possibly just public relations, the evidence suggests otherwise. In her introduction to the very first issue of its national newsletter in 1952, Planned Parenthood President Eleanor Pillsbury spoke of the "pressure of expanding population upon depleted resources." But perhaps nothing showed Planned Parenthood's ecological sensibility better than the organization's choice of national director in 1951: William Vogt.[24]

As national director from 1951 to 1961, Vogt mixed ecological messages about scarcity and degradation into Planned Parenthood publications and speeches. One brochure, for instance, stated, "With each turn of the earth around the sun, the world takes on 85,775,000 new babies while it loses soil, forests, food, and water. Half of the world's people go to bed hungry every night. This tidal drift of birth and depletion, new life and destruction—causing personal suffering, national unrest, international tension—must be stopped before it's too late."[25]

Reaching Limits on the Global Frontier: Biologists, Conservationists, and Population Growth

Conservationists, naturalists, and biologists did not leave all the population work to Planned Parenthood and the Population Council. They pushed a brand of environmental Malthusianism, emphasizing overconsumption, resource depletion, limits, the drawbacks of industrialization, and environmental degradation, often drawing from ideas of carrying capacity developed by Pearl, East, Chapman, and Leopold, and extended by Vogt and Osborn. But not all environmental Malthusians thought alike; a great diversity of opinions swirled around. Some envisioned a catastrophic crisis, others were less worried. Some called for authoritarian solutions; others, even though they saw large problems approaching, insisted that voluntary measures would suffice. In general, compared with later in the 1960s, there was less alarmism and fewer extreme remedies.

Fairfield Osborn's Conservation Foundation promoted population awareness and research, although nothing on the scale of the Population Council. The organization, which Osborn founded in March 1948, adopted a different approach than earlier conservation organizations such as the Sierra Club, the Audubon Society, and the Wilderness Society. In addition to wildlife concerns, it promoted ecologically informed research about issues such as pesticides and population growth, issues that would eventually dominate the new environmental agenda. A 1956 mission statement, for instance, included "To assess population trends and their effect upon environment." The Conservation Foundation sponsored studies by demographers Kingsley Davis, George Roberts, Judith Blake Davis, and J. Mayone Stycos. Osborn also wrote and gave talks frequently about population. As time went by, his concern about population grew. "The inordinately rapid increase of populations in this world," he wrote in *Our Crowded Planet*, a 1962 collection of essays he assembled on population, "is the most essential problem that faces everybody everywhere." The Conservation Foundation became a think tank and advocacy organization involving many environmentalists who would play key roles in the late 1960s environmental policy, such as the political scientist Lynton Caldwell, architect of the National Environmental Policy Act, and Russell Train, President Nixon's director of the Environmental Protection Agency.[26]

In *Fundamentals of Ecology* (1953), the standard ecology textbook for several decades, Eugene Odum argued that ecology was an increasingly important, holistic form of biology, and that within ecology, population was a crucial unit of analysis. In fact, Odum used human population growth to frame the entire book. Humanity had entered a dangerous new stage due to population growth, he emphasized in the book's first paragraph, and thus a much more ecologically sophisticated understanding of the "environment" was needed: "It is more necessary than ever for mankind as a whole to have an intelligent knowledge of the environment if our complex civilization is to survive, since the basic 'laws of nature' have not been repealed; only their complexion and quantitative relations have changed, as the world's human population has increased."[27]

In addition to other subjects, *Fundamentals* described population-level analysis, Royal Chapman's conception of carrying capacity, and the Kaibab deer irruption. Odum devoted his final chapter to "human ecology," the application of ecology to human society, and especially to population growth, "perhaps the most 'violent' topic in human ecology." He noted that human populations usually followed the same "S" curve that Raymond Pearl sketched thirty years earlier. True, because of its adaptability, culture, and dominance, humanity could raise the carrying capacity. But no firm predictions could be made about the future. Citing Malthus and William Vogt, Odum pointed out that rapid collapse was equally possible: "We cannot be sure that human populations will not exceed the carrying capacity, as did the Kaibab deer when adequate checks were removed." Industrial societies had been able to reduce population because of voluntary birth control, but "in China and India—perhaps a more 'simple ecosystem'?—such action seems less effective; density continually rises to or above the carrying capacity of the environment, food chains become shorter, and such progressive philosophies as democracy mean little to the hungry people!" Reiterating the impossibility of reliable predictions, he called for more study.[28]

Harrison Brown's more pessimistic book, *The Challenge of Man's Future* (1954), shows the lengths to which some scientists feared societies had to go to combat overpopulation. A Manhattan Project geochemist who wrote of the dangers of nuclear weapons (and later a leader of the Pugwash disarmament movement), Brown argued that humanity was in a "very precarious position" because of population growth, which he had seen firsthand during lengthy stays in Jamaica, and massive and growing resource consumption. He devoted most of the book to innovative and rigorous data-filled analyses of these subjects.[29]

Brown also mused about possible solutions, including world government and regulating population size with artificial inseminations and forced abortions.

He recognized the potential costs to individual liberty, but reminded readers of the status quo, where "abortions must be obtained frequently on kitchen tables, usually at great expense and under circumstances where the victims have the 'freedom' to choose between giving birth to unwanted children and endangering their lives by subjecting themselves to illegal operations under insanitary conditions." Regulating births, he added, would have a eugenic effect: it could prevent the "deterioration of the species." He admitted the possibility of "individual fluctuations," but he worried that "the feeble-minded, the morons, the dull and backward, and the lower-than-average persons in our society are outbreeding the superior ones at the present time." Indeed, the average intelligence quotient of Western populations "is probably decreasing significantly with each succeeding generation." Although "pruning" was not feasible because genetic knowledge was not precise enough, he believed that priority for births "could be given to healthy women of high intelligence whose ancestors possessed no dangerous genetic defects." Conversely, he said, priorities for abortions could be given to "less intelligent persons of biologically unsound stock."[30]

Concern about population growth in more moderate form was conspicuous at the June 1955 conference "Man's Role in Changing the Face of the Earth," which historians often credit for laying a foundation for the 1960s environmental movement because of its global, ecological focus on human beings as a dominant force. "Human population growth occupied, understandably, a large share in the symposium," observers for *Geographical Review* noted. Indeed, a sense that population growth had made the modern day unlike any seen before helped spark the idea for the conference in the first place. "With more people on the earth than at any time in the past," wrote Dr. William Thomas, the chief planner at the Wenner Gren Foundation, the conference sponsor, "this is really a most extraordinary period in which to be living." Noting humanity's capacity for "vigorous reproduction," geographer and conference chairman Carl Sauer identified the "growth of population" as one of the main factors determining human impact on nature.[31]

Many of the participants were well-known Malthusians: Marston Bates, F. Fraser Darling, Charles Galton Darwin, Alan Gregg, Lewis Mumford, Samuel H. Ordway, and Warren Thompson. Population growth came up in almost every discussion, in part because of the global focus of the meetings. Sauer believed that conservation had grown on the American frontier "where the depletion of lately virgin lands gave warning that we were drawing recklessly on a diminishing natural capital" and, following Leopold, Vogt, and Osborn, warned of the spread of industrialization into the new global frontier. "We present and recommend to the world a blueprint of what works well with us at the moment, heedless that we may be destroying wise and durable native systems of living with the land." To this, Harrison Brown added, "if industrialization spreads, and populations continue to grow, denudation will take place on a scale which is difficult for us to comprehend." But the United States also figured into the discussion, with Fairfield Osborn wondering how the country would deal with "250 to 300 million people . . . with their tremendous demands for raw materials, social amenities, and education." (The U.S. population in 2010 was over 310 million.) Paul Sears noted that "American culture is inclined to value continual expansion: some is good, more is better; more and more are both bigger and better." Osborn called for a harmless, effective, simple, and inexpensive check to human fertility. There was little or no talk of eugenics and coercion.[32]

Although most conference participants seemed concerned about population, some dissented. Fritz Heichelheim stressed that capital resources, not population, drove environmental change. Pierre Gourou favored studying local changes and local conditions, arguing that it was not "scientifically practicable to be preoccupied with world populations." Historian James Malin argued that "Malthusian doom might be postponed or even canceled out without conscious intervention of human planning."[33] At the end of a discussion called "Limits of the Earth," the economist and environmentalist Kenneth Boulding summarized his ambivalence with a poem jotted down while listening to the discussion:

A CONSERVATIONIST'S LAMENT
The world is finite, resources are scarce,
Things are bad and will be worse.
Coal is burned and gas exploded,
Forests cut and soils eroded.
Wells are dry and air's polluted,
Dust is blowing, trees uprooted.
Oil is going, ores depleted,
Drains receive what is excreted.
Land is sinking, seas are rising,
Man is far too enterprising.
Fire will rage with Man to fan it,
Soon we'll have a plundered planet.
People breed like fertile rabbits,
People have disgusting habits.

Moral:
The evolutionary plan
Went astray by evolving Man.

THE TECHNOLOGIST'S REPLY
Man's potential is quite terrific,
You can't go back to the Neolithic.
The cream is there for us to skim it,

Knowledge is power, and the sky's the limit.
Every mouth has hands to feed it,
Food is found when people need it.
All we need is found in granite,
Once we have the men to plan it.
Yeast and algae give us meat,
Soil is almost obsolete.
Men can grow to pastures greener,
Till all the earth is Pasadena.

Moral:
Man's a nuisance, Man's a crackpot,
But only Man can hit the jackpot.

Boulding would emphasize limits in later writings but not call much attention to population growth.[34]

In an essay in *Scientific American* in 1956, former UNESCO director general Julian Huxley combined near apocalyptic worry about population growth with a sense of social justice. Pointing out that the world added the equivalent of a good-sized town, more than 90,000 people, every day, he cautioned, "If nothing is done to bring down the rate of human increase during that time [the next thirty or forty years], mankind will find itself living in a world exposed to disastrous miseries and charged with frustrations more explosive than any we can now envision." Huxley also lamented the "alarming differentials in consumption between different regions and nations," which he found "intolerable" and "grotesque." He called for "a new and more rational view of the population problem" as well as "a large-scale concerted program of research" related to oral contraceptives.[35]

The biologist Raymond Dasmann distilled many environmental Malthusian ideas in his innovative 1959 textbook, *Environmental Conservation*. Even more than others, Dasmann shows the complicated interplay between the international and the domestic at this time. After seeing combat in the Pacific in World War II, Dasmann (1919–2002) studied zoology at the University of California with Carl Sauer and Starker Leopold, Aldo Leopold's son. His dissertation research examined deer irruptions in northern California. After writing *Environmental Conservation*, Dasmann spent two years as a Fulbright fellow studying range management in Rhodesia (later Zimbabwe) before publishing another book on global conservation, *The Last Horizon*, in 1963. In 1965, he published one of the earliest and most influential books on overpopulation within the United States, *The Destruction of California*.[36]

Dasmann dedicated *Environmental Conservation* to Aldo Leopold. He saw a need, he explained in the preface, for a "text written from a biological standpoint, which would take the long view of conservation problems by considering the history of human populations in relation to natural resources." Thanking Carl Sauer, whose inspiration launched the project, he adopted a global approach, drawing heavily upon "Man's Role in Changing the Face of the Earth," whose proceedings were published in 1956. Dasmann's theme was familiar to anyone familiar with *Our Plundered Planet*, *Road to Survival*, or "Man's Role": growing world populations, combined with the spread of industrial civilization, threatened human society. The "engulfing flow of western industrial civilization" was destroying both nature and human culture. Society, Dasmann wrote, was engaged in "war with nature." Population growth posed a "critical problem." It added to social and political problems and would "intensify all resource conservation problems."[37]

Leopold's influence was clear. Dasmann used the Kaibab deer irruption story to explain not only the concept of carrying capacity but also human population growth. A "spectacular increase in the deer population," he wrote, "resulted in the carrying capacity of the habitat being surpassed, the food supply being destroyed, and finally in a die-off of the deer to a new low level." In just six years, starvation had killed 80,000 deer. Human population growth, Dasmann noted, often mushroomed after medical supplies and sanitary techniques spread in agricultural societies, such as India. India's population had been stable, but after the British introduced Western technology, medicine, and transportation, the population began to grow rapidly. With industrialization "a new balance can be expected," but how large the population will grow in the meantime "is disturbing to contemplate."[38]

Although some Malthusians focused exclusively on the third world, Dasmann also worried about the United States. "Anglo-America" was an "island of prosperity" threatened by a "frontier" faith in inexhaustible resources and unlimited population growth. "In our reproductive rates we have shown an irresponsibility and disregard for the future," he lamented. "We still believe that more people means more hands to work and more customers to buy. We look forward to an ever-expanding business and industrial economy, often without critically examining the resource base that must support that economy." He warned about the "catastrophe that unbridled growth must bring." In *The Last Horizon*, like Brown, he added a concern about "a decrease in the quality" of the population.[39]

Dasmann called for "the highest level of resource conservation," and for "more effective, and more generally available and acceptable, methods of birth control." In *The Last Horizon*, he added an idea that would grow in importance in the ensuing decade, calling for new social roles for women and criticizing conservative traditions that defined them just as mothers. Far too often, education for girls stressed "social development in place of academic achievement," thus fostering "an early courtship and marriage pattern upon the young." And those "who do break loose and enter a career," he lamented, "still incur distrust and dislike from their home-bound sisters."[40]

Dasmann offers a good point of comparison. Unlike David Potter and even some in the population movement, he had a strong sense of limits and urgency, even applying Leopold's Kaibab model to human populations. Yet unlike some of the more strident Malthusians of the late 1960s, he held hope for voluntary measures. If people understand the problem and have the means, he believed, they will limit their population: "The need must be faced now, the machinery put into operation, the research performed, the education begun which will make voluntary family limitation within the reach of everyone in the near future. When this has been accomplished, there is every reason to believe that the rate of population growth will decline as it has in Europe in recent years." In *The Last Horizon*, Dasmann also pointed to Japan, where an intensive program of public education, the legalization of abortion, and widespread availability of contraceptives had brought about a dramatic fertility decline. He emphasized, "There was no coercion."[41]

Quality of Life and the "Environment"

As the population movement of the 1950s strove to change the world, it remade postwar American life in fundamental ways. For instance, driven by concerns about international overpopulation, the population movement during the 1950s and early 1960s created many of the reproductive rights and technologies that later generations would take for granted—including the birth-control pill, the intrauterine device (IUD), and sterilization procedures. All four individuals central to developing the "pill"—Margaret Sanger, philanthropist Katherine McCormick, biologist Gregory Pincus, and physician John Rock—worried deeply about global overpopulation. An oral contraceptive, Rock told a Planned Parenthood audience in 1954, "would serve to help avert Man's self-destruction by starvation and war. If discovered in time, the H-bomb will never fall." Rock also blamed communist aggression in Vietnam on overpopulation: "The frightful pressures on the last few years of French-held Dienbeinphu can truthfully be said to reflect the pressure of Chinese overpopulation on the resources of its weaker neighbor." Similarly, Malthusian concerns also motivated the Population Council doctors who developed the IUD. Concerns about population growth also appeared in the 1965 Supreme Court case *Griswold v. Connecticut*, which struck down bans on contraception for married people.[42]

The population movement also helped develop an important concept that changed American life, especially the emerging environmental movement, in a more subtle way: the idea of "quality of life." Thinking about the proper balance of food and people led those in the population movement to the question of existence at what standard: mere survival or something more? Besides food and shelter, what were the characteristics of a good life? The "quality of life" concept first gained wide currency in an international context, but Malthusians increasingly used it to understand an American population expanding because of the baby boom. Eventually discussion turned toward such things as freedom, creativity, clean air, space, and solitude, all of which were seen to suffer as populations grew. The concept became important to an "environmental" viewpoint, as well as to 1960s liberalism.

John Rockefeller III, founder and funder of the Population Council, was among the first to explicitly articulate quality-of-life ideas. He first described them at the Williamsburg meetings, and in September 1952, not long after the conference, he wrote a lengthy memo on the subject that became the guiding document for the Population Council. Rockefeller felt his approach to population was much more positive than what the terms "birth control" and "population control" connoted. To him, the issue went beyond "the relation of the food supply to the number of mouths to feed." Physical needs must be met, of course, but "more important still are the needs which differentiate man from animal." Human beings do not "live by bread alone." Consideration must also be given to "mental and emotional well-being—to education, religion, art and other forms of expression which enable man to achieve self-realization."[43]

Rockefeller was, in effect, redefining the idea of "quality" common in demographic discussions. Whereas in the eugenics movement, the term had referred to maintaining a genetic "quality of the population," the postwar population movement spoke increasingly about improving "quality of life." The opening lines of a 1965 in-house history of the Population Council makes this clear: "The world has several great problems: to maintain peace and international order, to extend political freedom, to improve standards of living, to promote health and well-being, to advance literacy and popular education, to spread art and culture. *Underlying all of these* and qualifying as a world problem in its own right is the population problem: to keep the quantity of human life from diminishing the *quality of human life.*"[44]

During the 1950s, quality of life was also a regular theme for conservationists, naturalists, and biologists worried about population growth. "The day is here when we Americans need to clarify," Fairfield Osborn wrote in 1953, "what we mean by a 'standard of living' and in so doing give greater recognition to the immeasurable values of social, and not merely economic, criteria." Osborn gave a talk called "The Quality of Life" to a Planned Parenthood group the next year. The power of their movement, he said, rests "not so much upon the physical or quantitative needs of human life as upon its social and even spiritual needs. The goal is the betterment of the 'quality' of life." Other conservationists made similar points. In *Road to Survival*, William Vogt had warned about dropping school quality, declining governmental capacity, and reduced services in museums, libraries, and colleges. Discussing overpopulation in *Fundamentals of Ecology*, Eugene Odum symbolically defined quality of life as being able to go fishing and occasionally eat steak. "We must reject the idea that the quantity of human beings," Julian Huxley wrote in 1956, "is of value apart from the quality of their lives." In 1962, naturalist Joseph Wood Krutch asked, "What is the optimum number [of people] from the standpoint of the possibility of a good life?"[45]

Harrison Brown added a political note to quality of life, emphasizing that population growth and overconsumption led to the "collectivization and robotization" of humanity. Vogt and Osborn had hinted at this connection to freedom and individuality in the late 1940s, but it became a regular topic for the environmental Malthusians in the 1950s and 1960s. "The first major penalty man will have to pay for his rapid consumption of the earth's non-renewable resources," Brown wrote, "will be that of having to live in a world where his thoughts and actions are ever more strongly limited, where social organization has become all-persuasive, complex, and inflexible, and where the state completely dominates the actions of the individual." In 1956, Fairfield Osborn warned about a "human anthill," and a decade later Lynton Caldwell of a "scientifically managed animal farm." Raymond Dasmann believed that humanity was on a "pathway to a world overpopulated, ground down to a uniformity, constricting of the individual, inimical to liberty."[46]

Quality-of-life concerns reconfigured postwar environmental politics. As Samuel Hays and others have argued, what separated prewar conservation politics from postwar environmental politics was the latter's emphasis on quality-of-life issues, especially environmental amenities such as clean streets and parks. "The most widespread source of emerging environmental interest," Hays wrote in his classic history of the environmental movement, *Beauty, Health, and Permanence*, "was the search for a better life associated with home, community, and leisure." Explaining the emergence of this new way of thinking, he explained that in the postwar years "a new emphasis on smaller families developed, allowing parents to invest their limited time and income in fewer children." This idea came despite, or perhaps in reaction to, the great baby boom of the 1950s and early 1960s.[47]

One of the best places to see the importance of the quality-of-life concept to emerging environmental thought is in Dasmann's *Environmental Conservation* (1959). "Conservation of natural resources," Dasmann wrote in the introduction, "is here defined as the use of natural resources to provide the *highest quality of living* for mankind. It must aim toward both a material and a spiritual enrichment of life for man on earth, now and in the future." In each chapter of the book, Dasmann described what quality of life meant in different sectors of American life, including cities, farms, forest and range lands, and wilderness areas. At the end of the book, Dasmann reiterated that the goal of conservation was "improved quality of living for mankind."[48]

For Dasmann, the idea of quality of life went hand in hand with a concept of environmental quality. This idea, in turn, went back to Aldo Leopold, who had argued that the overproduction of a single resource such as deer not only reduced the quality of the deer herd's lives but also, because of consumption chains, degraded the surrounding environment, especially soil. Substituting people for deer, Dasmann devised the concept of environmental conservation. "Because the living resources of an area and its human populations are closely tied together," he stressed, "a one-sided approach to conservation which emphasizes one resource and ignores the others will lead to difficulties or failures." For Dasmann, the very concept of environment came from thinking about human population growth. "A broad, environmental approach to conservation is a necessity because physical environment, biotic resources, *and people* form a whole."[49]

The quality-of-life idea had a political ambivalence built into it. Was the idea to spread a higher standard of living to poorer groups and poorer nations or to defensively protect one's own? Were the families, social groups, and nations with high population growth rates some sort of menace, or were they to be helped in improving their lives? Critics have sometimes attacked postwar environmentalists, especially environmental Malthusians, for protecting class privilege with their quality-of-life concerns, for hoarding the best resources— now increasingly defined in terms of quality of life—for themselves. In the Malthusian emphasis on environmental quality, they also see a reincarnation of the eugenic movement's elitist emphasis on genetic quality. In both cases, the objective was to control other people's behaviors, which were defined as unhealthy or environmentally unsound or just inappropriate. Moreover, critics also argue that environmentalists wanted to have their environmental cake and eat it too—that is, they sought "quality" environments for themselves but failed to see how their own reckless consumption contributed to environmental degradation. At its worst, this "not in *my* backyard" attitude ("nimbyism" for short) involved pushing the environmental consequences of consumption—the pollution, the degraded environments—away from wealthy neighborhoods, often onto less powerful poor and nonwhite communities, then blaming those communities for their irresponsibility. Taken together, Malthusian arguments about population growth amounted to a globalized mix of eugenics and nimbyism.[50]

In his exposition of an environmental conservation built around "quality of living," in addition to a brief reference to reform eugenics and his willingness to apply wildlife models to human societies, Dasmann expressed aesthetic views that could be seen as part of a larger nimby approach. Cities, for instance, should no longer be "ugly work centers" but should have more scenic beauty and recreational space. Farming areas should show "rural beauty and attractive home sites" and more opportunities for recreation and culture. Wilderness was important "for people seeking to keep contact with primitive values"—mostly an elite concern—but was threatened by the "recreation-seeking hordes."

Yet those who accuse environmental Malthusians of oversimplification should be careful not to oversimplify them. To Dasmann, those "ugly work centers" were also places where "the economically unfortunate are forced to live a closed and barren life." In forest and range lands, he also called for maintaining water and soil quality, as well as wildlife diversity. "The environmental approach to forest land and rangeland must consider all the values and potentials of these areas," he wrote. "They must be regarded, in the future, not just as sources of wood and forage, but as part of living space for all of the people, urban as well as those who work locally." Wilderness was needed for recreation, but also as a benchmark for scientific comparison and as a refuge for wild plants and animals. Moreover, echoing other environmental Malthusians, Dasmann repeatedly criticized urban overconsumption. Although not without flaws, this was no simple defense of middle-class privilege.[51]

During the 1950s, the larger population movement reflected this ambivalence. Many of those in the population movement stressed the threats to their own quality of life. Often the mentality was crisis-driven, the diagnosis oversimplified, the tone defensive, and the remedies repressive. But some saw population limitation as a way to share the benefits of a higher standard of living. "The resources of the world," Starker Leopold told the Sierra Club in 1959, "should be used to raise the living standards of individuals, rather than to support more individuals."[52]

Environmental Quality and 1960s Liberalism

Perhaps the most famous explication of quality-of-life concerns came from Harvard economist and advisor for the Democratic Party John Kenneth Galbraith in his 1958 bestseller, *The Affluent Society*. As with Rockefeller, Osborn, and Dasmann, contemplating abundance in a world of poverty led Galbraith (1906–2006) to novel ideas about the good life. During the mid-1950s, exploring a new book project, Galbraith traveled to India to investigate international poverty, but ultimately decided to write about the United States instead. Visiting India helped him realize that American society faced unique challenges. The poor man, he wrote on *The Affluent Society*'s opening page, "hasn't enough and he needs more." The rich man, though, can imagine "a much greater variety of ills." "Until he learns to live with his wealth," the rich man will have a tendency "to put it to the wrong purposes." As with individuals, Galbraith said, so too with nations: American society had reached a stage where, because basic subsistence was rarely a problem, the new challenge of government was to improve the quality of life for each individual from cradle to grave.[53]

As historian Adam Rome has noted, Galbraith's arguments helped shape the agenda of the Democratic Party in the 1960s, including that regarding environmental affairs. One of Galbraith's most famous lines described pollution and open space problems:

> The family which takes its mauve and cerise, air-conditioned, power-steered, and power-braked automobile out for a tour passes through cities that are badly paved, made hideous by litter, blighted buildings, billboards, and posts for wires that should long since have been put underground. They pass into a countryside that has been rendered largely invisible by commercial art.... They picnic on exquisitely packaged food from a portable icebox by a polluted stream and go on to spend the night at a park which is a menace to public health and morals. Just before dozing off on an air mattress, beneath a nylon tent, amid the stench of decaying refuse, they may reflect vaguely on the curious unevenness of their blessings. Is this, indeed, the American genius?

Galbraith connected these problems to U.S. population growth. Elsewhere in the book, he described the problem "of a burgeoning population and of space in which to live with peace and grace."[54]

In the early 1960s, quality-of-life concerns reached the highest levels of American government. In his well-known 1963 book *The Quiet Crisis*, Stewart Udall, the influential secretary of the interior in both the Kennedy and Johnson administrations, helped chart an agenda for the emerging environmental movement by describing the problems of increasing pollution, shrinking open space, and disappearing wilderness. He framed these problems in terms of population growth. "The one factor certain to complicate all of our conservation problems," he wrote at the end of his most important chapter, "is the ineluctable pressure of expanding populations." When the U.S. population doubled by 2000, he wrote, there was no guarantee that "life in general—and the good, the true, and the beautiful in particular—will somehow be enhanced at the same time." Indeed, fear of out-of-control population growth helped Udall clarify the core goal of postwar liberalism—improving the quality of life of each individual. "It is obvious," he concluded, "that the best qualities in man must atrophy in a standing-room-only environment. Therefore, if the fulfillment of the individual is our ultimate goal, we must soon determine the proper man-land ratio for our continent."[55]

The juxtaposition of "quantity-of-life" concerns with "quality-of-life" concerns—of ideas about poverty and abundance—also framed in the public statements of President Lyndon Johnson, including his famous State of the Union Address in January 1965. In the first half of the speech, devoted to international affairs, Johnson became the first president to warn directly about population growth and resource scarcity, speaking of "the explosion in world population" and "the growing scarcity in world resources." In its second half, he described in greater detail than ever before his Great Society programs related to poverty, education, health, civil rights, and conservation. At the heart of this slew of programs was the quality-of-life concept—Johnson used the term "quality of life" at least three times and argued that the first test for a nation is the "quality of its people." The pairing of international quantity of life with hopes for improving domestic quality of life was more than coincidental. In 1965, a core idea of Johnson's liberal vision—helping each individual not just survive but live a rich, meaningful life—seemed threatened by overcrowded, unplanned societies around the world and at home. In a time of baby booms around the country and around the world, managing population size became central to Johnson-style liberalism. "So long as we are concerned with the quality of life," one Johnson administration official wrote in 1965, "we have no choice but to be concerned with the quantity of life."[56]

American prosperity in a sea of poverty was a theme that engaged not just environmental Malthusians during the 1950s and 1960s. Especially during the Johnson Administration, concern about international and domestic population growth reached new proportions. When they did, they embodied the same ambiguities and ambivalences that characterized the population movement overall.

4

"Feed 'Em or Fight 'Em"

Population and Resources on the Global Frontier during the Cold War

"It may be that the greatest menace to world peace and decent standards of life today is not atomic energy but sexual energy."

–James Reston, *New York Times*, 1961

As Lyndon Johnson's concern about the "explosion" in population and "growing scarcity" of world resources illustrates, by 1965 the U.S. government had shifted emphasis from the early 1950s, when the Paley Commission had rejected Osborn and Vogt's ideas about overpopulation and resource depletion. Johnson offered a similar warning in his 1967 State of the Union address: "Next to the pursuit of peace, the really great challenge of the human family is the race between food supply and population." From 1965 to 1967, Johnson oversaw a revolution in federal policy toward birth control and population planning. Together with a supportive Congress, he made the U.S. government the largest provider of birth control around the world.[1]

How did this evolution come about? A curious article on population by a former U.S. army general in 1966 in *National Parks Magazine* suggests the intertwined domestic and international problems driving heightened concern about population growth. In "Parks—or More People?" General William Draper pointed to domestic problems that he attributed to overpopulation, but then warned of an even graver international problem—one that "jeopardized the existence of civilization as we know it." Massive overpopulation was causing food shortages and hunger in nations such as India. Fueled by public health measures and food aid programs, population growth was outstripping food production. Americans, Draper made clear, had something to fear from this.[2]

Draper was an expert in American foreign policy and economic aid. He had become aware of the population issue as chairman of a U.S. military committee, later known as the Draper Committee, set up by President Dwight D. Eisenhower in 1958 to assess U.S. foreign aid programs. The committee, he explained, quickly realized "that neither food production nor over-all economic development could keep pace with populations." By 1966, Draper wrote, the problem had only expanded. Food production was increasing 1 percent a year, but populations were climbing at twice that rate. The result was widespread hunger and malnutrition. India, Draper stressed, faced "the prospect of a real famine this fall" and the likelihood of "mass starvation."[3]

Many Americans shared Draper's concerns. By the time of President Johnson's 1965 reference to population growth and resource scarcity, ordinary citizens had become aware of the sea of poverty that surrounded the few prosperous nations of the world. In the 1960s, world hunger became the stuff of newspaper headlines, church charity campaigns, and college campus organizing. Humanitarian concern for their fellow human beings drove American concern in part, but the Cold War—as a general's interest in population suggests—was also crucial. In its struggle with international communists from Moscow and Beijing, the United States could not afford to lose the allies, strategic ground, and resources of the third world. "Today's struggle does not lie here," President John F. Kennedy announced in Europe in early 1963, "but rather in Asia, Latin America, and Africa." Overpopulation in these places, it was believed, created poverty and poverty created communism. At a time when Americans were beginning to send their sons, brothers, and husbands to Vietnam in increasing numbers, this was no idle worry.[4]

Although the Cold War's shift to the third world, bolstered by important domestic concerns, formed the backdrop for the increased governmental concern about population growth and resource scarcity, there was nothing inevitable about the links between the Cold War in the third world and concern about population growth, much less environmental Malthusianism. William Vogt and Fairfield Osborn had warned that overpopulation in India and other nonwestern places threatened U.S. national security in the late 1940s and early 1950s, but they were mostly ignored. At that time, population, sex, and birth control were taboo subjects. Moreover, even those who did worry about third-world poverty had faith that technology-based economic development programs could work wonders. Indeed, in 1959, President Eisenhower did not act on the recommendations of the Draper committee. President Kennedy highlighted the problems of third-world poverty, but he, too, did little about population growth.

All this had begun to change by the mid-1960s. Not only had third-world poverty emerged as a key national security issue but problems in Vietnam and especially India had also revealed that most international development programs, especially those celebrated by Kennedy's "New Frontier," were not living up to expectations. Problems in India, General Draper warned, threatened civilization. As economic development programs stumbled, concern about population growth mounted, and natural resource scarcities became linked with national security.[5] Population growth—although still conceived of by some as a global issue transcending East-West divisions—had become a top national security concern for American strategists.

Increased governmental attention to population and resource matters came at a crucial time for the environmental movement—not long after Rachel Carson's *Silent Spring* (1962) and not long before the first Earth Day in 1970. By focusing so bright a spotlight on population and resource problems, Johnson

FIGURE 8 According to this 1950 Herblock cartoon, the economic development programs of the United States were competing against global population growth. By the mid-1960s, after several years of great hope, this "race" seemed increasingly headed toward a bad ending for Uncle Sam.

1950 Herblock cartoon, copyright by The Herb Block Foundation.

gave a tremendous push to what was becoming an important strand of the fledgling environmental movement: the population limitation movement. But there was no simple relationship between governmental policy and environmental thinking. Even as they emphasized scarcity and population growth, people such as Draper and Johnson pushed solutions—mostly technological remedies such as "green revolution" hybrid seeds and birth control programs—that struck critics like Paul Ehrlich, drawing from early environmental Malthusians, as not sufficiently environmental. As the federal government took up population programs, the fault line between those Malthusians who saw hard limits and those who did not widened.

The First Population Bomb: Environmental Balance as National Security

Although some resisted linking the population issue with the Cold War, others aggressively connected the two. Leading the way was Hugh Moore, a quirky millionaire-turned-foreign-policy-activist who ranks with Vogt and Ehrlich as a passionate and ear-catching salesman for population planning. A businessman who had founded the Dixie Cup company in the 1910s, Moore became a population activist by way of an interest in world peace, which he felt could only be secured by rationally planning natural resources on a global scale. The failure to do so, he believed, had caused World War II—in which two of his sons fought—and now threatened another global conflagration. In 1948, he read Guy Irving Burch and Elmer Pendell's *Population Roads to Peace or War* and Vogt's *Road to Survival*, leading him to believe that human consumption fueled by population growth drove the competition for resources, which in turn sparked aggression and warfare. He realized that the population issue lent itself to just the sort of transnational planning that he had long advocated. He credited Vogt and Guy Irving Burch for "really waking me up!"[6]

In 1954, impatient for more action, Moore wrote a pamphlet entitled "The Population Bomb" to alert the "leaders of American opinion," especially businessmen, of the problem. Like many in the population movement, he believed that working through elites was the most effective way to get things done. By 1967, the year before Paul Ehrlich's bestselling book by the same name, "The Population Bomb" had gone through thirteen editions for a total of 1.5 million copies.[7]

Moore's "Bomb" combined Vogt's argument that population and resource imbalances threatened American national security with apprehension about the rise of communism in the third world. Moore cribbed Vogt's logic but, even more than Vogt, highlighted the underdeveloped world—Latin America, Africa, and especially Asia. Medical discoveries and advances in sanitation in these areas had improved health and lowered death rates, while not reducing birth rates.

The result was a "population explosion," which, because little hope existed for greater food production, threatened economic and political instability. Like Vogt, Moore argued that this risked the outbreak of war. "Hunger and poverty drive men to war," he wrote. "The struggle for 'lebensraum'—living space—has long been a justification of conquerors—for example Hitler, Mussolini and Tojo of most recent memory."[8]

Moore worried that no one stood to gain more from overpopulation-driven third-world hunger than communists. "As long as two thirds of [the underdeveloped countries] go to sleep hungry every night," he elaborated, "the odds favor Communism." He emphasized his point with a chart conflating population growth with rising communist strength around the planet. Moore did not invoke communism merely for rhetorical leverage; his anxiety was sincere. "We are not primarily interested in the sociological or humanitarian aspects of birth control," he wrote to John Rockefeller III. "We are interested in the use which Communists make of hungry people in their drive to conquer the earth."[9]

Moore's choice of metaphor to encapsulate the issue—a bomb—reinforced his view of population growth as a military threat. He used the bomb to dramatize the "deadly triangle" of population growth, communism, and war. Indeed, he often claimed that population growth threatened to create an explosion "as dangerous as the explosion of the H bomb." The population bomb, however, differed from nuclear weapons in one crucial regard: while the United States and the USSR stockpiled their nukes, the fuse of this bomb was "already lighted and burning."[10]

Arguments like Moore's helped make sense of the new links and vulnerabilities that came with the shift of the Cold War toward the third world. They reinforced the "domino theory," the powerful metaphor of global interconnection that presidents from Eisenhower to Nixon invoked to justify American involvement in small, far-off countries. At best only a loose collection of ideas, this "theory" suggested that because countries—even small, remote places like Laos or Nicaragua—stood linked together, the fall of one to communism could ripple across a whole region and even topple major liberal capitalist states like Japan. Population and food imbalances helped fill in one of the domino theory's chief logical weaknesses—how one domino was connected to another. Pointing out that food imbalances in one nation could quickly destabilize an entire region, one 1951 report warned, "Should Burma, Thailand, and Indochina fall, with them would go a good part of the normal food supply of India, Pakistan, Ceylon, Indonesia, Japan, and the Philippines." Drawing from concern about population and resource imbalances, Moore's "Bomb" affirmed the fear that, if unchecked, the rising tide of communism in Asia could eventually spread to America's doorstep.[11]

Moore and other Malthusians found their biggest ally in General Draper, the kind of Republican—common then, but rare after Vietnam—who saw international development as a national security issue and therefore tolerated a modicum of governmental intervention in economic matters. Draper's perspective derived from his experiences in war and business. After serving in World War I, and then working as a banker for twenty years, he was tapped in 1945 to direct the Economics Division of the American Military Government in postwar Germany. Over the next eight years, in the army and later in the State Department, he oversaw the economic planning of first Germany and then Japan. In Japan, he saw how a nation could use birth control to radically reduce population growth rates. Because of his high-profile background, Draper made headlines in the late 1959 when he announced, as chair of the Draper Committee, that America's Cold War effort in the third world would fail without vigorous government-aided population planning programs.[12]

Draper had not given the population issue much thought until his second day as the committee's chair, when he received a letter from Hugh Moore, along with the latest edition of Moore's "The Population Bomb." In this letter and later ones, Moore laid out why the United States needed to think more strategically about population growth: American goals around the world would fail *"unless resources and population are brought into balance."* Communism, he explained, "travels on empty bellies." Draper requested twenty additional copies of "The Population Bomb" to distribute to committee members. Several years later, Draper recalled how Moore "practically forced the so-called Draper Committee to speak its piece on population problems." Draper was also influenced by Coale and Hoover's book and advice from Robert Cook, the head of the Population Reference Bureau, although Cook did not think that the government needed to get involved in population matters.[13]

By this time, the Cold War had become much more of an economic struggle. In the early 1950s, the Eisenhower administration had supported third-world allies mostly with military aid, not economic programs. But Soviet maneuvers showed the need for increasing economic aid. In 1956, the USSR helped Egypt with technical aid for its giant Aswan dam—aid that the United States had refused. In 1958, Soviet Premier Nikita Khrushchev intensified the economic competition, famously declaring an "economic offensive" in the third world: "Growth of industrial and agricultural production," he predicted, "is the battering ram with which we shall smash the capitalist system, enhance the influence of the ideas of Marxism-Leninism, strengthen the Socialist camp and contribute to the victory of the cause of peace throughout the world." The Sputnik launching in 1957 gave Khrushchev's swagger some credibility, as did the expanding Soviet economy, which had miraculously industrialized the country in the span of a generation. Because third-world leaders hoped to do exactly the same, many leading Americans believed that the United States had to dramatically step up its own economic aid programs. "Today . . . it is clear that the focus of concern has shifted to the economic field," the *New York Times* explained in late 1958.

"In particular, it has shifted to concern over the political gains which the Communist nations may make from their program of economic aid to underdeveloped countries." International poverty had become a national security problem.[14]

Draper went beyond merely emphasizing economic aid, becoming the first senior U.S. government official to publicly warn of the dangers of population growth. "The population problem," he argued before the Senate Committee on Foreign Relations in May 1959, "is the greatest bar to our whole economic aid program" and to "the progress of the world." Released in December 1959, the Draper Committee's final report made the first small step toward a U.S. population-control policy. It called for the United States first to assist nations that request assistance addressing their population problems, second to increase its assistance to local programs relating to maternal and child welfare, and third to strongly support studies and research regarding population growth. Population limitation, the report insisted, should be integral to the nation's foreign aid programs: "No realistic discussion of economic development can fail to note that development efforts in many areas of the world are being offset by increasingly rapid population growth." The report explicitly invoked the nation's Cold War struggle. Population limitation would decrease "opportunities for communist political and economic domination" in the third world. The Draper Report was, as Hugh Moore put it, "the firecracker that put the thing on the front page."[15]

Yet Draper's position found little public support from President Eisenhower. At a press conference on December 22, 1959, Eisenhower denounced its findings. "I cannot imagine anything more emphatically a subject that is not a proper political or governmental activity or function or responsibility," he said. Eisenhower's reasoning is not clear. Privately, as his conversation with President Truman seven years earlier demonstrates, he saw population growth as a problem and believed birth control programs might help. It is possible that by this time he had grown more aware of the political fallout, both internationally and domestically. But it appeared he meant what he said, that it was not the proper activity for government. He passed the report onto Congress without recommendation.[16]

Population Progressives

It would be wrong to think that everyone who lobbied for population limitation in the 1950s and 1960s did so primarily for Cold War reasons. Many approached the population issue as part of a larger social agenda. Many supporters of birth control and population limitation were liberals at the forefront of movements to protect civil rights, eradicate international poverty, and expand roles for women. The leader of the population planning movement in the U.S. Senate, for instance, was Ernest Gruening, a Democrat from Alaska, member of the NAACP since the 1910s, former writer for the *Nation*, anti-imperial activist, and leader in the anti-Vietnam war struggle. For Gruening, the birth-control and population-limitation movement was part of a broader social critique, both at home and overseas.

Gruening first became aware of population growth and birth control in the slums of south Boston. In the early 1910s, while in his third year of Harvard Medical School, he did obstetrical service in the area. "There," he later wrote, "the children came as fast as nature permitted," with both children and mothers suffering as a result. At the time, contraception was "undreamed of." It was not taught at Harvard. After graduating and becoming a journalist for a series of Boston papers, Gruening championed various progressive causes, such as academic freedom, good government, workers' rights, woman's suffrage, birth control, and racial equality. A strong supporter of Margaret Sanger, Gruening and his wife served on the Conference Committee for the first American Birth Control Conference, which Sanger organized in 1921. One of the topics discussed at the conference was overpopulation as a cause of World War I. Sanger was arrested for her advocacy of birth control on the third day.[17]

Gruening became an expert on Latin American affairs with the publication of a book on the Mexican revolution, *Mexico and Its Heritage* (1927), which defended the revolution and especially applauded its land reform program. From 1934 to 1939, he served as director of the Interior Department's Division of Territories and Island Possession, which oversaw government operations in Alaska, Hawaii, Puerto Rico, the Virgin Islands, the Philippines, and several smaller Pacific islands. At the same time, he also headed the Puerto Rican Reconstruction Administration, pushing for change and opening twelve birth-control clinics on the island. "Unless some steps were taken to slow down the population growth," he later wrote in his autobiography, "our efforts would be nullified." These clinics were closed in 1936 when President Franklin Roosevelt's political advisors caught wind of them, for fear of alienating important Catholic supporters.[18]

In 1960, the year after he was elected Alaska's first U.S. senator, Gruening delivered a long floor speech calling for civil rights legislation. He quickly became known as an expert on foreign aid, a critic of Vietnam policy, and a leader on the population and birth-control issues. In August 1963, he co-sponsored with Senator Joseph Clark of Pennsylvania a resolution calling for more federal research on birth control. His views grew from his long-standing convictions, but also because of his frustrations with foreign aid programs, especially after a study trip to the Middle East in 1962 and Latin America in 1965. "There has been great indifference in many of the recipient countries to the welfare of their masses and whose government have not taken steps to make our aid effective," he wrote in the *New York Times* in 1968. "The foreign-aid program is bound to falter ... so long as no substantial efforts are made ... to cope with the overwhelming population growth which, in itself, is largely nullifying any and all efforts at economic and social improvement." His approach was part of a larger critique of Cold War foreign policy. In the name of fighting communism, he wrote in his 1973 autobiography, American programs "have poured millions of dollars into the pockets of the oligarchies of these countries, and have supported in power the most repressive of right-wing dictatorships, both of which have prevented land reform, equitable taxation and all the other social and economic reforms that would truly benefit the people." Whereas some within the population movement harbored fearful visions of brown and black people overrunning the planet, Gruening saw his population advocacy as consistent with a lifetime of progressive activism.[19]

Moreover, many advocates of population limitation, in contrast to those who looked through the Cold War lens, thought in global, not national terms, and held a hope to transcend national borders. This idea bounced around the anti-nationalist, "one-world" atmosphere of the late 1940s. In *Our Plundered Planet*, Osborn stressed that global interdependence implied a single human fate: *all* human beings lived within interconnected, interdependent networks of economy and ecology. "The peoples of the earth," he wrote, "are bound together today by common interests and needs." Dependence on nature united humanity. For Osborn, population growth was an issue that could—and should—unite nations, including the United States and the Soviet Union. "I have the wild dream," he announced in 1950, "that conservation can be the harmonizer of discord between these two powerful nations." A reviewer echoed this sentiment: "The great enemy of America is not Russia; it is hunger. The great enemy of Russia is not America nor any capitalist country; that enemy is hunger."[20]

Such sentiments had grown stronger by the 1960s, particularly among environmentalists. "By shifting our attention from the now completely irrelevant and anachronistic politics of nationalism and military power," author Aldous Huxley wrote in 1963, "to the problems of the human species and the still inchoate politics of human ecology we shall be killing two birds with one stone—reducing the threat of sudden destruction by scientific war and at the same time reducing the threat of more gradual biological disaster." "The thought that we're in competition with Russians or with Chinese," Nobel Prize–winning biochemist George Wald announced in a 1969 speech, "is all a mistake, and trivial. Only mutual destruction lies that way. We are one species." Environmentalist John Fischer echoed this idea in *Harper's* in April 1970: "The differences between capitalism and Communism no longer seem to me worth fighting about, or even arguing, since they are both wrong and beside the point." Paul Ehrlich expressed the idea with characteristic bluntness: "If we want to continue the argument about capitalism and socialism into the next century we shall have to do something about saving all our hides in this century." Making one-world arguments in no way precluded a turn to coercion as a solution. Those who stressed the global were just as likely to recommend nonvoluntary population limitation methods as the Cold War Malthusians.[21]

Several prominent political leaders also saw population and hunger problems as bigger than political divisions. "Perhaps the necessity of confronting the population dilemma will finally usher in the brotherhood of man," Adlai Stevenson pointed out in the mid 1960s. "The members of the United Nations have perhaps ten years left," U.N. Secretary General U Thant announced later in the decade, "in which to subordinate their ancient quarrels and launch a global partnership to curb the arms race, to improve the human environment, to defuse the population explosion, and to supply the required momentum to development efforts." Otherwise, they faced problems of "staggering proportions." Such concern helped push the United States toward détente, the policy of rapprochement with the Soviet Union that developed during the 1960s.[22]

JFK, the Third World, and Modernizers

The population programs of the mid-1960s grew out of—and partially in response to—the antipoverty modernization programs of the Kennedy administration. Perhaps no American politician emphasized fighting international poverty more than President Kennedy, who saw the economic development of the third world as a great opportunity to score victories against international communism. A 1961 White House task force called for less military and more economic aid. A 1962 National Security Council report reiterated the threat posed by the Soviet economic offensive. As part of this, Kennedy proposed the Development Decade, with programs such as the Peace Corps and the Alliance for Progress aimed at winning friends in the third world through economic assistance. Such optimistic development programs, more than anything except perhaps the space program, helped define the New Frontier. "Man," Kennedy announced in his famously inspirational inaugural address, "holds in his mortal hands the power to abolish all forms of human poverty."[23]

The Kennedy administration illustrates the unusual political dynamics that later enabled U.S. governmental population planning to become reality. The overlap of the Cold War with an international war on poverty brought together a number of strange bedfellows—including realists who stressed defending American self-interest, as well as antipoverty humanitarians who tended to emphasize broad human needs. Antipoverty programs that otherwise would never have been passed by Congress were thus brought into reality, creating the beginnings of a global welfare state, as some historians have begun to call it. Indeed, combining self-interest and humanitarianism, Kennedy (and later Lyndon Johnson) simultaneously appeared as an aggressive cold warrior and compassionate liberal. In his inaugural address, Kennedy called for helping those struggling to break the bonds of mass misery "because it is right." But he then added, "If a free society cannot help the many who are poor, it cannot save the few who are rich."

To develop third-world economies, Kennedy turned to advisors who promoted "modernization" programs. Building upon David Potter's view of America's mission as expressed in *People of Plenty*, modernization theorists stressed technology-driven growth. They gained prominence during the late 1950s and 1960s, as new trends in the Cold War increased the need for economic strategies. Kennedy placed great weight in the ideas of modernizers such as Walt Whitman Rostow—who, clearly drawing from Potter, coined the phrase "new frontier."[24]

Modernizers like Rostow built their programs on three tenets. First, they attributed poverty to cultural deficiencies, not fixed racial characteristics. Mutually reinforcing backward traditions, they believed, prevented economic development in the third world. "The central fact about the traditional society," Rostow wrote in *The Stages of Economic Growth: A Non-Communist Manifesto* in 1960, "was that a ceiling existed . . . [because] the potentialities which flow from modern science and technology were either not available or not regularly and systematically applied." This emphasis on culture as opposed to race helped modernizers distinguish themselves, at least in their own minds, from the old imperial powers of Europe, who often saw poverty as a function of the racial inferiority of colonial peoples. Their emphasis on culture also informed their second main belief: progressive change was indeed possible. Societies, no matter how far behind, marched toward progress. This led to their third main belief: Western nations could accelerate the evolution from tradition to modernity with a judicious infusion of advanced technology. Because all societies followed a common path from impoverished "tradition" toward modernity, an infusion of technology at just the right moment could enable a backward society to leapfrog forward. Rostow, the prophet of modernization in both the Kennedy and Johnson administrations, advocated a massive influx of Western know-how and money at the right moment—the so-called "take-off" moment—to catapult third-world economies into advanced status. These ideas of racial equality, progress, and technological transfer drove ambitious antipoverty programs both at home and abroad.[25]

Compared with thinking about development later in the decade, Kennedy's economic development programs stand out because of their extreme optimism. "The tricks of growth are not all that difficult," Rostow said at the time. "Euphoria reigned," Kennedy advisor Arthur Schlesinger Jr. later wrote, "we thought for a moment that the world was plastic and the future unlimited." Because of their faith in human rationality and progress, modernizers are often considered the ultimate in Enlightenment optimists.[26]

This is not to say that modernizers ignored population growth. Rostow, for instance, believed that a transition to modernity and prosperity could not happen without fertility control: "The view towards the having of children," he wrote in *The Stages of Economic Growth*, "must change in ways which ultimately yield a decline in birth-rate." Children required heavy investments and ate up progress, Rostow believed, but with enough investments rapid population growth could be overcome, especially if fertility rates were reduced. Technology could also increase production, particularly in places untouched by modern methods. "It is, therefore, an essential condition for a successful transition that investment be increased and—even more important—that the hitherto unexploited back-log of innovation be brought to bear on a society's land and other natural resources, where quick increases in output are possible." Because of their faith in technology, modernizers were optimistic in spite of rapid population growth.[27]

Many within in the population movement shared the assumptions of modernization theory. They believed that technology and planning—especially modern contraception—could overcome the cultural shortcomings that mired third-world nations. They thought of poverty not as a fixed concomitant of race, but rather as a cultural phase that would pass. What made people primitive were surmountable cultural traits such as religious and economic traditions favoring high birth rates. These "reproductive modernizers" saw themselves as social progressives doing battle with outdated colonial worldviews. Of this group, some were as optimistic as Rostow, some less so. Many switched from optimism to pessimism as the 1960s unfolded.

Yet not all within the population movement accepted the premises of the modernizers, especially those Malthusians concerned about environmental limits and degradation. Modernizers tended to be the heirs to Truman administration Secretary of the Interior Julius Krug's lineage of conservationists and the Paley Commission of the early 1950s, who saw their mission as the utilization of underused resources, not the protection of ecological integrity. Greater efficiency and growth, they believed, could overcome any scarcities. "Above all," Rostow wrote in *The Stages of Growth*, "the concept must be spread that man need not regard his physical environment as virtually a factor given by nature and providence, but as an ordered world which, if rationally understood, can be manipulated in ways which yield productive change and, in one dimension at least, progress." The goal of modernization was large-scale economic growth; the pinnacle of Rostow's stages of growth was a "high mass-consumption" society. Modernizers saw the transfer of powerful technologies—roads, factories, and dams—as the remedy to poverty and the key to economic "take-off." But for those who thought in terms of natural resource limits and environmental degradation, these modernizing policies seemed increasingly distasteful.[28]

Combined with political factors, faith in modernization programs explains why Kennedy sidestepped the population limitation recommendations of the Draper Committee report, which came out as the 1960 presidential campaign was heating up. The other Democratic candidates quickly supported Draper, as did a number of liberal protestant theologians. Just two days after the National Catholic Welfare Conference rejected the report, Kennedy added his own disapproval. It was a tricky issue for Kennedy: he didn't want a forceful rejection to highlight his Catholicism, nor to spur controversy by accepting Draper's recommendations. As president, he avoided the issue until his last year in office, when he began to allow underlings to acknowledge a problem and call for population research.[29]

Eventually, however, the optimism of the New Frontier and the Development Decade ran into the problems of Vietnam and the less well-remembered problems of food crisis in India and other third-world countries. When it did, American officials and the public began to embrace the population limitation movement more broadly.

India, the United States, and Population Planning

All of the factors at work in the change from the Paley Commission to the policy changes of the Johnson administration played out in American policy toward India in the late 1950s and early 1960s: the shift of the Cold War toward the third world and economic development, the focus on narrow and overly optimistic modernization programs, the political success of mixing humanitarianism and geopolitcs, and the eventual turn to population limitation programs.

American strategic concern about India began in the mid-1950s but intensified after the 1960 election. In December 1955, *two million* Calcuttans lined the streets to greet Soviet Premier Nikita Khrushchev, who had visited India to promise Soviet aid for the Bhilai steel works, India's flagship development project. Soon thereafter, the United States began targeting India with economic aid. In 1957 and again in 1958, Eisenhower increased aid. In 1959, he visited the subcontinent. In March 1959, National Security Council memorandum 5701 reaffirmed the strategic importance of South Asia and—significantly—called for its economic aid. But Eisenhower's increasing attention to India paled in comparison to Kennedy's. As a senator, Kennedy twice sponsored resolutions highlighting the strategic importance of India. And as president, he geared many of his well-publicized Development Decade programs toward India, even dispatching a top advisor, John Kenneth Galbraith, to be ambassador. Under Kennedy, India became the largest recipient of United States economic aid in the world.[30]

U.S. policymakers considered India important to the United States for several reasons. It was the largest and most populous of the newly independent nations. As home to roughly one-quarter of the world's poor, it had also long earned sympathy from Americans. High-profile American ambassadors such as Chester Bowles and John Kenneth Galbraith drew attention to India, as did visits from Adlai Stevenson, Martin Luther King Jr., William O. Douglas, secretaries of state John Foster Dulles and Dean Rusk, vice presidents Richard Nixon and Lyndon Johnson, and President Eisenhower. Most important, India was a top Cold War prize. A leader of the decolonization movement, a supplier to the United States of several important minerals, and a nation strategically located vis-à-vis both the Middle East and China, India's support was sought after by both the Soviet Union and the United States. American strategists thought India's economic development to be crucial in this competition. "The extent of India's development," a 1959 National Security Council memorandum declared, "will have international ramifications. . . . Asia and Africa will be watching and comparing what the Indian and Chinese regimes are achieving for their peoples, in terms of rapid industrialization, as well as in terms of the impact on human freedoms and living standards."[31]

Although the entire third world provided a giant testing ground for "modernization," India was the showcase. Indeed, the most important of Kennedy's troop of social science experts had developed their theories with India in mind. Walt Rostow and Max Millikan began a study on India in 1958 and established a research center there in 1959. India offered a convenient—and English-speaking—place to show not only how the American-run postwar order differed from the imperialism of old but also how democratic capitalism differed from communism. As Rostow put it, the contest between democratic India and communist China constituted "a kind of pure ideological test of great significance." Moreover, Rostow and Millikan believed that India was poised for "take-off." India was only one of a few developing nations that had a governmental structure, a skilled labor force, a supply of trained administrators, and a network of basic transportation, communications, and power facilities. It was the perfect place to inject outside capital and technology to jumpstart self-sustaining growth. American officials were especially pleased that Indian Prime Minister Jawaharlal Nehru's policies emphasizing quick industrial growth appeared to mirror their own ideas.[32]

Following the advice of the modernizers, both Eisenhower and Kennedy pumped millions of dollars into India's industrial sector. In particular they favored giant "megaprojects." Aid from the Eisenhower administration went mostly toward railroad rehabilitation, hydroelectric and thermal power projects, and steel imports; Kennedy administration aid funded large-scale dams and machinery parts needed for industrialization.[33]

India exemplified Kennedy's optimistic New Frontier framework. Kennedy evoked India's struggles in his inaugural address, and administration officials openly predicted that their modernization programs would help the giant country attain self-sustaining growth within *ten* years. Decades later, Kennedy's ambassador to India, John Kenneth Galbraith, would write, "More seemed then to be possible than one could now imagine."[34]

Tragically, the India of the mid and late 1960s was far from the hopeful New Frontier predictions of the early 1960s. When monsoon rains in India failed in 1965 and 1966, an already chronically bad food situation threatened massive human disaster.[35]

Americans heard frequently about India's troubles. The director general of the U.N.'s Food and Agriculture Organization (FAO) predicted "a disaster of unprecedented magnitude." The world, Dr. Raymond Ewell, a former advisor to the government of India, announced, "is on the threshold of the biggest famine in history." Roger Revelle, director of the Harvard Center for Population Studies, testified before Congress in February 1966: "It is almost certain that tens of millions of people will starve." According to a *New York Times* editorial, "Millions of the world's peoples are fighting a grim and losing battle for mere survival and many millions more are on the borderline between hunger and starvation." Newspaper headlines also reported the growing problems: "India Will Ration Grain in Cities," "Fighting Famine in India," "45 Million in India Face Lean Fare," "The Malthusian Specter," "Danger of Upheaval in India Discerned by U.S. Food Team," "India Asks U.S. Aid as Famine Looms." Books spread a similar message, such as Ronald Segal's *The Crisis of India*.[36]

In 1965, Alaskan Senator Ernest Gruening, aware of the forecasts for trouble in India and cheered by Johnson's willingness to address the issue, launched a three-year series of hearings on the population crisis, which included both international and domestic components. "No single government undertaking," one observer noted not long afterward, "had greater impact in publicizing birth control as a legitimate public policy issue in the mid-1960s than the Senate subcommittee hearings chaired by Senator Gruening."[37]

If the decade had started with great optimism about new frontiers of development on the global periphery, such hopes were beginning to fade by 1965.

LBJ, the International War on Hunger, and Population Programs

President Johnson also focused tremendous attention on India's troubles, even after the 1965 State of the Union speech. Later that year he made a similar statement at the United Nations: "Let us in all our lands—including in this land—face forthrightly the multiplying problems of our multiplying populations." India, he said in early 1966, stood "on the threshold of a great tragedy." Unless Indian production is supplemented by external assistance, he pointed out, "more than 70 million people will experience near-famine." Declaring an "International War on Hunger" in early 1966, Johnson organized a relief plan—involving the largest flotilla of U.S. ships "since the allied forces crossed the English Channel on D-Day" according to Walt Rostow—to deliver grain to Indian ports. At the height of the effort, two American ships arrived in India every day. For two years, American grain was the main source of sustenance for over sixty million Indians. Roughly one-quarter of the annual U.S. wheat crop went there. Johnson spoke of feeding India as "America's job."[38]

Many Americans, of course, followed the human drama of the relief effort. But as with William Vogt during the late 1940s and Hugh Moore during the 1950s, they also fretted about what famine and political instability in India might mean for them. For over a decade Americans had been hearing politicians describe how poverty and political instability in one part of the world could spread around the globe. Now India, according to the *New York Times*, "a vast bulwark of liberty in threatened Asia," faced "widespread famine and increasing political unrest." News accounts sometimes set forth dire consequences. "Aside from the humanitarian aspects," wrote *Times* columnist James Reston, "the social and political considerations of [the food crisis] at home and abroad are likely to be considerable," concluding that in Asia, the United States had to either "feed 'em" or "fight 'em." The crisis, President Johnson warned, was one of the largest threats to the world: "No peace and no power is strong enough to stand for long against the restless discontent of millions of human beings who are without any hope."[39]

India's problems initiated a dramatic change in American thinking away from the high hopes of the Kennedy years. Each ship of grain sent by the United States suggested that the foreign aid programs of the early 1960s, which had promised quick and dramatic results, were failing to reduce third-world poverty. Although Walt Rostow had claimed in the early 1960s that "the tricks of growth are not all that difficult," by the mid 1960s, the famous Swedish sociologist Gunnar Myrdal was noting a deep pessimism and disillusionment among development professionals. Two historians of development summed up the change in mood: "The 1960s, which the Kennedy administration declared the 'Development Decade,' instead became the decade in which development programs and the technocratic optimism that motivated them, ran aground."[40]

One indication of the pessimistic shift in American policy circles can be seen in the trajectories of Robert McNamara and Willard Wirtz, key architects of the nation's optimistic development programs at home and abroad in the early 1960s. McNamara, the number-crunching defense secretary in the Kennedy and Johnson administrations, became president of the World Bank in 1968. Wirtz was secretary of labor in the Johnson administration and a key player in LBJ's Great Society programs. By the late 1960s, both had become prominent Malthusians. Speaking about foreign aid and global economic development in 1968, McNamara described a "mood of frustration and failure." He also announced a new direction in World Bank policy—birth-control programs spanning the globe to counteract the "crippling effect" of population growth. At the end of the decade, Wirtz had also become a leading activist for population limitation: "A Secretary of Labor spends his years in office trying futilely to fight unemployment by creating more jobs; and only later, freed of the inhibitions of office and politics' restraints, faces the truth that there are too few jobs because there are too many people."[41]

President Johnson was at first cautious about fully embracing Malthusianism. In a January 1966 editorial called "Johnson vs. Malthus," the *New York Times* attacked Johnson's lingering optimism about places like India: "The President seems to be suggesting that the poor countries, having achieved a substantial degree of development, are now poised for a great leap forward to higher living standards. Unfortunately, this is not the case." Johnson's dream of improving the standard of living around the world "cannot be realized in a world where the Malthusian specter, more terrible than Malthus ever conceived, is so near to being a reality."[42]

Within a few months, however, Johnson was directly blaming population growth and calling for family planning programs. "Most of the world's population is losing the battle to feed itself," he explained to Congress in 1967. "India's plight reminds us that our generation can no longer evade the growing imbalance between food production and population growth." He continued, "We know that land can be made to produce much more food—enough food for the world's population, if reasonable population policies are pursued. Without some type of voluntary population program, however, the nations of the world—no matter how generous—will not be able to keep up with the food problem." In case this wasn't clear enough, he made it clearer: "Developing nations with food deficits must put more of their resources into voluntary family planning programs."[43]

Johnson oversaw a revolution in U.S. policy toward population planning programs. In 1965, he allowed the U.S. Agency for International Development (USAID) to begin providing advice and technical assistance on population issues, and he established an informal White House Task Force on family planning. The biggest policy changes, however, arrived in 1967, when Congress passed two landmark bills vastly expanding federal population programs: Title X of the Foreign Assistance Act, which earmarked $35 million for USAID family-planning programs; and the 1967 Social Security Amendments, which allocated funds for domestic family-planning centers. From 1965 to 1969, U.S. government annual funding for domestic family-planning programs increased from $8.6 million to $56.3 million, and funding for programs in the developing world increased from $2.1 million to $131.7 million. From then on, birth control would form the core of the "basic needs" approach, which dominated American economic aid policy in the late 1960s and 1970s. These programs were a mix of Malthusianism and modernization: they recognized the problems of rapid population growth and resource shortages, but they also stressed the progress that could be attained by overcoming obsolete cultural traditions with technological solutions—in this case birth control.[44]

The Agricultural Revolution

The shift to population planning, however, was only part of the story. Population programs were part of a larger shift in American foreign aid programs in the Johnson administration away from industrial development and toward agrarian restructuring and rural development. A new, major focus became agricultural modernization, including new "green revolution" programs that promoted high-yield varieties of wheat, rice, and other staple crops. Norman Borlaug had overseen the initial research on high-yielding wheat varieties in Mexico in the early 1940s with funding from the Rockefeller Foundation. In later decades much of the progress, especially for rice, came from the Los Banos experimental station in the Philippines. President Johnson, a ranch owner who had pushed agricultural modernization in his native Texas and other parts of the U.S. South for decades, enthusiastically supported green revolution programs. The group of critics coming to be called environmentalists, however, observed this new policy initiative with growing alarm.[45]

In his 1967 message to Congress on food aid, Johnson highlighted the new focus on agriculture, lauding an "air of change" and "innovation" around the world. This "agricultural revolution" was crucial, he said, because "long-term economic growth is dependent on growth in agriculture." Earlier U.S. aid efforts had not always understood this; agriculture must now receive "a much higher priority." Developing nations "can no longer take food supplies for granted, while they concentrate on industrial development alone." The United States could help usher in these changes. This year's economic aid program, Johnson said, "makes agricultural development a primary objective." Key to the agricultural revolution was exporting "American equipment and know-how. . . . Fertilizer, seed, and pesticides must be provided in much greater quantities than ever before." He specifically mentioned the "new varieties of rice introduced from Taiwan" and "high-yielding wheat seed imported from Mexico." He added a note of optimism: "We know that an agricultural revolution is within the capacity of modern science."[46]

In ensuing years, the "green revolution" received a tremendous amount of attention. "Many scientists are confident that the needed revolution in food production will be forthcoming," Jane Brody wrote in the *New York Times* in 1969, citing "grains that contain high levels of protein and that respond well to fertilizers." According to another article from the time, an agricultural revolution was under way "that may prove as important to mankind as the Industrial Revolution of the early nineteenth century." In 1970, Norman Borlaug won the Nobel Peace Prize for his green-revolution research. The green revolution only grew in reputation during the 1970s.[47]

In some ways Johnson's population and agricultural policies represented a shift from earlier modernization programs, but in other ways they were no shift at all. The programs reproduced one of modernization theory's core blind spots, its disregard of social and political contexts in favor of technological solutions. By shifting attention to population growth and new technologies, Johnson drew the focus away from the numerous social and political factors involved in India's troubles: that the 1947 partition had left India with 82 percent of the subcontinent's people but much less productive land; that the wealthiest 20 percent of the rural population, who controlled more than 50 percent of the cultivated area, cared mostly about rent revenues, not increased production; that Prime Minister Nehru had done little about these inequities and instead pushed industrial development; and that the massive American food aid program in place since the 1950s, designed to keep American farm state voters happy, had created disincentives for grain production. In sum, then, although the Cold War brought the world hunger and population problem a tremendous amount of attention, it also narrowed the population movement's political vision. Instead of addressing core political problems such as poor governance or disparities between the rich and poor, American population and agriculture programs emphasized technological deficiencies and solutions.[48]

It was against these technological programs that environmental Malthusians like Paul Ehrlich would react so strongly in the late 1960s. Ehrlich and others saw the same trouble brewing in India and the third world as did the Malthusians in the Johnson administration, but believed that Johnson's technological solutions could not overcome the underlying problem, which they, drawing from a strand of Malthusianism emphasizing hard limits, defined not as cultural but as biological and environmental. In doing so, though, they, like the Johnson administration before them, tended to overlook the social and political factors that were an important part of the problem. In other words, even environmental arguments deeply critical of the Johnson administration's technological solutions mostly operated within the same narrow Cold War political parameters, at least at first.

5

The "Chinification" of American Cities, Suburbs, and Wilderness

[Because of population growth, Americans will face] "increased crime, gambling, sexual promiscuity, riots, air and water pollution, traffic congestion, noise and lack of solitude. More and more there will be 'no place to hide.'"

—Irving Bengelsdorf, *Los Angeles Times*, 1965

"Chinification has already begun right here at home."

—William and Paul Paddock, 1971

As international concerns about population growth were developing in the mid 1960s, concerns were also mounting about population growth within the United States. In the same 1966 article in which he stated that population growth in India and other third-world nations "jeopardized the existence of civilization," General William Draper also blamed population growth for two pressing problems within the United States. The first was poverty, especially urban poverty. In many American cities, poverty and public welfare had become "a way of life" because both white and black migrants from the countryside "brought higher birth rates which are characteristic of all rural families." As with India, Draper suggested that there was something to fear: studies showed, he said, that "delinquent boys were found to come from larger families." Additionally, Draper blamed population growth for a problem that was seemingly unrelated to poverty: the crowding of national parks and other outdoor recreation areas. "Soon the day may come," he wrote, "when we will have to make a reservation five years ahead even to drive in a massive traffic jam through the gates of a national park." Many causes might be put forward, but "the real pressure on our irreplaceable natural resources," he emphasized, came from "the pressure of people." During the 1950s and especially the 1960s, Americans began to "feel" crowded not just globally but within the United States, as well.[1]

During these decades, the United States, like many third-world countries, experienced unprecedented population growth. This growth came as a surprise. When William Vogt and Fairfield Osborn wrote their books in 1948, the great "baby boom" had already started, but no one knew that it would last, which is ultimately what made it distinctive. The massive increase in birth rates did not peak until 1957, and ended only in 1964. During this "fertility splurge," Americans of all races and regions had more children on average, starting at a younger age. All told, the boom added nearly ninety million children, during a period of little immigration. In 1950, the U.S. population stood at 152 million; in 1960, at 181 million; in 1970, at 205 million—a one-third gain in twenty years.[2]

This postwar population growth, unique among industrial powers, changed American culture: in the 1950s, according to one historian, it made the United States into "a child-centered nation." In later years, as this huge generation became teenagers and then college-aged, American life shifted according to their interests and concerns.[3]

At the same time, other demographic trends also reshaped the United States. Ethnic neighborhoods dispersed, and "ethnics" turned into "whites." African Americans joined other rural residents moving to cities, sending urban and semi-urban populations dramatically upward. In 1945, more than half of all Americans lived on farms or in towns of fewer than 10,000 people. By 1970, nearly 75 percent lived in or near communities much larger than that—many in gargantuan metropolitan areas like New York, Chicago, and Los Angeles. Much of this urban growth was uneven: African Americans clustered into inner cities, but most growth occurred in new suburbs on the outer ring of cities. Some urban centers grew, but by 1970, roughly half were losing numbers. By this time, the United States had become one of the world's most decentralized industrial societies.[4]

In these years of rapid growth and change, among the loudest critics of population growth were the environmental Malthusians, an important strand of a group coming to be called environmentalists. Downplaying other often more moderate explanations, they increasingly saw population growth as the core problem behind the threats to American "quality of life": the poverty and violence in inner cities and the crowding and dysfunction in suburbs, parks, and wilderness areas. "Underlying it all," two environmentalists wrote in 1967, was humanity's "breeding fervor." In these arguments, the international often overlapped with the local. Many environmental Malthusians, for instance, worried about the "Chinification" of the United States—the loss of quality of life because of too much quantity of life. In the years following Rachel Carson's attack on chemical pesticides in *Silent Spring* (1962), domestic concerns about "too many Americans" combined with international concerns about overpopulation in India and the third world to create a snowballing sense of environmental crisis.[5]

Failed Modernization in American Inner Cities and the Turn to Population

In the early 1960s, poverty was not just a top international issue but a top domestic issue, too. Michael Harrington's *The Other America* (1962) put the issue in the headlines, and Lyndon Johnson's ambitious "War on Poverty" kept it there. Applying Malthusian ideas to America's domestic context followed a pattern similar to the Indian example: many people turned to overpopulation as part of a general attack on poverty, especially after more optimistic culturally based modernization programs seemed to fail. After riots in Watts, Detroit, and hundreds of other American cities during the second half of the decade, many saw the underlying problem as demographic. As with India, they did so for different reasons—some from altruism, some from class and racial fear, some for both reasons. They also did so with different degrees of sophistication—some incorporated population into a broader analysis of poverty, but others focused narrowly on population. Indeed, it was in these years that the phrase "blaming the victim" first emerged.[6]

Domestic poverty—and people of color—had been a focus of groups like Planned Parenthood for decades. Margaret Sanger had started opening birth control clinics to reach women who, like her mother, lived in poverty and struggled to find the resources for children. "Right here at home, in the United States, and in our possessions," William Vogt announced in 1951, "we have thousands of little Indias. Go into the slum sections of Washington or New York or Los Angeles or Chicago . . . and you will surely find woman after woman with five, six, seven or eight children, who cannot be decently taken care of on the family income. Rather than mope in dingy tenements many of them range the streets and fall into juvenile delinquency." In the 1950s, Planned Parenthood took the lead in providing services in poor neighborhoods in the United States, but even the Population Council occasionally focused on domestic population growth. "Excessive fertility by families with meager resources," a 1958 Population Council report explained, "must be recognized as one of the potent forces in the perpetuation of slums, ill health, inadequate education, and even delinquency."[7]

Launched in 1964, Lyndon Johnson's War on Poverty provided a golden opportunity for the population movement. Aiming to extend the welfare state to those Americans not helped by the 1930s New Deal programs, in particular Americans who worked outside of factories, Johnson's program included the Model Cities and Head Start programs, legal services for the poor, the Community Action Program, as well as Medicare and civil rights legislation. Incorporating many of the same assumptions as international antipoverty programs, these programs were also extremely optimistic. The "conquest of poverty," the Council of Economic Advisors announced in 1964, "is well within our power."[8]

Where there's awareness of poverty, Malthusian ideas are often close behind, especially when public money is involved. From 1964 to 1967, historian Donald Critchlow writes, the population movement conducted a major lobbying effort to capitalize on the political mood to cut welfare costs and out-of-wedlock births among the poor. Hugh Moore helped organize the Population Crisis Committee, which ran ads in major newspapers to draw attention to the overpopulation of inner cities and foreign nations. Planned Parenthood continued with its campaigns. Population Council founder John D. Rockefeller III also associated inner-city poverty with overpopulation.[9]

More liberal views about sexuality and birth control helped their cause. Since World War II many Americans, but especially younger people, had become increasingly comfortable with a wider expression of sexual behavior and with new birth control techniques, including the "pill," introduced into the U.S. market in 1960. The changing views of Catholics were also crucial, especially to Democratic politicians such as Lyndon Johnson, who relied on support from large blocs of white ethnic voters. In 1959, 50 percent of Catholics supported making birth control information more available; in the early 1970s, over 80 percent did. For a while in the mid 1960s, even the Catholic Church itself showed signs of liberalizing.[10]

The federal government began to act in 1965. Lyndon Johnson's concern about global population growth opened the door for more action, within the United States as well as overseas. Senator Ernest Gruening devoted a chunk of his Senate hearings to population growth within the United States, especially among the poor. Advocates repeatedly argued that the United States could not call for other nations to institute population and birth-control services that it did not offer at home. Later that year, Johnson established an informal White House task force on family planning. Federal funds, although limited compared to those in subsequent years, began to flow.

One indication of the concern about overpopulation within the Johnson administration came from Labor Department official Daniel Patrick Moynihan. Little noticed amid the controversy over his August 1965 report "The Negro Family: The Case for National Action," which generated fierce criticism for calling attention to single-parent African American families, was that Moynihan had also stressed that the poverty "of Negro Americans [is] compounded by the present extraordinary growth in Negro population." Moynihan contended that large families could not provide the resources needed to keep their children from falling into a lifetime of poverty. He believed that the problem was only getting worse.[11]

Michael Harrington, the man most responsible for placing poverty on the national agenda in the first place, agreed. "Now one of the most important developments in the twenty or so months since the President declared the war on poverty," Harrington told a Planned Parenthood group in October 1965, "is that we have made a more precise and sophisticated description of who the poor are and, above all, we have related poverty to family size.... Millions of children, many of them unwanted, are basic to the culture of the other America."[12]

Just a week after the Moynihan report came out, an event happened that added momentum to the call for population programs: a riot ripped apart the predominantly African American Watts neighborhood of Los Angeles. It was the first of many. In 1966, thirty-eight disturbances rocked the nation's cities, resulting in millions of dollars more in property damage and many deaths and arrests. In 1967, 164 disturbances erupted, including 33 requiring the state police and 8 the National Guard. The most destructive of these—the July riot in Detroit—left 43 people dead, 7,000 arrested, 1,300 buildings destroyed, and 2,700 businesses looted. The next year, 1968, would see even more unrest.[13]

The roots of this violence ran deep. In Detroit, historian Thomas Sugrue has recently argued, African Americans who came from the South starting in the 1940s found themselves trapped in a substandard housing market, and in a job market governed by discrimination. In the 1950s, the situation worsened, as automation, decentralization, and relocation eliminated thousands of good working-class jobs from the city, leaving African Americans with low-paying, dead-end positions. On top of this, governmental programs favoring suburban development, such as home mortgage guarantees and interstate highway programs in effect subsidized urban decay. At the same time, racial divisions began to harden. A powerful, well-organized white homeowners' movement emerged to block black migration to white neighborhoods, and African American civil rights activists showed a new militancy. Perhaps it was only a matter of time before violence broke out.[14]

The riots left deep frustration among the larger American public, even among liberals. "The Great Society," Robert Lekachman, a liberal Keynesian economist, wrote in January 1967, "has ground to a halt." It has "lost its token bouts with racism and poverty." The unrest in Detroit, the worst urban violence of the decade, was particularly frustrating because it ravaged a city with a Democratic mayor and model urban renewal and antipoverty programs. If such efforts could not lift the poor out of poverty and prevent violence, then what could? "By the time the war on poverty lost its political glamor," historian James Patterson writes, "its limitations had intensified the country's doubts about the ability of experts to diagnose, much less cure, the ills of the poor."[15]

Notably, the riots erupted during the same years as India's food crisis and the war in Vietnam were undermining the high hopes of modernizing the third world. Although most experts on inner-city poverty in the early 1960s believed that, as historian Clayborne Carson has put it, "modernization and assimilation ... were universal and inexorable social trends," Watts, Detroit, and other riots showed that nothing was inexorable.[16]

As with India, when other programs stumbled, family planning programs gained support. In part as a result of the riots, the federal government passed its most important birth-control legislation in 1967. The Social Security Act of 1967 required state welfare agencies to develop family planning programs. By 1968, the federal government, by way of the Office of Economic Opportunity, was supporting 160 birth-control programs in thirty-six states.[17]

Historians have disagreed in their interpretation of these federal programs. Several scholars, viewing the population movement as an outgrowth of the pre–World War II eugenics movement, have emphasized the race and class fears behind them. Derek Hoff, on the other hand, acknowledges the racism that animated some supporters of family planning, but argues that the "ghost of eugenics" arguments overlook many of the progressive forces supporting family planning. He points to the well-meaning and sophisticated approaches of the main backers of family planning legislation in the U.S. Senate, especially Senators Ernest Gruening, Joseph Clark (Pennsylvania), and Joseph Tydings (Maryland). Others have stressed that, whatever the intentions, the general narrowness of the War on Poverty continued with these programs. "By linking federal family planning to the War on Poverty," Donald Critchlow writes, "the Johnson administration implicitly accepted a perspective that a technical solution of having fewer children could be found to those deeper social problems of welfare dependency, out-of-wedlock births, and urban decay."[18]

There is evidence supporting each of these positions. At the time, many minority groups, particularly African American men, distrusted government-sponsored birth control. Although birth control had wide acceptance among the African American middle class and civil rights leaders such as Martin Luther King, skepticism about the intentions of birth control advocates ran deep, especially in Black Nationalist organizations and poorer communities more generally. As early as 1962, Malcolm X met with Planned Parenthood to discuss his concerns that family planning programs were racist. The issue heated up after the Watts riots, especially in response to Moynihan's controversial claims about "pathology" in "the" black family. "It is a plot rather than a solution," members of the Student Non-Violent Coordinating Committee (SNCC) declared at a Southern Christian Leadership Conference (SCLC) meeting in 1966. "Birth control is just a plot just as segregation was a plot to keep blacks down. . . . Instead of working for us and giving us rights—you reduce us in numbers and do not have to give us anything." A March 1968 article in *Ebony* asked, "Is birth control just a 'white man's plot' to 'contain' the black population? Is it just another scheme to cut back on welfare aid or still another method of 'keeping the black man down'?"[19]

Yet Daniel Patrick Moynihan showed a deeper understanding of population than he is often given credit for. Although little noticed, in his 1965 report he identified the biggest problem facing urban black residents as the disappearance of secure, well-paying jobs. Michael Harrington also offered a complex view. "Yes, the figures are overwhelmingly clear . . . we have to have federal support of family planning," he told Planned Parenthood in 1965, "but we must avoid the danger of thinking that it provides a simple way out of the poverty problem. It is one element in a program against poverty, but not a total program." However, not everyone who supported population programs at this time was as informed as Moynihan and Harrington.[20]

The reality of how programs actually played out in local communities was extremely complicated and unexpected, as historian Beth Bailey emphasizes in a study of family planning and population programs in the mid 1960s in Lawrence, Kansas. In 1963, Kansas passed a law allowing public agencies to distribute information about contraceptives. In 1965, driven by concern with population growth, Dr. Dale Clinton, the head of Lawrence's public health clinic began distributing contraception. "The ever accelerating wild rate of population growth," he wrote in the Lawrence newspaper in early 1965, was the world's "most serious health problem." According to Bailey, Clinton did not target the poor, but in addition to students from the nearby universities, his clinic happened to serve mostly lower-income residents of Lawrence, many of whom were African Americans. Those who could afford a private doctor already had access to contraception.

In 1966, concern about population growth inspired a group of Lawrence residents, mostly women associated with the University of Kansas, to organize a Planned Parenthood group. The group consisted of "crusading volunteers" who displayed no particular racial agenda, but wanted to make sure poor women got access to contraception. Unlike Clinton, who just handed out pills, the Planned Parenthood clinic had a broader, but more involved approach. They instructed clients about the dangers of population growth (often using films on population ecology and planned families), conducted examinations, gave counseling, and only then gave prescriptions. They had difficulty sustaining their clinic, and it lasted only from 1967 to 1970.[21]

In the early 1970s, Dr. Clinton was forced to resign by women upset that he refused to do exams and laboratory tests before handing out prescriptions. Research had revealed that such background work led to better individual treatment. Dr. Clinton argued that he was a public health officer, not a doctor for the public. Just as he might give inoculations but not treat broken bones, he distributed pills but did not offer clinical service. He was replaced by a doctor who provided the required exams.

Bailey notes the change as positive, but does not condemn Clinton completely. Even in the mid-1960s, she points out, most Americans disapproved of premarital sex, and many doctors refused to give contraception to unmarried people. "I give prescriptions to married women and to girls who come in and show me they are getting married," one female private doctor told a reporter in

1966, "But to unmarried women? Heavens no!" But in Dr. Clinton's clinic they had few troubles. Marital status, he said, was not a medical criterion. "His passion was population control," Bailey writes, "but because of that passion he provided birth control pills to many young women" without requiring "the demeaning act of lying about an impending marriage." Young people had their own interests, Bailey writes, and often "found practical allies among adults who saw contraceptives as a tool in their struggle against the population explosion."[22]

Historian Johanna Schoen has emphasized similar themes. Writing about twentieth-century birth-control and sterilization programs aimed at poor and nonwhite Americans, she criticizes their often racist and misguided rationales, but stresses that recipients were often able to wrest benefits from the programs nonetheless. "Although clients were not blind to the race and class prejudices that underlay many family planning programs," Schoen writes, "they valued contraceptive information, took advantage of the services offered, and bargained with authorities over the conditions of contraceptive advice." In some cases, this meant the difference between repression and liberation. "Some women," Schoen notes, "suffered negative consequences from their participation" in population programs, but others "found that birth control introduced them to a new way of life—one unthreatened by frequent pregnancy and childbearing." Programs designed with one goal in mind could sometimes be redirected by implementers or recipients for their own purposes.[23]

Reaction of black women to a federally funded Planned Parenthood clinic in Pittsburgh supports Schoen's position. In 1968, a clinic in the Homewood-Brushton area closed down after a male African American resident, equating family planning with "black genocide," reportedly threatened violence if "anyone tries to operate a birth control project in the area." But an informal group of seventy area women, most of them also African American, pushed to reopen the clinic. "We're getting tired of these statements by Mr. Haden. We should make the decision ourselves," said one of the women, adding, "Mr. Haden is only one person—and a man at that—and he can't speak for the women of Homewood." A second woman put it bluntly, "Why should I let one loudmouth tell me about having children?" Pointing to an additional aspect of the problem, another said, "The press has created this monster. I've seen him [Haden] on television more than Bullwinkle." In other cities, other groups of black women, including poor black women, expressed similar views.[24]

In this complicated political terrain, some scientists, conservationists, and environmentalists offered their own ostensibly scientific interpretation of urban poverty. A common theme was that inner-city crowding created social pathologies. One of the first to make this argument was physician Alan Gregg, vice president of the Rockefeller Foundation, in an oft-cited 1955 article from *Science*. As part of an extended (and often borrowed) analogy between rapid human population growth and rapidly multiplying cancer, Gregg singled out inner cities as the most diseased. "At the center of a new growth," he wrote, "necrosis often sets in—the death and liquidation of the cells that have, as it were, dispensed with order and self-control in their passion to reproduce out of all proportion to their usual number in the organism. How nearly the slums of our great cities resemble the necrosis of tumors raises the whimsical query: Which is the more offensive to decency and beauty, slums or the fetid detritus of a growing tumor?" In population arguments, science was sometimes used, often by very privileged individuals, to strip ordinary human beings, especially the poor and nonwhite, of their basic humanity—and portray them as useless, unproductive, and even downright dangerous.[25]

At the 1955 conference on "Man's Role in Changing the Face of the Earth," the urban planner Lewis Mumford associated urban crowding with crime. "Already there are metropolitan bathing beaches and 'wild' recreation areas, where, on a Sunday afternoon in summer, the sign 'Standing Room Only' describes the facilities available. Perhaps some of the perversity and criminal mischief exhibited in our cities, particularly by the more muscular types, may be due to this very constriction of space."[26]

In the 1960s, several biologists began to test Thomas Malthus's claims that high population caused vices. The best known was John Calhoun, a biologist at Johns Hopkins University who studied Norway rats. Under high population densities, Calhoun reported in a 1962 article in *Scientific American*, many female rats were unable to carry pregnancy to full term or survive delivery, and many male rats showed behavioral disturbances including sexual deviation and cannibalism. Under such conditions, some males occasionally went "berserk, attacking females, juveniles and the less active males, and show[ed] a particular predilection—which rats do not normally display—for biting other animals on the tail." It is obvious, Calhoun concluded, that the behavioral patterns of the Norway rat "break down under the social pressures generated by population density." Ominously, he alluded to "analogous problems confronting the human species."[27]

Other researchers followed. In 1964, Hudson Hoagland reported in the *Bulletin of the Atomic Scientists* (and *Time*) that overcrowded female flour beetles destroy their eggs and turn cannibalistic, while males lose interest in females. In 1965, popular nature writer Sally Carrighar wrote in *Wild Heritage* about communities of lemmings who, because of overcrowded conditions, raced into the sea and drowned themselves en masse. In a March 1970 synopsis of his book *The Human Zoo*, Desmond Morris, the curator of mammals for the London Zoological Society, suggested that overcrowding of animals can cause "social chaos" and "the collapse of the animals' social structure." Researchers such as Morris did not hesitate to extrapolate to human crowding. Humans create overcrowded conditions in cities, he pointed out, "And then we wonder why there is so much violence in the streets. What is surprising is that there is not much more violence."[28]

In the late 1960s, environmentalists, too, began to point to urban violence as an example of U.S. overpopulation. Borrowing John Calhoun's logic in their book *Moment in the Sun*, for instance, Robert and Leona Rienow blamed the recent riots on overcrowding. In 1968, Paul Ehrlich would make similar arguments. Urban poverty and violence were part of a growing sense among environmentalists that modern society had grown far out of balance.[29]

In contrast to out-of-control urban areas, many Americans held the assumption that suburban life was normal and good. "What every *normal* man wants for himself and his family," an ethnologist quoted in an October 1970 *Ladies' Home Journal* article on "people pollution" explained, "is a detached house and an adequate garden." Some environmentalists favored this ideal, but others found its postwar American application deeply flawed.[30]

Losing the "Good Life": Suburbs

If Malthusian anxiety about inner city poverty seemed to parallel worries about the third world, concern about overpopulation in middle-class spaces—suburbs, national parks, and wilderness—presented a paradox. These were hardly places of starvation and civil unrest. Nonetheless, deep and growing concerns about overcrowding in these areas pervaded the 1950s and 1960s. During these years, more magazine articles about population invoked crowding in mostly white middle-class spaces like suburbs than any other concern.[31]

Suburbs were nothing new in the United States. As the United States industrialized in the nineteenth century, wealthier Americans began to link cities with dirty and noisy factories, immigrant slums, and machine politics, and began to see spaces on the edge of cities not as desolate wastelands or productive space for food, but as quiet, safe havens from the city. Trolley lines reached these semi-rural, semi-urban spaces in the 1880s. By the turn of the century, historian Jennifer Kalish writes, "the suburban, single-family, detached dwelling lay at the heart of the American dream."[32]

Suburban development after World War II, however, was new in scale and type. During the 1950s alone, over 15 million new suburban homes sprouted up; each year developments consumed a million acres, an area the size of Rhode Island. In 1950, 36 million Americans lived in suburbs; by 1970, the year of the first Earth Day, 74 million did, more than in any other form of community. An urban housing shortage and cheap land made newly available by cars drove the boom, as well as interstate highways and guaranteed mortgages subsidized by the federal government. Family considerations were also important. Almost all new suburbanites moved, James Patterson emphasizes, "because they were seeking a more satisfying family life." Race also played a role: some whites moved to suburbs to get away from the rural, poor African Americans who had moved to urban areas in the thousands.[33]

In another example of how war shaped American environments, two of the technologies that gave postwar suburbs their distinctive feel—bulldozers and assembly-line house construction—came out of World War II. Bulldozers were used to clear jungle islands in the Pacific, and assembly-line house construction methods, where workers specializing in one skill moved from site to site, had been developed to build military bases quickly. These methods were most famously adapted by the Abraham Levitt and his sons, who did for houses what Henry Ford had done for the car, and Ray Kroc for the hamburger. In Levittown, New York, their assembly-line techniques built thirty houses a day, for a total of 17,000 houses and 82,000 residents.[34]

These new methods made houses cheaper than ever, but often at the expense of the natural environments. Bulldozers remade wetlands, hillsides, and other places Americans had rarely built on before. Developments gobbled up tremendous amounts of rural space, especially as average lot size increased, often damaging watersheds and destroying natural beauty. In 1950, urban and suburban areas covered 5.9 percent of the United States; in 1970, 10.9 percent. Postwar suburbs had few parks and outdoor recreation spaces, but lots of private lawns, which required heavy doses of chemical fertilizers and pesticides.[35]

In response to suburban development, an "open space" movement sprang up in the late 1950s and early 1960s, led by William Whyte, author of *The Organization Man* and editor of *Fortune* magazine. Whyte lamented the loss of prime agricultural and recreational spaces to bulldozed, parkless, monotonous subdivisions. He was joined by environmental and—significantly—population-planning organizations. "Anybody who has seen fields, forests, orchards, and meadows obliterated to provide sites for ticky-tacky housing," the Sierra Club board of directors wrote in 1965, "should realize that overpopulation is today's problem as well as tomorrow's." A 1966 Planned Parenthood report specifically targeted "the space eaters." Malthusian arguments, many drawn from an overseas context, added new steam to the open-space movement, and the problems of sprawl helped suburbanites better understand the issue of population growth, which had mostly been framed internationally.[36]

One of the earliest and most influential of the Malthusian arguments about suburbs came from the biologist Raymond Dasmann in *The Destruction of California* (1965). Dasmann's chapter titles reveal his mood: "A Not So Golden State," "The West Ends Here," and "Once There Was a Place Called California." Dasmann especially decried how housing and industry had despoiled the land on the edge of cities, engulfing farm and forest, marsh and pasture with "no end in sight except the dismal one of a gigantic, disorganized megalopolis." He blamed population growth, opening the book by describing how upset he was the day he saw a sign on the Bay Bridge celebrating California overtaking

New York as the nation's most populous state. "The greatest reality in California today is the population problem," Dasmann wrote. "It touches on every facet of the land and its life, from the conservation of wilderness and national parks to the restoration of meaning and pleasure to life in the cities." Drawing from the language of wildlife biology, Dasmann even spoke of a "human population irruption that may well end tragically both for the people and for the land." He also warned of a moral decline. Mindless growth, he said, had fed a "spiritual rot."[37]

Environmental Malthusians often mixed environmental critiques with cultural laments about mass society. Such laments were common at the time, especially by highbrow cultural critics such as Whyte, David Reisman, and John Kenneth Galbraith. They were insightful but sometimes condescending. To these critics, the uniformity of mass-produced housing epitomized the monotony, conformity, and general lifelessness of a mass society organized by soulless corporations and regimented bureaucracies. Suburbs seemed the postwar equivalent of the army base, physically and culturally. What would such structures mean for individual freedom and human creativity—for culture with a capital C? "Mass culture," Dwight Macdonald wrote in 1953, was "a cancerous growth on High Culture." Is this what progress had brought? Perhaps no one summarized these complaints better than folksinger Malvina Reynolds, who sang of the new suburbs as "little boxes made of ticky tacky . . . little boxes all the same."[38]

Not long before writing *The Destruction of California*, Dasmann had written about global overpopulation in *Environmental Conservation* and *Last Horizon* (1963) and had spent two years researching in Africa. His extensive international experience and writings suggests that the sense of crowding within middle-class America and the concerns about overpopulation in the third world were not altogether separate concerns. Indeed, it's possible to see them as structurally complementary, where the thought of the one often helped define the other. What middle-class Americans hoped to create in the suburbs—a high quality of life—was the opposite of what they feared held sway in the overpopulated third world. The converse was also true: when Americans expressed concerns about overpopulation in the third world, they were often implicitly comparing them to idealized views of the family residential patterns epitomized by the suburbs. The same could be said about the relationship between the inner city and suburbs. When students and practitioners read Dasmann's line in *Environmental Conservation* about "Anglo-America" as an "island of prosperity," did they think about the global context that he intended, or about urban poverty and "white flight" within the Unites States?

For most in the open space movement, the problem of suburban crowding could be solved by dispersal or concentration. William Whyte, for instance, adopted a conservative approach: Americans, he believed, could solve the open-space problem without infringing on the free market in land and individual property rights. Instead of regional planning agencies and zoning, Whyte favored voluntary measures such as conservation easements. "If there is to be any hope of having open space in the future," he wrote in his 1968 classic *The Last Landscape*, "there is going to have to be a more efficient pattern of building."[39]

But for others, Whyte's conservative approach did not go far enough. They scorned his failure to challenge the nation's pro-growth mentality. Some proposed the governmental regulation of land use, such as zoning, which Whyte had rejected. Others, though, wanted quicker and deeper social change. Rejecting the incremental approach of many antigrowth groups, people such as Sierra Club director David Brower, according to one observer, "increasingly believed that an environmental Armageddon was taking shape, a battle between the forces of darkness and light." Brower and his allies zeroed in on what they believed to be the underlying problem: uncontrolled population growth. "There are many . . . propositions being formulated for alleviating our open-space crisis," Robert and Leona Rienow wrote in *Moment in the Sun*, their bestseller from 1967, but "they are cures of symptoms, not of the disease. They all strive toward accommodating the agglomerating masses. . . . Our 'cures' for congestion thus far are all aspiring for the headaches, while the disease spreads in the nation's body untreated." The true disease, they said, was biological: human overpopulation. Human fertility, they cautioned, was a "monstrous thing."[40]

Critics like Dasmann, Brower, and the Rienows saw new babies as the driving force behind an environmentally devastating, consumption-based growth economy centered in the suburbs. By the mid 1960s, the economy that Aldo Leopold, William Vogt, and Fairfield Osborn had warned about had grown tremendously. "It is now difficult after twenty-five years," economist John Kenneth Galbraith wrote in 1981, "to give an impression of the commitment of economists and the official intellectual establishment in the 1950s to the absolute importance of maximizing the output of goods—of the increase in the GNP as a measure of social excellence." Presidents Kennedy and Johnson especially emphasized consumption-based growth. Keynesian thinking was so dominant in these years that Keynes landed on the cover of *Time* in December 1965. The results were spectacular. The U.S. economy grew every single month from 1961 to 1969, a new record. From 1960 to 1973, growth averaged close to 4 percent per year.[41]

Environmental critics hated this growth. "The solution of conservation problems," Dasmann wrote, "demands a rejection of the expanding-economy concept." "An increased idolatry of production," the Rienows wrote, "has been sweeping the land, and its icon is the growth chart." They lamented what economist Kenneth Boulding in 1965 called the "cowboy economy," a Keynesian economic system, based on frontier ideas of limitless resources, measured more by the quantity of supplies it consumed rather than the quality of the products and experiences it produced.[42]

Like many others, environmentalists saw suburbs as central to this growth. Coming out of World War II, Adam Rome writes, American policymakers saw suburban growth as central to economic prosperity. In scholarly studies, magazine articles, policy conferences, business meetings, agency memos, and congressional hearings, he writes, a great range of people "argued that a full employment economy required a homebuilding industry capable of serving a mass market." New houses created construction jobs, but they also drove the consumer appliance boom. "The production of houses, automobiles, electrical devices, schools, and hospitals," Rome says, "went hand in hand."[43]

For Malthusian critics, suburban consumer culture also went hand in hand with the production of babies. Indeed, babies often drove the whole system. "The twentieth century is not likely to be the Century of the Common Man," the sociologist David Reisman noted in 1959, 'but the Century of the Child.'" Many of these children grew up in suburbs, often new suburbs. Babies were so prominent in Levittown, the huge subdivision outside of New York City, that people started referring to it as "Fertility Valley" and "the Rabbit Hutch." Writing about suburban California, historian Kevin Starr has noted that "in each home—or at least in the myth of each home—was a crypto-theological, even overtly theological, concept of marriage as a most sacred and most necessary institution, based on committed sexuality, procreation, the nurture and education of children. So, at least, went the most accepted story line of the postwar era." Dasmann and the Rienows would have agreed with historian Steve Gillon when he wrote that the baby boom "transformed America into a nation of schools, suburbs, and station wagons."[44]

Environmental critics were responding to the vocal pro-population growth advocacy of many politicians, businessmen, and economists, who viewed babies as engines of economic growth. "Your future is great in a growing America," read signs in the New York subway. "Every day 11,000 babies are born in America. This means new business, new jobs, new opportunities." A study of postwar magazine articles on population found that a large portion favored increased growth. "Babies Equal Boom," *Reader's Digest* proclaimed in 1951. A 1958 cover story in *Life* called kids a "Built-in Recession Cure." "Listen!" a caption in *Minneapolis Star Tribune* said next to a picture of a crying baby, "You can hear the population explosion in the Twin Cities' suburbs. . . . Dad may disagree (especially at 3 A.M.), but to a marketing man a baby's cry is sheer music."[45]

Environmental Malthusians, however, saw children through the lens of the resources they would gobble up. Although a "disarming little thing," the Rienows pointed out, each new baby in the Keynesian economy will become a voracious consumer, beginning with 26 million tons of water, 21,000 gallons of gasoline, 10,150 pounds of meat, 28,000 pounds of milk and cream, 9,000 pounds of wheat, among other foodstuffs. He will need $5,000 to $8,000 for school building materials, $6,300 worth of clothing, $7,000 worth of furniture. They concluded, "We can no longer revel in abundance." Children also required a lot of space. In San Francisco, Dasmann noted, the middle class fled the city "hoping to find a lawn, a garden, and a quiet street where children could grow in the open air." Children also remade house structure, car size, and, indeed, much of the postwar economy. They created suburban sprawl as much as the bulldozer and interstates.[46]

Environmental Malthusians often reduced consumption to a factor of population growth: as population went up, so did consumption. But some recognized that it was not just children, but middle-class ideas about the proper way to raise children, that shaped house and subdivision architecture and drove consumption. When Dr. Spock told parents in his hugely influential advice books that they should not let children sleep in parental beds, that required more space. Ideas about proper child rearing were central to the quality of life and the standard of living. Thus, curtailing this level of consumption was difficult. Part of why Malthusians targeted population was because they saw fighting this consumption as next to impossible. "Nothing is sacred or too remote," a Sierra Club member wrote in the club's magazine in 1964, "when people multiply too rapidly and excessively and need more and more and more housing plus all its ancillary structures, facilities and space." As a way to reduce overall economic growth, it was easier to fight the production of children than the consumption Americans felt each child deserved. "The pressures of consumption," the Rienows concluded, "are relentless."[47]

In retrospect, the environmental Malthusian view of suburbs in the 1950s and 1960s seems mixed. Blaming suburban sprawl and consumption on overpopulation was overly simple and even misanthropic. Population growth alone did not create suburban sprawl and consumption. Many cities in other countries have large populations but few American-style suburbs. And American cities such as Portland, Oregon, grew in numbers in the postwar years but managed to minimize the negatives of sprawl through better planning. Moreover, consumption during these years actually far exceeded population growth. From 1950 to 1980, for instance, the U.S. population increased by 50 percent but the number of automobiles increased by 200 percent. This was especially the case in suburbs. Indeed, some of the population movement's documents reflected this. "Americans are growing fast," a 1966 Planned Parenthood document stressed, "but they are gobbling up space even faster." On top of this, the Malthusians fell in a little too easily with the highbrow critics, making them seem like snobby "not in my backyard" types. That said, despite their flaws, Dasmann and the Rienows and many other Malthusian environmentalists called attention to the significant environmental problems caused by suburban sprawl and overconsumption—and did so in a way that contributed to the growing environmental awareness of the 1960s.[48]

Overcrowded Parks and Wilderness

Concern about overpopulation also influenced early postwar American ideas of wilderness. At first this might seem surprising. What did poverty and food shortages in India, inner-city political instability, and suburban sprawl have to do with snow-covered mountaintops, vast canyons, and ancient forests? Yet many Americans tended to think of the country's national parks and wilderness areas as places without people. Overcrowded places represented the diametric opposite of wilderness—the conceptual flipside.

By the 1960s, American national parks and wilderness seemed increasingly at risk. Unusual problems began to plague America's parks, such as traffic jams and even smog. Many observers blamed population growth. "National Parks are created for people," William Catton Jr., argued in a 1964 *National Parks Magazine* article, "but people—as their numbers increase—are a threat to national parks." Speaking of a "human tide," Catton identified two ways in which people threatened the parks: by creating pressure to open parkland for resource development, and by the vast numbers visiting parks. Catton even wondered whether he, himself, had the right to have a fourth child. "We live in a finite world," he wrote, "whose human population is already so large that it has become a significant ecological factor."[49]

In another *National Parks Magazine* article, the biologist Bruce Welch added a cultural dimension to the problems that Catton identified. With continued population growth, Welch noted, "there will be a limit to human tolerance, the advent of social and cultural stagnation, the disappearance of freedom—and compassion—and sensible morality, the reign of an artificially tranquilized and emotionless sub-animal existence." Although many conservation-minded biologists strove to show how the human species was governed by the same laws that governed other animals, Welch tried to prop up those boundaries, to prevent humans from sinking to an animal level. "If we hold wilderness and natural beauty to be important for man's most meaningful habitation of the earth," he concluded, "then we must act immediately to curb his uncontrolled increase in numbers."[50]

Concerns about population growth, international war, domestic poverty, and race added momentum to the wilderness protection movement during the 1950s, culminating in the run-up to the Wilderness Act in 1964. The "real threat to wilderness," Welch wrote in 1963 at the height of the campaign for the Act, was population growth. The core problem was not saving and protecting untouched land, but "stopping the ominous, uncontrolled expansion of our population." In 1964, a headline in the *Population Bulletin* summed up the problem: "Outdoor Recreation Threatened by Excess Procreation."[51]

Concern about overpopulation as a threat to wilderness went back to at least the late 1940s and early 1950s. Harvey Broome, a founding member of the Wilderness Society, often wrote about population growth in his hiking journal, reprinted in *Living Wilderness* in the early 1950s. "I have reservations about the ever expanding dynamism of the present day," he wrote one day, taking a break from the trail. "It must rest upon definitely circumscribed resources, and lands. It springs from an expanding population which must perforce bear an increasingly unfavorable relation to resources." Echoing William Vogt and Fairfield Osborn, Broome connected his concerns to geopolitics: "Our very advances have brought closer the dreadful clouds of war." He continued, "Do the people of the earth—or rather the people of these snow-balling economies of the United States and Russia—lack the ability to assimilate all the potentialities of their headlong rush? Or is the rush in itself so intoxicating that there is no turning back?" In the 1950s, it seemed, international affairs affected Americans even during quiet moments walking around the woods.[52]

Discussions about population growth in the Sierra Club in the late 1950s and early 1960s show the range of ideas circulating among traditional conservation groups as the postwar environmental movement coalesced. Perhaps the first formal mention of population came from David Brower, the Sierra Club's executive director, at the 1957 Wilderness Conference: "A serious problem, upon which no conservation organization I know of has adopted a policy, is the population problem—an especially touchy cat to bell." At the same meeting, Lowell Sumner spoke of "more people, more complications, more stress." Sumner, a career Park Service biologist greatly influenced by *Man's Role in Changing the Face of the Earth*, helped introduce the Sierra Club to ecology.[53]

The Sierra Club first took up the issue seriously at its 1959 Wilderness Conference. In a special presentation, the UCLA ecologist Raymond Cowles argued that population was the root cause of many threats to natural beauty. Cowles, a South African who had moved to southern California in 1916, pointed to the "subtle corrosion of our standard of living" as a result of population growth. The "amenities of life," he explained, "are dwindling away rather rapidly because of overcrowding." Examples were rife: the pressure for lake fronts, stream sides, beach frontage, mountain property—"anywhere there is beauty." "Underlying everything," he stressed, "is our population size, again and again." The U.S. population, he said, would double in forty years to 340 million.[54]

Cowles also added an international dimension to the Sierra Club discussion. "We, as organisms, have the capacity to continue reproducing . . . until we will resemble China, India, and other such countries." More specifically, he likened the United States to his homeland, South Africa, which he had recently visited for the first time in decades. In South Africa, Cowles saw "an alarming future," in which "man's needs and his numbers will increasingly jeopardize the wilderness and himself." Using biological concepts to explain racial competition, he noted that because South Africa's blacks and whites belong to "nearly identical subspecies of an organism," the "inevitable biological solution is war between the subspecies." Whites, blacks, and the mixtures are having a "race to see who outpopulates the other." He specifically tied wilderness protection to white control: "I am convinced that preservation of South Africa's wildlife and wilderness areas, at any time beyond the next generation, can continue only so long as there is White domination."[55]

Cowles saw the potential for instability around the world, wherever the "backward people, impoverished people, or wealthier people with a steadily diminishing supply of food . . . are apt to become restive." He noted the Cold War consequences: "This is particularly acute to us today, because we [Americans] are having to win the friendship of the so-called backward people, or the underdeveloped countries, throughout the world." The United States was directly tied to these places through its food shipments. Because South Africa and places like it were dependent upon U.S. surplus grains, what would happen if the U.S. population doubled and it ran out of food?[56]

Cowles concluded by spelling out the consequences for American national parks and wilderness. "If man reproduces unchecked, what consequences will there be to the remaining small bits of uninfected land, the islands, the wilderness areas, the sanctuaries, the parks, and so on?" His answer was no surprise: "Unless we know how to hold our wilderness areas against the rising needs of an increasing abundance of humanity there does not appear to be much point in establishing them, since they are ultimately to be ravaged by would-be exploiters supported by a self-interested and unappreciative, or unwise, public."[57]

The ensuing discussion witnessed some dissent but a lot of support. It displayed the mix of conservative and liberal ideas that often came together in population politics in the 1960s. To the well-known biologist and conservationist Frank Fraser Darling, the real problem was America's "expanding economy," which he labeled "the greatest continuous illusion that we could possibly have." Then, pointing out that in Africa "uncivilized" people outnumbered the "civilized" ten to one, he called for a remedy that was far ahead of its time. "Education," he said, "is the only possible means by which to raise the standard of living and lower the birth rate." "Until you can bring the womenfolk of Africa and Asia along at the same speed as you educate the men, you cannot hope for any different attitudes." Wilderness lovers, he said, needed to consider these issues deeply. Educating women in Africa may seem "a long way from wilderness, but I am sure it's a very important part of it."[58]

Cowles himself suggested a far different set of remedies. He called for the development of birth-control technology that would be easy enough for simple folk around the world to use. Echoing William Vogt, he suggested giving financial bonuses for not having children at all. In addition to its other benefits, he said, this would "act on a eugenic basis." These bonuses, he said, again echoing Vogt, "would be most effective in the poorest and least educated and least forward-thinking part of our population: those that are usually dependent, those that require social security and unemployment compensation."[59]

Population growth, it appears, took on many meanings for conservationists concerned about wilderness. Cowles's presentation revealed a reductive focus on population as part of a larger defensive struggle, conceived of racially. Darling's comments, on the other hand, reflected a deep concern about unbridled economic growth, a common concern among Malthusian environmentalists.

At the end of this 1959 meeting, the Sierra Club adopted a resolution warning of overpopulation and urging that the government give it "urgent attention." The resolution read, "As wilderness is one of the first of the earth's important natural resources to come into short supply as a result of world-wide human 'population explosion,' the final destiny of all wilderness may hinge on this trend. This Conference . . . recommends that . . . social, governmental and other appropriate agencies give immediate and urgent attention to the development of desirable population controls." Several members voted no, signaling a divide on the issue that would last at least until the end of the 1960s. One person particularly impressed with Cowles and the larger discussion was the man who invited him to speak, David Brower, who would write of the discussion's influence on him in his memoirs thirty years later.[60]

As Raymond Cowles's views suggest, postwar problems such as suburban sprawl, national park crowding, and even wilderness protection were domestic concerns never entirely distinct from the outside world: the larger international context often added layers of meaning to these spaces. A rich visual example of this global/local mix is provided by the well-known photo and essay book by Ansel Adams and Nancy Newhall from 1960, *This Is the American Earth*. Flip through this huge coffee-table book, which was part of the Sierra Club's campaign for wilderness protection, and you will find predictable images of people-less landscapes: rugged mountain ranges bathed in light, storm clouds hovering over conifer-covered valleys, flowing streams blanketed in snow. Among these pictures of American wildness, however, is a surprising image: a crowd of thousands of Indians bathing along the banks of the Ganges River. The picture underscored the problems of overpopulation, as did a second picture of India depicting an anguished Indian woman holding a baby and labeled "famine." Two-thirds of the population of the world, the text read, "find want and hunger multiplying like themselves." Together with two other unusual pictures for a wilderness book—one of a dirty, turn-of-the century American slum and the other of a postwar subdivision with seemingly endless rows of cookie-cutter houses—these images warned of the "Chinification" of the United States due to population growth: "Headlong, heedless, we rush . . . to blast down the hills, bulldoze the reeds, scrape bare the fields to build predestined slums, until city encroaches on suburb, suburb on country." The American Earth, it seemed, could not be understood fully without understanding the rest of the world.[61]

In the early and mid-1960s, the population issue came up frequently at Sierra Club annual meetings. One frequent advocate of population control was Daniel Luten, a chemist who spent two years in Japan as a resources advisor to General Douglas MacArthur. "Does a wilderness program, a wilderness policy, without a population policy make any sense?" he asked at the 1961 meetings. Luten also specifically attacked America's growth fetish: "If we believe, with Lord Keynes, that a capitalist economy must grow to survive, then we will perhaps seek a society with an ever-growing population." Letters to the editor of the club's magazine also often addressed population growth. "For two decades now we have seen the effect of 'population explosion' on our several resources, particularly the loss of scenery, and recreation areas," one member wrote in 1964. "Conservationists should be the leaders in initiating a move toward population increase control."[62]

Another member, writing in 1965, showed how the population issue wrapped together the war on poverty, the welfare of women, and the conservation movement. "Records show," he wrote, "that there are thousands of poor women who do not have access to medical help who would greatly appreciate relief from the unending bearing of children. Provision for this group, so important in relation to the anti-poverty program as well as to conservation, should be made by the dissemination of birth control information." Although showing a hint of sympathy and understanding, the author also displayed a darker condescension and defensiveness. He ended by quoting a judge from a local town: "It is the desperately poor, the ignorant, the constitutionally inadequate, the culturally deprived, the mother supported by welfare, who have neither the knowledge nor the means to use contraceptive devices. From these same persons, in these same families, are bred our criminals, our delinquents, our dependent children and our mental defectives."[63]

Not every Sierra Club member approved of the focus on population growth. Club member and *Scientific American* publisher Gerald Piel offered a strongly worded dissent at the 1961 annual meeting. "More than once in the publications of the Sierra Club one finds the dour shade of Thomas Malthus presiding in the chair," he wrote. "Misanthropy on those occasions becomes explicit. We hear the good old days recalled when 'viruses, bacteria and starvation [curbed] the growth of population.'" He also took issue with likening of human populations to a cancer and the anti-egalitarian tone of the "No Trespassing" protections around wilderness areas. The men and women "who dreamt the American dream not quite two centuries ago had a different vision." Piel's comments earned an asterisk in the text and a note from David Brower, as well as a specific rebuttal in a speech by Fairfield Osborn at the 1963 meeting. Brower argued that the Sierra Club devoted a large part of its program to encouraging the multitude to get into the wilderness, and that, "Were the H-bomb to eliminate pressures man failed to reduce voluntarily, new days wouldn't be good, either." Osborn denied that concern about new health measures was in fact misanthropic.[64]

In March 1965, the Sierra Club again passed a resolution on population similar to its 1959 resolution. The population "explosion," it said, had upset the ecological relationship between mankind and his environment, caused a scarcity of wilderness and wildlife, impaired the beauty of whole regions, and reduced the quality of living. It called for increased education. Other population organizations also sometimes took up the wilderness cause. In a 1966 Planned Parenthood pamphlet, August Heckscher, director of the Twentieth Century Fund, wrote, "The result of technological forces—combined with increased numbers of people, increased wealth, increased mobility and increased leisure—is to threaten the existence of every geographical place which is separate and distinct, every integrity which gives to the individual the possibility of standing apart and meeting the world on his own terms."[65]

By 1967, as inner-city and suburban problems intensified as well as problems in India, calls mounted within the Sierra Club for a much stronger stance against population growth. In the spring 1967 Wilderness Conference, Robert Rienow made an impassioned plea for more action: "Today we cannot save the land unless we are ready to stir the political mud." He called for a "full-scale mobilization of political power." David Brower, the leader who had over the previous fifteen years helped turn the organization from mostly an outing club to a national advocacy organization, led a faction calling for more urgent and broader action.[66]

In early 1968, Brower, a master of public relations, found an opportunity to do so: he met an outspoken Stanford biologist who shared his worries about overpopulation—Paul Ehrlich—and, together with Ian Ballantine of Ballantine Books, convinced him to write what would become *The Population Bomb*, published in May 1968.

Brower's worries about population, expressed in the foreword of Ehrlich's book, signal a transition in American environmental politics. Conservation organizations like the Sierra Club, he wrote, have been "much too calm about the ultimate threat to mankind." Protecting wilderness was crucial, of course, but so was the "nine-tenths or so of the earth that had already felt his [Man's] touch." Conservationists needed to think more about broader threats to nature in cities and suburbs. They "could start with Manhattan, or Los Angeles" or wherever "growth itself was growing." Brower wanted to fight quality-of-life problems wherever they existed.

For Brower, international issues such as war overlapped with concerns about wilderness and out-of-control domestic growth. "Man can undo himself," he wrote in the first line of the foreword, "with no other force than his own brutality." Echoing Aldo Leopold in "A Biotic View of Land" and Osborn in *Our Plundered Planet*, Brower, a veteran of combat in Italy during World War II, lamented the "new brutality" scourging civilization. "The hand that hefted the axe against the ice, the tiger, and bear," he said, quoting Loren Eiseley, "now fondles the machine gun." Overpopulation, Brower believed, drove this violence. "The roots of the new brutality . . . are in the lack of population control."[67]

In 1968, all of these complicated and seemingly unsolvable international and domestic issues—India, war, the inner city, and middle-class crowding—appeared headed for crisis. Combined, they created the "feel" of overpopulation just as the environmental movement was taking shape in the years before the first Earth Day in 1970.

Nobody brought together these domestic and international concerns as powerfully as Paul Ehrlich in *The Population Bomb*.

6

Paul Ehrlich, the 1960s, and the Population Bomb

"By coincidence, my field of expertise (population biology) has overlapped many of the world's most pressing problems."
—Paul Ehrlich, 1982

"There can be little doubt that cultural factors strongly influence the biologist's view of the structure of nature."
—Paul Ehrlich and Richard W. Holm, *The Process of Evolution*, 1963

In the opening of *The Population Bomb*, Paul Ehrlich famously wrote, "The battle to feed humanity is over. In the 1970's the world will undergo famines—hundreds of millions of people are going to starve to death in spite of any crash programs embarked upon now." Ehrlich rejected Lyndon Johnson's solutions to the world food crisis as too little, too late. Because the experts had overlooked environmental concerns, he argued, even their worst-case scenarios were overly optimistic. Massive food shortages had already started and even worse shortages were on their way.

Ehrlich called for some measures that struck many as long overdue, such as sex education, expanded birth control, and abortion rights, but also authoritarian measures that seem shockingly draconian in retrospect. At home, he fudged on the issue of coercion, not calling for it directly but not renouncing it either. "We must have population control at home," Ehrlich wrote, "hopefully through a system of incentives and penalties, but by compulsion if voluntary methods fail." Internationally, he pulled few punches, endorsing scaled-back food aid for the starving, and even forced sterilization. "We must use our political power," he wrote, "to push other countries into ... population control." Appearing in 1968, *The Population Bomb* sold over two million copies and made Ehrlich into the best known of the environmental Malthusians.[1]

Three puzzles stand out about the environmental Malthusians of the late 1960s and early 1970s. First, why did they respond to population growth in India and elsewhere so differently than the Johnson and Nixon administrations? What was lacking in the remedies that many other Malthusians believed in, such as voluntary birth control programs and the "green revolution" in agriculture? Second, why were some environmental Malthusians, including some liberal-progressives, willing to turn to such drastic remedies such as food hoarding and coercive population control? What was their relationship with the eugenics movement and how should we understand them politically? And finally, why did the issue of population seem to resonate with so many Americans during the late 1960s? Why did Ehrlich and other environmentalists suddenly seem like prophets, especially to America's middle-class youth?

Answering these questions, and understanding Ehrlich, requires revisiting some predictable places, such as postwar biology, Rachel Carson's fight against pesticides, India's food shortages, and Johnson's "agricultural revolution," but also events that historians rarely associate with environmentalism: the civil rights movement and inner city riots, the Vietnam war, the youth rebellion, and the "counterculture." In the most tumultuous years of the late 1960s, Ehrlich came to see biology, especially population biology, as a master key that could unlock the nation's most intractable problems, from suburban sprawl and race riots to famine in India and international communism. His revolutionary ideas regarding the nation's economic growth policies and the social conservatism governing sexual and birth control matters contained great insights and had tremendous appeal at the time, yet, like much of 1960s liberalism, often showed a shallow understanding of poverty and a lack of understanding of the potential abuses of government power.[2]

The Evolution of an Evolutionist

Ehrlich was fascinated by biology from a young age. Born in 1929 to liberal Jewish parents, he grew up in Maplewood, New Jersey, a suburb outside Newark, "chasing butterflies and dissecting frogs." As a high school student, he liked to visit the biologists at the American Museum of Natural History in New York. He studied biology at the University of Pennsylvania, and then pursued a Ph.D. in entomology and population studies at the University of Kansas, graduating in 1957. While an undergraduate, Ehrlich read William Vogt's *Road to Survival* and Fairfield Osborn's *Our Plundered Planet*.[3]

At mid-century, American biology was undergoing a revolution. Few people realize that not until the 1940s and 1950s did most biologists, even those who enthusiastically accepted evolution, fully accept Darwin's idea of natural selection, where genetic advantages are passed along to subsequent generations. Within a decade or two of the 1859 publication of *On the Origin of Species*, most had accepted Darwin's ideas about the variation of species, but for decades they debated the exact mechanism by which evolutionary change occurred. "Well into the twentieth century," historians of science Gregg Mitman and Ronald Numbers have written, "the overwhelming majority of evolutionists in the American scientific community relegated natural selection to a secondary role in the evolutionary process."[4]

All this changed in the middle decades of the twentieth century with what is called the Darwinian Synthesis, which combined advances from genetics and field biology.[5] During the 1920s, geneticists began applying advanced mathematical techniques to entire biological populations, as opposed to lone individuals, and during the 1930s, Ukrainian-born evolutionary biologist Theodosius Dobzhansky combined these insights with observations of wild populations. The result was, as Mitman and Numbers write, a "consensus about the efficacy of natural selection in evolution." "The theory of evolution," the French biologist Jacques Monod wrote in 1971, "did not take on its full significance, precision, and certainty until less than twenty years ago."[6]

Ehrlich cut his teeth intellectually and professionally on the Darwinian Synthesis. In the late 1950s, he helped with one of the earliest investigations of natural selection at work in the wild—a study of Lake Erie's water snakes. Examining the changing proportion of snakes with bands and snakes without bands in different environments, Ehrlich believed he was seeing evolution at work before his very eyes. This research profoundly influenced him. "My early experience with that project greatly shaped my career," he later wrote. In his 1963 textbook on evolution, Ehrlich explicitly acknowledged his intellectual debts to the pioneers of the Synthesis: Theodosius Dobzhansky, Ernst Mayr, George Gaylord Simpson, G. Ledyard Stebbins, and Sewall Wright.[7]

Within the natural world, natural selection highlighted the importance of environmental scarcities, population growth, and reproduction. Not coincidentally, it was the part of Darwinian thinking that owed most to Thomas Malthus, whose ideas formed a framework Darwin used to reach his revolutionary conclusions.

The new emphasis on natural selection also shaped how Ehrlich and many other biologists thought about human society. First, it reinforced a sense that human beings, like all animals, had strong sexual drives. Others had made this point before, including Edward Murray East in the 1920s and Vogt and Osborn in the 1940s, but the Synthesis gave it more scientific backing. As it happened, this science coincided with the movement toward greater acceptance of sex then spreading through American society. Second, natural selection challenged the idea that human beings stood apart from—over and above—nature. Contrary to popular belief, Darwin's *On the Origin of Species* did not fully undercut faith in human specialness. Well after 1859, many Americans, including biologists, believed that humans had evolved from other species, but that they had evolved into something qualitatively different and unique. By demonstrating that human beings were just like other animals—bound to an interconnected world of plants, animals, and abiotic materials, where survival depended exclusively upon scarce resources—natural selection made this line of thought more untenable.[8] Finally, natural selection challenged the idea of social progress that had lived on well beyond *Origin*: natural selection emphasized that evolution came about by chance variation and adaptation, not by divine design. To Julian Huxley, one of the key theorists of the Synthesis, change became "a product of blind forces." The evolution of humanity was due to the same forces as "the falling of a stone to earth or the ebb and flow of the tides."[9]

These changing ideas shaped environmental politics. According to historian Susan Schrepfer, even after early twentieth-century conservationists recognized humanity's misuse of nature, they nonetheless maintained a faith in science and technology at least in part because they believed in a pre-Synthesis model of directed evolution that placed humans at the top of a hierarchical animal kingdom. But after World War II, Schrepfer writes, the Synthesis's rejection of directed evolution "undercut the hierarchical ecology that had placed man and his technology confidently at the apex of natural history." In the 1940s and 1950s, philosophically inclined scientists who popularized the new synthesis, such as Huxley, microbiologist Rene Dubos, and Loren Eiseley, labeled evolution "opportunistic," "random," "unconscious," "disorderly," and "lacking in design." Natural selection, Huxley pointed out, "does not ensure progress." In 1959, biologist Garrett Hardin was asked if progress was relevant to geology. No, he said. Geology cannot "be said to involve direction at all." The Synthesis, Schrepfer concluded, "proved just as conducive to the questioning of human dominance and material progress as did awareness of environmental deterioration. The synthesis underscored the chance course of natural history and the equality of species."[10]

The idea that change did not mean progress had special meaning for biologists worried about human population growth. Consider, for instance, how Raymond Dasmann combined the existential implications of modern evolutionary biology with Aldo Leopold's concerns about population irruptions in *The Destruction of California* (1965): "There are times," he wrote, "when the *change without apparent direction*, and the growth without control, give the appearance of socially acceptable madness, of a human population irruption that may well end tragically both for the people and for the land."[11]

Ehrlich also questioned "directed evolution" and progress. Devoting an entire section of his 1963 textbook on evolution, *The Process of Evolution*, to attacking human-centered logic, and criticizing the ideas of "purposive forces guiding evolution," "species chauvinism," and "the nearly ubiquitous tendency to use Homo sapiens as a standard," he emphasized that there is not "a shred of evidence ... that man is the ultimate goal of evolution." For an example, he noted that cockroaches were more likely to survive a thermonuclear war than human beings.[12]

As with East, Pearl, Leopold, Osborn, Vogt, and Dasmann, war was never far from Ehrlich's mind. As a youth, he followed World War II closely and read widely about the history of combat. In the early 1950s, during an animated conversation about the battle of Dunkirk, Ehrlich met his wife, Anne. At that time, the Ehrlichs later recounted, the war was still the "defining event of our lives and a great source of mutual interest." The year before Ehrlich published *The Process of Evolution*, the Cuban missile crisis had created the tensest showdown between the United States and the Soviet Union during the entire Cold War. If Ehrlich needed evidence that human society did not inevitably tend toward progress, few subjects could have worked better.[13]

In 1967, a year before he published *The Population Bomb*, Ehrlich articulated a variant of this rejection of biological progress in an article on the "balance of nature." Citing population crashes, he wrote, "The notion that nature is in some sort of 'balance' with respect to population size or that populations in general show relatively little fluctuation in size is demonstrably false." Catastrophe was just as likely an outcome of change as progress.[14]

Although others had emphasized sexual urges and questioned progress long before the Darwinian Synthesis, the postwar years saw a remarkable overlap between modern evolutionary biology and emerging environmental thought. Whether the science drove these moral views or vice versa is perhaps unanswerable. There's little doubt, however, that biologists such as Ehrlich believed that scientific truths undergirded their judgments about the world around them. Advances in biology gave them unbounded confidence when trying to make sense of the problems facing modern society, which appeared to be mounting in the late 1950s and 1960s.

Ehrlich, Numerical Taxonomy, and Race

Few social issues were as pressing as race relations. The civil rights movement, in which Ehrlich was more involved than most Americans, profoundly shaped his thinking and activism. In 1959, as a young postdoctorate researcher, Ehrlich helped organize an antisegregation protest at a restaurant in Lawrence, Kansas, moved to action after a visiting Jamaican scientist was forced to subsist an entire weekend on dispensing-machine candy bars because restaurants refused to seat him. No doubt growing up in a Jewish family during World War II played a role in Ehrlich's sense of racial justice. But Ehrlich's own biological research as a graduate student and young professor also confirmed and extended his views.[15]

During the 1950s and early 1960s, at precisely the time when civil rights activists were attacking racial hierarchies, Ehrlich was attacking the traditional system of categorizing species within biology. Disturbed by the artificiality of the categories used to classify butterflies, his main research subject, he titled one of his articles, "Has the Biological Species Concept Outlived Its Usefulness?" Elsewhere, he concluded that scientists had wrongly assumed the existence of "well defined clusters called species" which, in his view, simply did not exist.

Not all plants and animals fell easily into well-defined categories, he argued; some fell in between. Biologists often confused intellectual frameworks for tangible biological reality. "The reality of such concepts as 'fish,' 'bird,' 'mammal,' 'conifer,' and 'grass,'" Ehrlich wrote, "is rarely questioned." The term *Tribolium confusum* used for a kind of beetle "is not an entity, it is a taxonomic concept." The old system also lent itself to "fruitless arguments over whether or not a species or subspecies is 'good'" and highlighted a specimen's superficial features, not their far more important systemic roles.[16]

Instead of the old taxonomic system, Ehrlich suggested a "numerical taxonomic" system focusing on "the entire 'population' level of biological organization"—the core insight of population biology. This approach, he stressed in a journal article, emphasized *relationships*: "In populations, variation, growth, genetic equilibria, selection, behavior, and so on are not 'things' but relationships. . . . What is of interest in population biology is the pattern in which organisms are related in space and time." To Ehrlich, numerical taxonomy represented a move from a descriptive science that understood each part of nature as an isolated entity to an analytical science that saw each piece as part of something much larger. This was a move from a system based on imagined origins to one based on observed functional interrelationships. In a word, it was more *ecological*. Ehrlich felt he was helping to unravel some of the core mysteries of the world. Numerical taxonomy, as he put it in 1964, could help reveal the "ecological structuring of the biotic world."[17]

Ehrlich saw two additional benefits of numerical taxonomy, both of which shed light on his later population activism. First, because numerical taxonomy relied on numbers, it was "relatively objective" and had a "high predictive value." Ehrlich had great faith in numerical analaysis. The computer, he wrote in 1963 in another reference to contemporary race relations, "does not have a deep fear of miscegenation." Second, numerical taxonomy "transcends biology" and allowed biologists to push beyond the borders of their subfields, even outside of biology. "The techniques being developed today by population biologists," Ehrlich wrote in 1965, "will find employment in the future in such diverse fields as medicine, linguistics, history and economics."[18]

In 1964, Ehrlich publicly attacked the artificiality of *human* race categories in an article called "A Biological View of Race," in a book edited by Ashley Montagu, an anthropologist well known for his civil rights activism. "Taxonomic glasses have warped our view of the structure of nature," Ehrlich wrote to Montagu, "we find the term 'race' about as useless and unfortunate as any in the biologists' vocabulary." In the article, Ehrlich explained: "The concept Negro has much in common with the concepts bird and flower. Sociologically, Negro is defined differently in the United States and Brazil. In the southern United States anyone who is not 'pure white' is a Negro. In Brazil, anyone who is not 'pure black' is a Caucasian. Biologically the concept Negro has even less unity. Heavy skin pigmentation may be associated with a wide variety of other characteristics." Ehrlich would continue to publish attacks on the misuse of racial categories in public affairs for the rest of his career, including a 1977 book attacking the ideas of Nobel laureate William Shockley about racial differences in human intelligence.[19]

In making such arguments, Ehrlich followed in the footsteps of Theodosius Dobzhansky, who had penned an early popular treatise on the biological basis for racial equality, and Julian Huxley, who helped write the 1950 UNESCO Statement on Race that famously refuted scientific justifications of racial hierarchy. But Ehrlich, because of the direct relevance of his research, his family background, and personality, took a much more active role in civil rights than most biologists of his generation. His views about biology and human races are important: seeing the lessons that modern biology held for society helps explain what pushed him—unlike other biologists—to jump out of the "ivory tower" and into public affairs. They also explain why, once he did, he stressed the "we are all brothers under the skin" racial universalism of Fairfield Osborn, with all its benefits and flaws.[20]

Moreover, Ehrlich's research about biological categories also offers a window into his attitude in the early 1960s about knowledge, tradition, and change. Writing in *Science* in 1962, he attributed the faulty taxonomic system to the outdated "bonds of tradition" within biology. To a friend, he criticized those biologists who clung to the current taxonomic system out of habit: "I am not in favor of the taxonomists who, with their heads in the sand, have tried to ignore the opportunities for improving their methods."[21]

Ehrlich viewed his research as part of an inevitable Kuhnian revolution. Sometime in 1962 or 1963, Ehrlich read and admired Thomas Kuhn's paradigm-shifting history of modern science, *The Structure of Scientific Revolutions* (1962). One of Kuhn's main points was that scientists often placed their faith in a given theory not just because of its explanatory power but also because of what intellectual generation they belonged to. Young scientists at the beginning of their intellectual careers, noticing the marginal bits of data that failed to fit neatly into reigning scientific explanations, were much more likely than older scientists to devise and embrace new explanations. Borrowing Kuhn's term, Ehrlich referred to his own work in taxonomic categories as part of the "taxonomic revolution." He saw himself as part of a new generation jettisoning old, outdated ideas and ushering in new paradigms of truth. In just a few years, he would apply the same sense of self-righteous tradition-busting to environmental politics.[22]

Bay Area Population Biology and the Evolution of an Environmentalist

In 1959, not long after the anti-segregation protest in Kansas and his research on natural selection of Lake Erie snakes, Ehrlich and his wife, Anne, moved to Stanford University in Palo Alto, California, where they would live and work over fifty years. At the time California and especially the San Francisco Bay area were becoming fertile territory for environmental Malthusians, including Berkeley demographer Kingsley Davis and many future stalwarts of Ehrlich's organization Zero Population Growth. Raymond Dasmann, who that year had published *Environmental Conservation*, taught at Humboldt State University, north of San Francisco. Garrett Hardin, the most strident of all, taught at the University of California in Santa Barbara.[23]

A former frontier extraction zone, California in the postwar decades was moving quickly from being a landscape of abundance into one of perceived scarcity. Like much of the West, it was urbanizing remarkably fast. Its population, fed by wartime growth and Cold War industries, especially in northern Californian counties like Santa Clara, was growing faster than anywhere in the country, reaching twenty million people in 1967, over 85 percent of whom lived in cities. City and suburban growth in Western cities, moreover, looked different from that in other regions; cities were more dispersed, more decentralized, and more sprawling than anywhere else. Historian Mark Findlay attributes this distinctive Western pattern not to poor planning or profit-seeking developers but to the child-centered, consumer-oriented middle-class family structure that emerged in postwar America.[24]

Ehrlich could see many of the symptoms of the population problem from his Palo Alto doorstep: tremendous economic and suburban growth, threatened national parks and wilderness areas, and troubled inner-cities. Oakland and San Francisco were close by, and Los Angeles, the site of the decade's first urban riots, not far to the south. At the same time, the Golden State was also home to many of the technological innovations, especially in agriculture, that gave modernizers hope. As the nation's largest and most advanced agricultural producer, California epitomized the agricultural revolution that Lyndon Johnson wanted to export to around the world. In combination with his personal history and biological research, this cultural and material milieu shaped Ehrlich's growing urgency.

At Stanford, Ehrlich created one of the nation's strongest programs in population biology and built a network of scholars who would over the years provide ideas as well as political support. In the early 1960s, he used a grant from the National Institute of Health to bring noted population scholars to Stanford, such as Dobzhansky and Hardin, and to organize the first conference for West Coast population biologists. He also convened a weekly seminar on population issues that attracted faculty from around the university, especially from the medical, mathematics, and genetics departments. Five years later, in the acknowledgments to *The Population Bomb*, Ehrlich thanked numerous colleagues at nearby universities as well as a dozen colleagues in the Population Biology Group at Stanford for commenting on the manuscript.

On the Stanford campus, this group was surrounded by, and sometimes at odds with, the engineers and scientists who helped constitute the nation's "military-industry-university" complex during the Cold War. Like MIT and other research universities in the 1950s and 1960s, Stanford was transformed by World War II and especially the Cold War. In the 1930s, large research universities had maintained a critical distance from industry and government, but during the war and afterward, they began partnering with both. "Ballistic missiles, guidance systems, hydrogen bombs, and radar," historian Rebecca Lowen has written, "required the expertise of highly trained scientists and engineers." For this research, the federal government gave out a billion dollars a year during the Cold War, of which half went to just six schools, Stanford among them. By the 1980s, Lowen says, Stanford had become "the exemplar of the cold war university." Later, when Ehrlich argued for U.S. policymakers to consult less with engineers and physicists and more with ecologists, he was taking issue with some of his Stanford colleagues.[25]

As they studied population dynamics, Ehrlich and the Population Biology Group believed they saw the effects of overpopulation not just around the world but also in their neighborhood and workplaces. All around them were the growth problems that Raymond Dasmann decried in *The Destruction of California* (1965). A favorite hobby—flying light aircraft—gave Ehrlich an unusually comprehensive view of the Bay Area and the entire state. "A moment's reflection, preferably in an airplane over the sprawling suburbs of the Los Angeles basin," he wrote in 1967, "brings one to the basic cause of this deterioration. There are simply too many people in this state, and in the world in general." California was quickly losing the natural beauty that had cured Ehrlich's homesickness when he first arrived. "You would not believe what a nice place California is," he wrote to a friend in 1961, "you really ought to come out and see it sometime!" By the winter of 1965, however, Ehrlich was writing of being "very depressed about the growing urbanization in the Bay Area with the increase in smog, etc." He voiced these misgivings just about the time he first publicly spoke out about population growth.[26]

California's growth also infringed on Ehrlich's research area at Stanford's 735-acre Jasper Ridge Biological Preserve, a hilly area not far from campus. In 1963, a *San Jose News* headline warned of approaching problems: "Jasper Ridge—Untouched by County Growth—As Yet." In subsequent years, as a result of local development, Ehrlich's research teams would record the disappearance of almost a dozen butterfly communities. Eventually, Stanford's biologists felt forced to enclose Jasper Ridge with a fence. Years later, Ehrlich attributed much of his passion about overpopulation to what he saw near his research areas. "You have a powerful emotional reaction," he said, "when you see people building freeways and shopping malls on top of precious habitat and farmland. I have seen most of the places where I have done fieldwork destroyed in my lifetime. That gets you emotionally."[27]

Studying the population dynamics of checkerspot butterflies at Jasper Ridge in the early 1960s, Ehrlich made a discovery that would make his reputation as a scientist and transform his thinking about humans and their food supply.

The breakthrough came after Ehrlich and colleague Peter Raven—a plant biologist who would also become a famous environmental activist—noticed a peculiar pattern. Certain plants at Jasper Ridge contained chemicals that, like a potent insecticide, discouraged all insects—with one exception: the checkerspot butterfly. Ehrlich and Raven concluded that the butterflies and these plants had evolved in relationship with each other. "Coevolution" became an important branch of evolutionary and ecological theory. For Ehrlich, it also reinforced the central tenet of ecology: that organisms are bound to each other. Coevolution, he wrote a few years later, "provided some of the most compelling evidence that there is only a single web of life on earth."[28]

What coevolution taught about the relationship between plants and the animals led Ehrlich to conclude that the human food supply faced a grave threat. Because of evolutionary forces, the success of any synthetic chemical—the kind of chemical upon which California agriculture and increasingly the rest of the world depended—could only be temporary. Eventually, the original pest would adapt to the chemical and return in more menacing form, or a new pest would take advantage of the new abundance. Either way, after an initial rise in production, the agricultural crop—the corn or wheat protected from pests—would decline, perhaps precipitously. Ehrlich realized that the human food system could crash almost overnight.

When Rachel Carson's *Silent Spring* appeared in 1962, what struck Ehrlich was not so much what she said as how she said it. By this time, Ehrlich had long been frustrated by the misuse of pesticides. During the 1940s, as a teenager in New Jersey, he had seen how the "miracle" pesticide DDT had damaged the insects he loved collecting. As he later recounted, he "often found it impossible to raise caterpillars on local plants because of overspraying with pesticides." A few years later, as a graduate student at Kansas, he studied how fruit flies developed resistance to DDT. In 1957, while a postdoctoral researcher in Chicago, Ehrlich joined the Chicago Society for Exterminating Exterminators, a group that fought—unsuccessfully—against the U.S. Department of Agriculture's aerial spraying of fire ants in the U.S. South. His research on the coevolution of animals and the plants they ate only intensified his concerns.[29]

Silent Spring showed Ehrlich that something could be done about the problem. Since the failed fire ant campaign, Ehrlich had struggled to figure out how scientists could make their voices heard, especially in Washington. But while a phalanx of scientists had tried and failed to make pesticides a public issue, Rachel Carson had triumphed through her popular book. Ignoring many critics, including allies who argued that scientists should not be public activists, she boiled down complicated scientific arguments into easily comprehensible language for a general readership.[30]

Ehrlich gave his first major public address about population growth and environmental problems to a group of Stanford alumni in January 1965, a few weeks after Lyndon Johnson had mentioned population growth in his State of the

Union speech and Dr. Dale Clinton had decided to distribute contraception at his clinic in Lawrence, Kansas. In his speech, called "The Biological Revolution," Ehrlich highlighted the recklessness of humanity's growing domination of nature as well as the problem of overpopulation. Unlike many environmentalists, however, he argued that the key turning point in environmental history was not industrialization but the rise of agriculture. Modern agriculture particularly worried Ehrlich because it often degraded land quickly. He had seen this problem in California. In order to "support an ever increasing population," he warned, "uncoordinated, though often well-meaning attempts to modify the face of the earth" could have "disastrous consequences."[31]

Surprisingly—given what he would be saying about India just a few years later—Ehrlich downplayed the threat of food shortages. Advances in public health, agriculture, and transportation, he pointed out, had greatly *reduced* the threat of famine. Instead, the real threat was "improved technology," which had increased the potential of war "as a population-control device." Indeed, technologies such as nuclear weapons had given humans "the means for self-extermination."

Ehrlich concluded the speech with a call for scientists—especially ecologically informed biologists—to speak up in policy debates, to "come out of our ivory towers." "Following Rachel Carson's lead," he implored, "we must fight abuses wherever they occur."[32]

Applying Natural Selection to India

In hindsight, perhaps the most interesting aspect of Ehrlich's "Biological Revolution" speech in 1965 was its relative optimism. Compared to *The Population Bomb*, which came out only three years later, "Biological Revolution" displayed little sense of immediate crisis. Humans would reach the Earth's limits within decades, not within a few years, as he argued in *The Population Bomb*. Ehrlich even showed some hope in exactly the programs the Johnson administration would stress: expanded birth-control programs and more efficient agricultural methods.

Ehrlich's hope would soon evaporate. By late 1967, he was warning of catastrophe within years. In *The Population Bomb* from early 1968, he predicted disasters within nine years.[33]

What brought about this change was the mid-1960s food crisis in India. Ehrlich used the South Asian giant throughout *The Population Bomb* to illustrate every major aspect of the population problem: the extent of the third-world "population explosion," how this explosion threatened political unrest and even war, how the "green revolution" and other solutions actually made the situation worse, and why the United States needed to go beyond voluntary birth-control programs. Elsewhere, he called India "the ultimate convincer."[34]

India's food problem in the mid-1960s also alarmed a number of prominent environmentalists, a surprising number of whom traveled to or wrote about India: Lester Brown, future founder of Worldwatch; Garrett Hardin, author of "The Tragedy of the Commons" and "Lifeboat Ethics"; and future Zero Population Growth (ZPG) and Sierra Club director Carl Pope, who had conducted family planning work as a Peace Corps volunteer in India. Members of this group did not think identically, but they all agreed that India illustrated why Americans needed to think about human society more ecologically. The overriding goal of many population-minded environmentalists, historian Donald Fleming noted in 1972, was "no more Indias."[35]

Ehrlich witnessed India's troubles firsthand. During the summer of 1965, just a few months after his "Biological Revolution" speech, he started a year of research in Australia, New Guinea, Malaysia, Thailand, Cambodia, India, and East Africa. Ehrlich had high expectations for Asia, but the trip shattered his notions of tropical paradise. No place seemed more "spoiled" than India, where Ehrlich, his wife, and daughter visited in late June and early July 1966, right in the middle of the food crisis. Immediately on arrival in Delhi, they got a shock. "Our first scene outside of the airport was a man defecating in the street," Ehrlich wrote in a letter to friends in the States. "That sets the tone for India." The Ehrlichs spent much of their three weeks in Kashmir, the beautiful alpine province of India that had recently been the site of a war with Pakistan. They stayed in a dirty hotel, then moved to a Dal Lake houseboat whose owner tried to overcharge them. "Pay it; Americans care nothing for money," the man reportedly said, enraging Ehrlich. The butterfly collecting was little better. Ehrlich found "the fabled high-altitude meadows" of Kashmir "biologically barren, grazed to within a fraction of an inch above the ground." Of all his travels in Asia, Ehrlich wrote a friend later that year, Kashmir stood out as "the big disappointment."[36]

Ehrlich made sense of India's problems through biological theories about wildlife and food supply, but he also drew from expertise in related fields. After reading a gloomy 1967 article by demographer Kingsley Davis, he lost faith in the efficacy of birth-control programs in India. Davis, a preeminent population expert, helped establish the modern field of demography. His 1951 book on South Asia had done much to convince demographers of the effectiveness of birth-control programs in the first place. But revisiting his intellectual stomping grounds fifteen years later, Davis found that India's family planning program, one of the most touted in the world, had accomplished little if anything. No voluntary program, he emphasized, would work if Indians wanted large families. Birth control did not necessarily mean population control: "By stressing the right of parents to have the number of children they want, it [family planning] evades the basic question of population policy, which is how to give societies the number of children they need. By offering only the means of couples to control fertility, it neglects the means for societies to do so." Ehrlich would quote this line in *The Population Bomb*.[37]

Another book mostly about India—*Famine 1975!* by William and Paul Paddock—helped Ehrlich see how sluggish the world's food production trends were. Written by two brothers from Iowa—one a tropical agronomist, the other a Foreign Service officer—*Famine 1975!* combined agricultural and foreign affairs expertise to show the inevitability of massive famine and its potentially tumultuous political consequences. Methodical and packed with statistics, the book derided as wishful thinking one technological remedy after another—synthetic foods, desalinization, food from the ocean, agricultural research, fertilizers, and irrigation. It also rejected land reform, government regulations, private enterprise, and other political-economic responses. The green revolution showed promise, they believed, but it would clearly not raise production levels soon enough.[38]

Famine 1975! deeply influenced Ehrlich. Whereas in the "Biological Revolution" speech in 1965, he expressed faith that human beings could triple their food supply by the end of the century, by the end of 1967 he was claiming that the battle to feed humanity was over. He began to refer to *Famine* as "a massively documented book" and to quote from it extensively.[39]

These insights, combined with Ehrlich's concern about modern agriculture and what modern biology taught about reproductive imperatives and the dynamics of population explosions, prompted him to conclude that the technological remedies Lyndon Johnson was suggesting would fall on their face.

1968

The Darwinian Synthesis, the civil rights movement, Ehrlich's butterfly research, the runaway development of the Bay Area, Rachel Carson's attack on DDT, and India's food crisis reveal a great deal about the origins of *The Population Bomb*. But a growing sense of crisis in the United States—stemming from varied yet simultaneous crises that all seemed beyond the problem-solving capacity of the system—also fueled Ehrlich's pessimism and helped drive the appeal of the book.

In the late 1960s, the vast changes that had so transformed American life in the previous two and a half decades—economic growth, suburbanization, civil rights, the women's movement, the Cold War, Vietnam, student protest—came to a boil. The year Ehrlich wrote *The Population Bomb*, 1968, was particularly tumultuous. Each month brought a new crisis. In late January, the Viet Cong's Tet Offensive showed that the forces of international communism were on the march, not in retreat. In late March, President Johnson announced that he would not seek reelection, throwing the fall campaign wide open. In early April, after James Earl Ray shot and killed Martin Luther King Jr. in Memphis, scores of cities erupted in a new round of racial violence. In April and May, the nation's universities witnessed a wave of protests, including a strike that shut down

Columbia University, highlighting what some called the "youth quake." Summer saw the publication of William Whyte's *The Last Landscape* lamenting the cultural and environmental problems of the suburbs. In August, antiwar protestors and Chicago police came to blows at the Democratic National Convention, signaling the breakdown of normal political processes. In November, Nixon's paper-thin margin of victory highlighted the country's divisions. "The world was exploding," one young man later recalled. "It wasn't changing. It was exploding. We literally were afraid the world was coming apart. That we were going to have anarchy or civil war. It was frightening."[40]

In the midst of this upheaval, Ehrlich realized that the biological lens that had revealed India's underlying problems could also shed light on America's many problems. The nation's problems seemed so intractable, he believed, because conventional solutions focused on surface cultural factors, not their underlying biological causes. An ecological approach stressing population dynamics, however, could cut through the superficial issues to the core causes. Ehrlich applied this approach to a score of problems, from suburbs to pollution to national parks to India, and particularly to Vietnam and inner-city unrest.

In the late 1950s and early 1960s, Vietnam was to be, like India, a model of how American modernization programs would transform a strategic third-world country bound to poverty because of traditions into a prosperous modern state friendly to the West. By late 1967, though, many Americans had realized that efforts to build a modern South Vietnam were failing. Patriotic claims of America's inevitable victory seemed way off the mark.

Ehrlich began offering an environmental interpretation of American travails in Vietnam in late 1967, a year after he returned from India, arguing that the United States could not hope to guide the fate of third-world places without addressing the core biological dynamics driving their political instability. The war, he declared, represented "a symptom" of the larger problem of uncontrolled population growth. "Steadily increasing population pressures continue to deny man the breathing space he requires to alter his socio-political systems," he explained. "Until such an opportunity can be provided, the chance of finding a sane solution to world problems seems nil." He sketched grim possibilities: "War in Vietnam will inevitably be followed by war elsewhere—perhaps in Thailand, the Philippines, Korea, or California. Bad as world tensions are today, consider what they will be like when almost everyone is hungry and nations are competing for increasingly scarce food resources."[41]

Escalating racial tension also formed a backdrop for *The Population Bomb*. In an October 1967 speech, just months after the devastating riot in Detroit, Ehrlich blamed urban troubles, including "riots and increased drug usage," on overpopulation. In January 1968, he described riots as "symptoms of mankind's serious disease of overpopulation." In February, he predicted "increasing riots" if population problems were not addressed. That same month, the Kerner

Commission submitted its direct and divisive report, blaming the unrest in Detroit and other cities on white racism.[42]

Ehrlich wrote *The Population Bomb* in March and April 1968. On April 4, 1968, an assassin's bullet killed Martin Luther King, sparking hundreds of riots. That very day, Ehrlich told a Utah audience that the United States was an overpopulated nation in which "riots tear cities apart." At different points while Ehrlich was working on his manuscript, the U.S. Army was quelling riots in Chicago and Baltimore, and federal troops in Washington were setting up machine guns on the steps of the Capitol, right across from the Supreme Court.[43]

National problems coincided with a handful of personal problems for Ehrlich. "My life has just been a shambles," he reported to an Australian friend in May. In addition to his overcrowded work schedule, a student sit-in crisis had almost "dissolved" the university. Moreover, the draft situation had him "on the ropes." As graduate director for Stanford's biology department, he had to decide which students were eligible for deferments, and who had to serve and potentially die in the military. Finally, on top of all this, a good friend died. "Will this year ever end?" he lamented. But the situation only got worse. In July he wrote his Australian friend that, because of budget cuts, 1968 was "the most depressing time that I can personally remember in the history of American science."[44]

As these and other crises unfolded, Ehrlich yearned for a way to influence the national debate. The November presidential elections, he realized, offered a chance—perhaps the last chance—to head off disaster. The election, he told a group of dentists in January, "may well be the most important political event in the history of the world. . . . We must do everything possible to make population policy a major campaign issue." It dawned on Ehrlich, however, that lectures to dentists might not be enough to spark the crusade he hoped for. Not long afterward, the Sierra Club's David Brower suggested he distill his ideas into a paperback. In March 1968, with substantial help from his wife, Anne, he revised his notes and lectures, and, after getting feedback in April from academic allies at Stanford and Berkeley, put the finishing touches on the manuscript of *The Population Bomb*, which Ballantine Books published later in the spring. American environmentalism would not be the same again.[45]

The Population Bomb

Ehrlich used the book's first section, "The Problem," to provide a demographic snapshot of the world in 1968. He stressed the doubling time of the world's population, which was then just thirty-seven years. It had taken eighty years for the previous doubling to occur, two hundred years the time before that, and a thousand years before that. In particular, Ehrlich stressed rapid growth in the world's undeveloped countries (UDCs): in Kenya, the doubling time was twenty-four years; Nigeria, twenty-eight; Turkey, twenty-four; Indonesia, thirty-one; Philippines, twenty; Brazil, twenty-two; Costa Rica, twenty. "Think of what it means," Ehrlich wrote, "for the population of a country to double in 25 years." Food, buildings, roads, power, and trained doctors, nurses and teachers must be doubled. One of the "most ominous facts": roughly 40 percent of the undeveloped world was under fifteen years old.[46] Population growth in developed countries (DCs) also worried Ehrlich, although doubling times there ranged from fifty to two hundred years. Highly urbanized, most DCs could not feed themselves and suffered from a host of growth problems. The United States had numerous "headaches" caused by speedy population growth: "not just garbage in our environment," but "overcrowded highways, burgeoning slums, deteriorating school systems, rising crime rates, riots, and other related problems."[47]

To explain the origins of the population problem, Ehrlich emphasized the lessons of evolutionary biology as reinforced by the Darwinian Synthesis, especially human sexual drive: "Facet number one of our bind—the urge to reproduce has been fixed in us by billions of years of evolution." It was "the key to winning the evolutionary game." Cultural values reinforced the drive for reproduction, the desire for sex shaping "our self-esteem, our choice of friends, cars, and leaders." Indeed, many of the titles that Ehrlich considered for the book stressed reproductive potency: Breed Now Starve Later, Breeding and Oblivion, Are Your Neighbors Still Breeding?, Sex and Starvation, Breed Now Pay Later.[48]

As had Malthusians such as William Vogt, Ehrlich also blamed modern medicine, which had brought about a major decrease in death rates. Ceylon (Sri Lanka) provided his best example. In the late 1940s, spreading DDT to control malaria there had halved the death rate to ten per thousand in 1954 and sent the population skyrocketing.[49]

Ehrlich also emphasized how food supply lagged behind population growth. Following World War II, food production per person had kept pace with population, but in the late 1950s, "the stork passed the plow" and DCs started shipping food to UDCs in massive quantities. In 1965–1966, the situation deteriorated dramatically: the world population climbed by seventy million, yet food production plateaued. India faced the worst straights. That year, the United States shipped to India nine million tons of wheat—one-quarter of total U.S. production. And India's population was booming. Within four years, India would have to feed an additional fifty to seventy million people, in the next thirteen years, two hundred million more. Increasing food production by that much, Ehrlich wrote, was simply impossible. With other parts of the world just as bad, he saw calamity ahead. "If the optimists are correct, today's level of misery will be perpetuated for perhaps two decades into the future. If the pessimists are correct, massive famines will occur soon, possibly in the early 1970's, certainly by the early 1980's."[50]

Ehrlich wanted Americans to confront this poverty. "Most Americans either do not know or choose to ignore the true depths of the misery and despair," he wrote. Although the book often seemed about protecting American quality of life, Ehrlich was not without sympathy for fellow human beings: "We must never forget as we contemplate our unprecedented problems—that in all the mess of expanding population, faltering food production, and environmental deterioration are enmeshed miserable, hungry, desperate human beings."[51]

Up until this point, little distinguished Ehrlich from other Malthusians, including those in the Johnson administration, except perhaps his emphasis on reproduction and his pessimism. But then, in a section called "A Dying Planet," a title he considered for the entire book, Ehrlich highlighted an angle of the problem that was "almost universally ignored": the problem of environmental deterioration. Efforts to increase production in the short term were eroding the world's capacity to produce food in the long term. Soil erosion, which had destroyed whole civilizations in the past, would "accelerate as the food crisis intensifies." Worse, "use of synthetic pesticides, already massive, will increase." Pesticides were not only toxic but also "one of man's potent tools for reducing the complexity of ecosystems." Drawing from new ideas about genetic resistance as well as his own research on coevolution, Ehrlich expounded on the dangers of ecosystem oversimplification: one of the "basic facts" of population biology was that "the simpler an ecosystem is, the more unstable it is." In addition to requiring ever more pesticides, simplified ecosystems increase the possibility of disease. "It is difficult to predict," Ehrlich wrote, "the results of another 25 years of application of DDT and similar compounds, especially if those years are to be filled with frantic attempts to feed more and more people."[52]

At the core of Ehrlich's concern lay an often overlooked connection between population growth and synthetic pesticides. The "intimate relationship between pesticides on one hand and the population crisis, food shortage, and environmental deterioration on the other," Ehrlich noted repeatedly, "is often not recognized." Modern agriculture, he stressed in *The Population Bomb*, caused "striking" environmental alterations. Worse, the world's mushrooming population depended on exactly this system. Yet very few people, not even other Malthusians, connected the two issues. Many spoke of resources, but few emphasized environmental deterioration and limits. Ehrlich stressed shortages, but he also rang the alarm about the damage from pesticides, fertilizers, and other high-production methods. In doing so, he combined the concerns of Edward East and William Vogt with those of Rachael Carson.[53]

Ehrlich's critique of modern agriculture distinguished his brand of environmentalism not only from other Malthusians but also from other conservationists. By the 1960s, many conservationists saw their job as finding new technologies to make better use of resources to increase production. One Ehrlich supporter even described a war between conservationists and environmentalists. "Many conservationists," he wrote, "are in the enemy camp, having spent the last two generations telling us how pushbutton farming and food for all are just around the corner."[54]

To the problems of agriculture, Ehrlich added a catalogue of environmental concerns: smog, water pollution, lead poisoning, and even climate change. *The Population Bomb* was one of the first general primers on environmental studies. With each problem, Ehrlich made the same point: that even experts missed below-the-surface ecological relationships. The "subtle ecological effects," he said, may be "much more important than the obvious features of the problem." He emphasized unforeseen risks: "What do we gain by playing 'environmental roulette'?"[55]

Ehrlich attributed all these environmental problems to population growth. Although in other parts of the book he mentioned other factors, in this section, he reduced these complicated matters to a function of just population. "The causal chain of deterioration is easily followed to its source. Too many cars, too many factories, too much detergent, too much pesticide, multiplying contrails, inadequate sewage treatment plants, too little water, too much carbon dioxide—all can be traced easily to *too many people*."[56]

Within this mix, Ehrlich included racial unrest. As proof of our "deteriorating psychic environment," he pointed to "riots" and "rising crime rates"—his second mention of riots in the text. Such riots, he implied, were an environmental issue. Elsewhere, he borrowed from the research of John Calhoun and others who directly linked violence among animals to high populations and crowding. "We know all too well that when rats or other animals are overcrowded, the results are pronounced and usually unpleasant," Ehrlich wrote. "Social systems may break down, cannibalism may occur, breeding may cease altogether. The results do not bode well for human beings as they get more and more crowded." Americans, he added in early 1969, must learn to view each new baby as "a user of irreplaceable resources, an increaser of crowds, and even a potential rioter."[57]

Ehrlich called for a new kind of conservation. He knew *The Population Bomb* did not dwell "on the themes that typically characterize the pleas of conservationists." He had said little about redwoods, passenger pigeons, bison, California grizzlies, nor about national parks and wilderness areas. Instead, he focused on problems that "seem to bear most directly on man." This was a conscious strategy. In spite of all the efforts of conservationists, the conservation battle "is presently being lost." Why? Because "nothing 'undeveloped' can long stand in the face of the population explosion." When it came to nature for its own sake, "most Americans clearly don't give a damn." Ehrlich's assessment showed the shifting priorities of environmental politics in the late 1960s, and suggests a factor contributing to his human-centric, and indeed, America-centric rhetoric: he may have done so not out of fear of the world's poor but in order to stir Americans out of their apathy. Although effective in the short term, this strategy led to sharp criticisms down the road.[58]

In *The Population Bomb*, Ehrlich tried to convince Americans to "give a damn" with two scenarios in which famine ignited apocalyptic wars. In the first, war in Thailand and famine-sparked food riots in China prompt the Maoist behemoth to escalate involvement in the Vietnam War, which spurs the United States to bomb Chinese air bases. The Chinese respond by detonating five "dirty" bombs off of the West Coast, killing 100 million Americans. In the second scenario, years of famine across Asia, Africa, and South America open the door to a sharp increase in Russian involvement in Latin America, especially Mexico, leading to a nuclear war with the United States that leaves the northern two-thirds of the Earth uninhabitable.[59]

War was a common theme of the book. "Population pressures promote wars," Ehrlich warned, whereas "diminishing population pressures will reduce the probability of war." He did not specifically mention the Vietnam War but, as with race, in the supercharged context of the late 1960s, he didn't have to.[60]

Much of this military context was captured by the book's title—the same as Hugh Moore's pamphlet—chosen late in the writing with encouragement from Ballantine Press. As with Carson's use of fallout imagery in *Silent Spring*, the bomb metaphor evoked the overriding concern of the era—war and destruction. Bombs suggested the drawbacks of modernity, how man-made advances in science and technology could create horrors. They also had a temporal quality that Ehrlich and other Malthusians stressed: they often lay unseen until, instantaneously and unexpectedly, they create massive destruction. The exponential aspect of population growth would creep up quietly, then suddenly wreak havoc.[61]

What Is Being Done

Bombs also demand immediate action, but, Ehrlich stressed, not much was getting done. The technological solutions that the Johnson administration and Congress had turned to the year before—birth control and "green revolution" programs—were no match for the laws of reproduction and scarcity that governed biology.

Voluntary birth-control programs in India, as Kingsley Davis had pointed out, were simply falling on their face. Birth control may work for individuals, as Davis said, "but it does not control populations." Ehrlich hammered this point home with his own inimitable language: "Well-spaced children will starve, vaporize in thermonuclear war, or die of plague just as well as unplanned children." Even families who used birth control still wanted large families. "The story in the UDCs is depressing the same everywhere—people *want* large families. Family planning is all too often used to lock the barn door after the horse is stolen."[62]

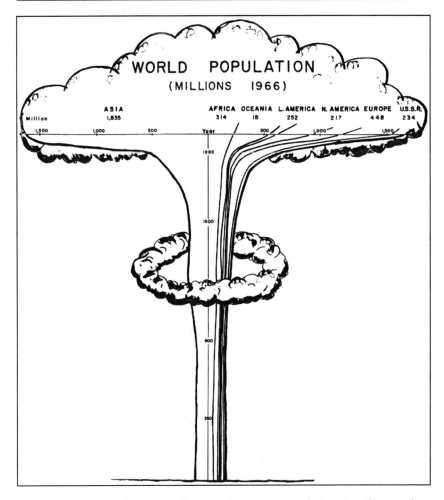

FIGURE 9 From the late 1940s through the 1960s, population growth was often conceptualized as a military problem that tapped into the apocalyptic fears of the atomic age. Swedish geographer Georg Borgstrom often wrote about population and hunger issues.

Illustration from Georg Borgstrom, *Too Many: An Ecological Overview of Earth's Limitations* (New York: Collier Books, 1971), x.

Echoing the Paddocks' *Famine 1975!*, Ehrlich rejected as unrealistic the "gee whiz" food production solutions that most "techno-optimists" pointed to, such as cultivating new lands (none available except in the tropics, where soil quality is poor), farming the sea (technology not available for decades), and culturing microbes on petroleum (limited by oil stocks and also unpalatable). He thought the green revolution was the most feasible option. Hybrid seed programs were already underway and required no great infrastructural or cultural change. But Ehrlich was cautious, again pointing to the ecosystem simplification: "We do not know how they will do under field conditions over the long run—how resistant they will be to the attacks of pests." Citing William Paddock, a plant pathologist, he wrote: "One wonders what unknown pathogens await." Moreover, he reiterated that the new seeds required prodigious amounts of fertilizer and pesticides. Technology had gotten the world into its current mess; it wasn't going to get it out.[63]

At the heart of Ehrlich's pessimism stood a core difference with Johnson's modernizers: whereas they saw poverty as essentially a cultural condition that could be overcome with the right technology, Ehrlich saw places such as India as beyond hope because of biological realities. India had out-bred its carrying capacity, just like the Kaibab deer, and disaster loomed. "As zoologists know well," he said in 1972, "animal populations often considerably overshoot the carrying capacity of their environment—a phenomenon invariably followed by a population crash." Efforts to intervene—to stretch the carrying capacity through measures like the green revolution—would only make matters worse by degrading environments.[64]

Ehrlich's critique also displayed the social lessons he drew from the Darwinian Synthesis: that humans were inextricably part of nature, and that change did not necessarily lead to progress. Whereas Johnson's modernizers believed in progress through technology, Ehrlich displayed a legendary pessimism. On virtually each page, he offered stories of limits, technology gone awry, inevitable misery, and mass death. Indeed, if the Kennedy administration's optimism about modernization during early 1960s made Americans think of "new frontiers," Ehrlich's worldview evoked the pessimism of limited resources. "We have now exhausted all of the world's frontiers," he announced in 1969.[65]

Arguing against the green revolution in the mid-1960s put Ehrlich up against two formidable forces. Many believed that new hybrid crops were the answer not just to world hunger but also to the Cold War. Ehrlich countered with his own hard-to-dispute approach: a preventative strategy. Why not prevent hungry populations from coming into existence in the first place? Then no environmentally destructive green revolution was needed.

Ehrlich's preventative strategy helped reconceptualize environmental politics. In addition to blaming new technologies of production—such as DDT or the green revolution—it also blamed demand. If population growth could be curtailed, then the demand for destructive technologies could be reduced. Even though his logic was flawed and reductive—it did not eliminate faulty technologies and often overlooked vast disparities in consumption—Ehrlich, like Osborn and Vogt before him, refocused attention on the consumption side of environmental problems.

This logic had important political consequences. Without necessarily losing sight of systemic drivers such as capitalism, it made environmental problems into something that not just governments could control but also individual Americans. It was a call for individuals to get involved. The personal was political.

Population Control and the Counterculture

Envisioning imminent collapse, Ehrlich in *The Population Bomb* called for much stronger measures than the Johnson administration did. At home, Ehrlich sought "drastic" programs in population planning. He considered the idea of placing chemical contraceptives in the water supply but decided it was not feasible. More likely, he wrote, were less drastic measures, such as reversing the government's system of encouraging reproduction, and creating financial disincentives with taxes on children, cribs, and diapers. He called for a Department of Population and Environment, with duties such as writing a federal law "guaranteeing the right of any woman to have an abortion" and providing sex education "at the earliest age recommended by those with professional competence in this area—certainly before junior high school." Ehrlich remained vague about coercive measures in the United States.[66]

He called for broad social change in the United States. He saw dysfunction and stagnation everywhere: "Many of our institutions no longer function properly." He wanted "extremely fundamental changes" in American approaches to sex, religion, and the economy. At the time, antiestablishment feeling was swelling in the United States and the "counterculture" was emerging, especially among young people.[67]

As Ehrlich's ideas about abortion and sex education suggest, he especially wanted Americans to rethink their notions of sex, and base new ideas upon the lessons of modern biology, not the teachings of conservative religious institutions. Sex, he said, was inevitable and irrepressible. To deny that, as Ehrlich felt calls for self-restraint did, was to deny nature itself. Societies must move away from "the anti-human notions that have long kept Western society in a state of sexual repression." Moreover, sex must be separated from reproduction. "It is now imperative," he said, "that we restrict the reproductive functions of sex." Liberal access to birth control and abortion allowed for just such a possibility, if societies could cast away their traditions reinforcing reproduction. But, Ehrlich pointed out, the "conservative elements in the Church hierarchy . . . still resist change." A "rational atmosphere" might lead to what Ehrlich hoped for: a population "thoroughly enjoying its sexual activity, while raising smaller numbers of physically and mentally healthier children."[68]

This was part of a larger attack on traditional religion. Borrowing historian Lynn White's famous argument from 1967, Ehrlich argued that Western attitudes toward nature originate in the Judeo-Christian tradition, which more than other cultures viewed man's basic role "as that of dominating nature, rather than as living in harmony with it." The Western tradition also fostered an exaggerated faith in science and technology, and an inclination to control. He applauded the counterculture for adopting religious ideas from the non-Christian East.[69]

Ehrlich also wanted the nation to rethink its obsession with economic growth. The "whole economy," he said, "is geared to growing population and monumental waste." Drawing from Robert and Leona Rienow, he singled out new babies. "Each new baby is viewed as a consumer to stimulate an ever-growing economy." Like them, he listed the vast supplies that each baby would require in a lifetime: 26,000,000 gallons of water, 21,000 gallons of gasoline, 10,000 pounds of meat, 28,000 pounds of milk and cream, over $5,000 in school building materials, $6,300 worth of clothing, and $7,000 worth of furniture (1967 dollars). "It's not a baby," Ehrlich concluded, "it's a Superconsumer."[70]

Here, too, he wanted to revisit basic assumptions, including the entire point of material progress. Echoing quality-of-life arguments, he thought that Americans should "ask exactly what our financial prosperity is for.... Do we want more and more of the same until we have destroyed ourselves?" He noted with approval that "Sizeable segments of our population, especially the young, are already answering that question: 'Hell, no!'"[71]

Ehrlich laced these arguments with impatient antiestablishment attacks. "People in positions of power," he said, "have either ignored the problem or have recommended solutions that are inadequate." Deploying both sarcasm and righteous indignation, he attacked conservative Catholics. "The Catholic Church recommends the rhythm method of contraception," he wrote. "Unfortunately, people who practice this method of contraception are commonly called 'parents.'" He called the rhythm method "Vatican roulette." He also aimed his sarcasm at the U.S. government, criticizing not only "entrenched bureaucrats" but also the military-industrial complex. The population budget of all the U.S. agencies combined, he pointed out, "would not buy more than a dozen sophisticated military jets. It is roughly the same amount as the government appropriation for rat control." He lambasted the USDA for its bungled campaign against fire ants. Showing how environmentalism could overlap with New Left anticapitalism arguments, he offered an extended attack, by turns irreverent, by turns sarcastic, of the Mississippi fish kill, in which a polluting corporation claimed that a rare disease was to blame for dead fish, not the gallons of toxic endrin it was dumping into the river. "The greed and stubbornness of industries, the recalcitrance of city governments, the weakness of state control agencies, and the general apathy of the American people," Ehrlich wrote, "have combined to keep progress discouragingly slow." It's not hard to see why young people, especially those yearning for change and suspicious of government and corporations, looked up to Ehrlich.[72]

If Ehrlich had limited his recommendations to the United States, he might have received less criticism. But India and the third world had propelled him to write *The Population Bomb*. Internationally, he was far more authoritarian. He advocated a system of "triage" to govern American foreign aid in which the United States should distribute food aid only to those nations not beyond hope. Nations too far out of balance, such as India, would only waste resources needed elsewhere. To Ehrlich, there was no other "rational choice." If this wasn't chilling enough, Ehrlich also wanted to see the U.S. support forced sterilization programs, such as a plan by the Indian minister Sripati Chandrasekhar that had been rejected by Indian authorities. "When he [Chandrasekar] suggested sterilizing all Indian males with three or more children, we should have applied pressure on the Indian government to go ahead with the plan," Ehrlich wrote. "We should have volunteered logistic support in the form of helicopters, vehicles, and surgical instruments. We should have sent doctors to aid in the program by setting up centers for training para-medical personnel to do vasectomies." Was this coercion? "Perhaps, but coercion in a good cause.... We must be relentless." As it turned out, the political tides would turn, and forced sterilization would become Indian state policy during in the mid-1970s.[73]

Describing the international context, Ehrlich elaborated on a metaphor that runs throughout the book: the idea of population growth as a cancer. "A cancer is an uncontrolled multiplication of cells," he wrote. "The population is an uncontrolled multiplication of people." He stressed getting to the root cause, not dealing just with symptoms. "Treating only the symptoms of cancer may make the victim more comfortable at first, but eventually he dies—often horribly. A similar fate awaits a world with a population explosion if only the symptoms are treated." Ehrlich's remedies for this cancer were very disturbing: "We must shift our efforts from treatment of the symptoms to the cutting out of the cancer. The operation will demand many apparently brutal and heartless decisions. The pain may be intense. But . . . only with radical surgery does the patient have a chance of survival." Here, he is talking not of malignant cells but real people, whose only fault was the bad luck of being born in poor countries. It seems that those who live in the West should be spared and those who live elsewhere "cut out." It's hard not to see a hierarchy and a moral judgment in this idea.[74]

Spaceship Earth and a Taxi in Delhi

Returning to the years between *Silent Spring* and the first Earth Day suggests that America's global context—and India's troubles in particular—transformed the thinking of Malthusian environmentalists such as Ehrlich in ways that shaped the fledgling environmental movement. These problems, combined with a growing concern about overpopulation at home, convinced him of the need for immediate and dramatic action.

Above all, Ehrlich wanted Americans to think more about the environmental consequences of their actions, especially their consumption. "We've got to change," he summarized in *The Population Bomb*, "from a growth-oriented, exploitative system to one focused on stability and conservation." He saw reducing population growth as the most feasible and effective way to do so. His ideas were a curious mix of liberal and conservative. Some ideas grew from a well-intentioned effort to address starvation in India and poverty in the inner city, to think globally instead of nationally or racially, and to advance the cause of birth control rights, abortion rights, and new more liberal roles for women. But other parts of the argument appeared to blame the poor for their own poverty, defined them as a threat to the American standard of living, and recommended harsh policies.[75]

Running through *The Population Bomb* were two metaphors that illustrate this ambivalence. One was the idea of "spaceship earth," an idea that a variety of well-known people, including former Democratic presidential nominee Adlai Stevenson, economist Kenneth Boulding, journalist Barbara Ward, and quirky inventor Buckminster Fuller, had pioneered, to highlight how modern human technologies had bound together previously far-flung peoples into a delicate web of interdependence. "This planet," Ehrlich pointed out in a 1967 speech, is "a spacecraft with a limited carrying capacity." In *The Population Bomb*, he wrote of the "good ship Earth," and even considered naming the book "Our Spaceship Is Bursting." For Ehrlich, the spaceship highlighted, even more than islands, another common metaphor for carrying capacity, the irreplaceable nature of resources and the precariousness of living in an artificial technological system that could, at any moment, collapse. The spaceship also stressed the united fate of all of Earth's residents, a theme running back to Fairfield Osborn's *Our Plundered Planet* and an important point for Ehrlich because of his belief about racial equality. Whereas on an island, residents starve one by one, when the high-tech life support systems give out on a crowded spaceship, all passengers perish together.[76]

At the same time, Ehrlich offered a second, more defensive and self-interested model with a striking, and soon to be famous, vignette on the first page of *The Population Bomb*. Like Spaceship Earth, this model also involved a transport metaphor, except in this case Ehrlich used a car and emphasized that the overpopulated masses were on the outside, not the inside. The emphasis on car and family evoked an American suburban context, but the story actually described a 1966 taxi ride in New Delhi that Ehrlich said had helped him understand the "feel" of the population problem "emotionally."

> My wife and I were returning to our hotel in an ancient taxi. The seats were hopping with fleas. The only functional gear was third. As we crawled through the city, we entered a crowded slum area. The temperature was well over 100, and the air was a haze of dust and smoke. The streets seemed alive with people. People eating, people washing, people

sleeping. People visiting, arguing, and screaming. People thrusting their hands through the taxi window, begging. People defecating and urinating. People clinging to buses. People herding animals. People, people, people.

"All three of us were," he concluded, "frankly, frightened."[77]

In just a few sentences, Ehrlich described the problems that he associated with overpopulation: fleas, heat, unhygienic conditions, begging, and conflict. Whereas modernizers stressed cultural and knowledge deficiencies, this story stresses the biological. Readers learn little about the political or historical factors that might help explain the squalid conditions; instead, they hear of eating, sleeping, screaming, defecating, and urinating—not much appears to separate India's poor from a herd of animals. By offering no other explanation for why these people live this way—why they were poor—the passage suggests that the answer lies in their very numbers. Moreover, the possibility of violence hung in the air, threatening a life of order, peace, health, technological comfort, and quietness. Quantity of life could undermine quality of life. All that separated Ehrlich and his family from this threat was a technological contraption likely to collapse at any moment. Fear, Ehrlich suggests, was the proper response. Nothing but strong measures could help.

The contradictory impulses of *The Population Bomb* grew out of 1968, an unusual and transitory moment in American political history. Dramatic problems that year had upended established political patterns, highlighting the need for bold new thinking. The old solutions seemed insufficient, yet the narrowness of the Cold War years persisted, and new ideas were not yet fully formed. *The Population Bomb* exploded in the urgency and amorphousness of the situation.

In the 1970s, a new political landscape, featuring a resurgent left and a "new" right that both found Ehrlich's biological approach reductive and authoritarian, would pick apart his contradictions. But for a few years in the late 1960s and early 1970s, especially from 1968 to 1970, the years leading to Earth Day, the power of Ehrlich's arguments and their ambivalences would attract to environmental Malthusianism many friends.

7

Strange Bedfellows

Population Politics, 1968–1970

> "Our world is no longer an endless frontier."
> —Senator Gaylord Nelson, 1969

Although Ehrlich's *The Population Bomb* is remembered as a classic of the environmental movement, few signs of a movement existed when it appeared in 1968. By the spring of 1970, however, Americans could hardly pick up a magazine or newspaper without seeing mention of ecology and the environment. Seemingly overnight, several concerns converged into a movement with the power to bring about new forms of thinking and even shape public policy. As historian Donald Fleming put it two years later, "an extraordinary diverse concatenation of impulses suddenly flashed together." Nothing signified the movement's arrival better than April 22, 1970, when upward of twenty million Americans participated in Earth Day, a national "teach-in" designed to call attention to the health of the planet. Growing concerns about population growth, both internationally and at home, contributed to the sense of environmental crisis.[1]

After the appearance of *The Population Bomb* in 1968, several factors energized and transformed Malthusian concerns: biologist Garrett Hardin's "The Tragedy of the Commons" appeared in *Science*; Pope Paul VI issued his famously controversial *Humanae Vitae* (Of Human Life); a strand of the women's movement started pushing hard for birth control and abortion rights; and President Richard Nixon issued a statement about population growth to Congress. In late 1969, Senator Gaylord Nelson, deeply worried about population growth and a host of environmental problems, called for a rally on environmental problems—what became Earth Day.

With support mounting, environmental Malthusians grew uncharacteristically optimistic. Writing to Hugh Moore in December 1969, Planned Parenthood funder Harold Bostrom expressed his sense that "the big breakthrough is coming. We will bring together the power of the conservationists, the birth-controllers, the anti-pollutionists, those against crime, war, poverty, insurrection and social decay—all societies under one roof. . . . Our voice will boom and the rich will listen, the poor will listen; the politicians will listen, and eventually, we will have the Pentagon and the Red army listening."[2]

That so many strange bedfellows—women's liberationists, Richard Nixon, and environmentalists—supported some form of population limitation suggests the multiplicity of meanings the idea could take on, as well as the sense of frustration and even crisis that gripped many Americans. It also suggests the narrowness of the moment—within a few short years attacks from both the left and the right of a quickly widening political spectrum would make Malthusian arguments much less appealing.

Garrett Hardin, the Tragedy of the Commons, and Coercion

As a sense of crisis grew, so did calls for strong governmental regulation of reproduction. Calls for coercion had been heard before, but as long as the government hesitated to get in the business even of family planning, such action seemed remote. By the late 1960s, however, political leaders in both major parties seemed newly willing to consider new roles for government. Ehrlich's *The Population Bomb* intensified the calls for strong action.

Strong action could take many forms, and population control meant different things to different people. For some, it meant reducing population growth through voluntary measures; for others, forced sterilization and abortions. In between these "voluntarist" and "compulsionist" positions was a "population control" position that avoided direct coercion, but called for socioeconomic measures, such as education, governmental propaganda for small families, and eliminating tax breaks for children. Ehrlich was a mix of the second two. He wanted noncoercive population control in the United States, but, in a 1970 textbook (co-authored with his wife Anne), pointed out that if such means should prove insufficient, laws "could be written that would make bearing a third child illegal and that would require an abortion to terminate all such pregnancies." In the developing world, he continued to consider coercive measures, as in *The Population Bomb*.[3]

Among prominent Malthusians, the person who called for the strongest forms of coercion was not Ehrlich but University of California–Santa Barbara biologist Garrett Hardin (1915–2003), whom one local journalist called a "more deadly version of Dr. Paul Ehrlich." In December 1968, Hardin published an essay in *Science* entitled "The Tragedy of the Commons," a philosophical defense of coercion so influential, especially in environmental circles, that it was called the "Magna Carta" of compulsory population control.[4]

Hardin had studied population dynamics under the well-known ecologist W. C. Allee at the University of Chicago in the 1930s. During World War II, while working on a government project to develop algae as a food source, he grew convinced that population limitation was a better solution to the world's food problem. He published a biology textbook in 1949 that cited both William Vogt and Fairfield Osborn and included a graph whose line representing population growth literally ran off the page. During the 1960s, in addition to writing articles about population, Hardin became a fierce early advocate for expanded birth control and abortion rights. His concerns about population overlapped with eugenics. In early 1969, he wrote to a supporter, "Forcing poor women to bear children they do not want (while richer women can avoid doing so) results in the poor outbreeding the rich. If poverty is even in part genetically caused (as it surely must be) such class discrimination in the availability of this form of birth control must have a dysgenic effect. Eliminating our anti-abortion laws and making abortion available to all at a very low cost would markedly improve the genetic trends of our times."[5]

Hardin opened "The Tragedy of the Commons" by pointing out that with the most dangerous and challenging problem of the day—nuclear weapons—Americans shrank from making hard choices and sacrifices and instead relied too much on technological fixes. He felt the same about environmental problems, especially population problems. Because population problems concerned how to distribute resources and prevent degradation in a resource commons, they ultimately required social solutions; technological solutions such as birth control could never really solve the problem. To explain his point, Hardin evoked a pasture open to all herders. Pasture herders, each out to maximize individual benefit, had no incentive to restrain herd size. "Each man," Hardin wrote, "is locked into a system that compels him to increase his herd without limit—in a world that is limited." The result was tragedy: "Ruin is the destination toward which all men rush, each pursuing his own best interest in a society that believes in the freedom of the commons."[6]

Hardin identified several environmental problems his logic helped illuminate: overgrazing in western ranges, fishing in oceans, and tourist visits in national parks. He also linked the commons problems to "quality of life" concerns, such as with the "reverse commons" that existed for pollution (where people added to, instead of extracted from, a commons) or with leisure activities, such as vacation spots. To solve the commons problems in national parks, Hardin proposed either selling them off as private property or strictly allocating the right to enter: privatize or regulate. Although his logic could work either way, in "Tragedy" he leaned toward community or governmental regulation, attacking Adam Smith's brand of laissez-faire capitalism for the flawed assumption "that decisions reached individually will, in fact, be the best decisions for an entire society." Ever since Smith, Hardin wrote, "invisible hand" ideas have interfered with "positive action based on rational analysis."[7]

With population growth, Hardin believed that technological answers such as green revolution hybrid wheat programs and voluntary birth control programs were ultimately insufficient, because the underlying social dynamics—the reproductive commons—had to be addressed. With or without birth control, individuals pursuing their own reproductive interests and desires would create a population so large that everyone suffered. Criticizing "laissez-faire in reproduction," Hardin warned that the "freedom to breed will bring ruin to all." Genetics, he believed, made the situation worse: those with little public conscience usually outbred those who acted more responsibly. Because of natural selection, he wrote, "conscience is self-eliminating." Furthermore, in welfare-state systems like that of the United States (which had expanding dramatically in the 1960s) the population problem was especially dangerous. Those who had too many kids were not punished but rewarded. "In a welfare state," Hardin asked, "how shall we deal with the family, the religion, the race, or the class . . . that adopts overbreeding as a policy to secure its own aggrandizement?"[8]

Only regulating reproduction could solve the reproductive commons, according to Hardin. He took issue with a 1967 United Nations declaration on the right to bear children, rejecting the implication that decisions about family size "must irrevocably rest with the family itself, and cannot be made by anyone else." Instead, he called for "mutual coercion, mutually agreed upon by the majority of the people affected." This position was, he later realized, just an elaborate way to describe ordinary "majority rules" democracy, forgetting that democracies generally also have provisions like the Bill of Rights that guarantee basic freedoms against the tyranny of the majority. To Hardin, a solution to the commons "need not be perfectly just to be preferable." Injustice, he insisted, is "preferable to total ruin."[9]

In subsequent years, calls for population control and coercion would grow among environmental Malthusians, as would the counterarguments. But for the most part, in 1969 and 1970, such calls were overshadowed by a strong pushback from conservative forces unhappy with *voluntary* birth control. Even as Hardin published his "Tragedy of the Commons," Ehrlich was engaged in a public dispute with the Catholic Church not so much about coercion as about condoms and the pill. This debate—which brought to the surface deeper differences about sex, nature, and social structure—helps to explain the energy surrounding environmental Malthusianism and especially Paul Ehrlich in 1969 and 1970. The pope became the "tyrant" against which the population movement rallied.

Pope Paul VI and the Environment

In July 1968, Pope Paul VI promulgated *Humanae Vitae* (Of Human Life), which, unexpectedly rejecting a liberalizing trend within the Church, reaffirmed its total ban on birth control. Today, that encyclical letter is mostly remembered as an attack on the birth control pill and on reproductive freedom. But at first it caused an uproar for related but different reasons: Americans understood birth control as crucial to controlling runaway population growth. "Thoughtful men around the world," the *New York Times* wrote, have reacted to the papal words "with astonishment and alarm" because they understand the world's "terrible Malthusian nightmare."[10]

Appearing just a month after *The Population Bomb*, the Vatican's encyclical quickly became entwined with Ehrlich's book. Sales jumped and Ehrlich was interviewed on *Face the Nation* and other network shows. *Humanae Vitae* warned of governmental misuse of birth control, but the core point of difference with Ehrlich and environmental Malthusians was the purpose of sex. Procreation, Pope Paul argued, should be the sole purpose: "Marriage and conjugal love are by their nature ordained toward the procreation and education of children." Each and every marital act "must of necessity retain its intrinsic relationship to the procreation of human life." Birth control was morally unacceptable because it was "specifically intended to prevent procreation."[11]

Like other Malthusians, Ehrlich had long viewed such logic as a recipe for uncontrolled population growth. "Progress in science," he argued in his 1963 textbook on evolution, "has far outstripped the ability of our extremely conservative religious and governmental systems to adjust to the changes." Because he wanted to remove any moral sanction against birth control, he fought the stigma attached to sex for pleasure instead of for procreation, attacking the pope's position as "anti-human and anti-sexual." Echoing Freud and the sexual revolution, he contended, as he often did, that the desire to have sex was unavoidable. "Night baseball," he joked, "is not going to replace sex." Moreover, as he emphasized in *The Population Bomb*, evolutionary biology had already shown that sex had not evolved primarily for procreation.[12]

Surprisingly, the debate revolved around ideas of nature. Pope Paul rejected the notion that humans should control reproduction, repeatedly citing what was "natural." Births should not be regulated by human "intelligence and will" but by "the specific rhythms" of human bodies, which were laws "written into the actual nature of man and of woman." These laws made unfettered procreation the "fundamental nature of the marriage act." "Artificial birth control," the pope contended, "obstruct[ed] the natural development of the generative process." The "rhythm method," on the other hand, worked "by nature."[13]

Indeed, deploying the rhetoric of environmentalism, Pope Paul criticized "man's stupendous ... domination and rational organization of the forces of nature." At the same time, however, the pope explicitly rejected the arguments of those who believed that modern biology showed the arbitrariness of life. "Marriage," the pope wrote, "is far from being the effect of chance or the result of the blind evolution of natural forces." Rather it was "His loving design." In another twist on environmental logic, the pope called for Catholics to acknowledge that they were not "the master of the sources of life but rather the minister of the design established by the Creator."[14]

Ehrlich flatly rejected the pope's insistence on leaving reproduction solely in God's hands. He had often argued that humans could and should "rationally" plan reproduction; in fact, he defined population control as the "conscious regulation" of reproduction. His calls for rational planning, however, presented a contradiction. Skeptical of technology, environmentalists like Ehrlich generally bristled at human interference in the ways of nature. If human nature and their technology could not be trusted in other realms of society, why would this be any different? Ehrlich's answer: place the responsibility for rational planning in the hands of scientific experts, especially ecologists. They alone understand the interrelations of nature, and only they can maintain the intellectual distance necessary to make hard choices.

In the debate with the Church, Ehrlich turned to fellow scientists for support. Along with Harvard professor Ernst Mayr and University of Puerto Rico biology professor Jeff Baker, Ehrlich organized a petition for scientists against the encyclical, hoping to get "a massive and unambiguous expression of scientific opinion." Mayr—one of the deans of postwar biology, and a popularizer of the Darwinian Synthesis—had often sparred with Ehrlich over biological theories, but they agreed about overpopulation. In April 1968, Mayr wrote Ehrlich that he hated "the almost universal stupidity that prevents more efficient population control." He was particular appalled by rumored Nixon administration programs that, he said, amounted to rewarding poor people for more children. By December 1968, twenty-six hundred scientists had signed the petition. Broadly comparing the pope to Hitler, it decried the "evil consequences" of the encyclical. By slowing efforts to halt world population growth, the pope's action "perpetuates the misery in which millions now live and promotes death by starvation of millions."[15]

Women's Liberation, Compulsory Pregnancy, and ZPG

The debate between environmentalists and the Catholic Church was not just about sex and nature. It coincided with a larger argument about the role of women in families and society. By 1968, the push for birth control rights by women had intensified greatly. Birth control rights formed part of both the mainstream women's movement as well as the more radical "women's liberation" movement that was just emerging, one strand of which, advanced by Lucinda Cisler and Shulamith Firestone, made birth control rights their rallying cry. Cisler was chair of the National Organization of Women's Taskforce on Reproduction, and Firestone was a co-founder of a series of New York–based radical feminist groups, including Redstockings and New York Radical Women. Both women saw the women's liberation movement and the ecology movement as allies. "The best new currents in ecology and social planning," Firestone wrote in *The Dialectic of Sex* (1970), "agree with feminist aims." Both movements not only advocated much greater access to birth control but also rejected ideas of womanhood centered around children. Because of the often bitter clashes between feminists and environmental Malthusians in the 1970s, many have forgotten that they were allies for much of the 1950s and 1960s.[16]

A 1969 essay by Cisler, "Unfinished Business: Birth Control and Women's Liberation," shows the growing emphasis on birth control and abortion access within the women's movement. Cisler argued that because nothing oppressed women more than their narrow role as mothers, nothing was more important for "liberation" than birth control access. "Different reproductive roles are *the* basic dichotomy in humankind," Cisler wrote, "and have been used to rationalize all the other, ascribed differences between men and women and to justify all the oppression women have suffered." Women have often wanted great things for themselves, she pointed out, but they had no choice but to be "reproducers." Birth control was crucial: "Without the full capacity to limit her own reproduction, a woman's other 'freedoms' are tantalizing mockeries that cannot be exercised. With it, the others cannot long be denied."[17]

But in 1969, Cisler saw many barriers blocking this liberation. Not only were abortion and sterilization services hard to find, expensive, and often illegal, but in some cases so was contraception. "Most people seem to think," Cisler wrote, "that there are no laws against contraception left on the books in this country, but that is not the case: only 40 percent of the states have no laws limiting the distribution or display of contraceptives." She called for individuals and organizations to fight "repressive religious, political, and medical forces." No doubt she had Pope Paul's encyclical in mind. But she also attacked Planned Parenthood for reinforcing women's roles as mothers and for being slow to push for abortion rights and the conceptive rights of single people. "'Family planning' and 'planned parenthood' carry a distinct connotation that you should get busy and plan a family—plan to be a parent—and if you aren't doing it, why aren't you."[18]

Cisler found many allies among environmental Malthusians. Over the years, few had lobbied harder for birth control rights. Twenty years earlier William Vogt, who directed Planned Parenthood for much of the 1950s, had pushed for more effective and more accessible birth-control technologies. More recently, Supreme Court justice and environmental Malthusian William O. Douglas had designed the "right to privacy" in *Griswold v. Connecticut*, the 1965 decision that established the constitutional protections for contraception for married adults. In *The Population Bomb*, Ehrlich called repeatedly for birth control rights.[19]

Moreover, at a time when even many leaders of the birth control movement were hesitant about abortion, environmental Malthusians such as Garrett Hardin pulled no punches, calling for complete repeal, not just reform, of abortion restrictions. Speaking in 1965, very early in the abortion rights movement, Hardin argued that "any woman, at any time, should be able to procure a legal abortion without even giving a reason." Hardin, in fact, coined the phrase "abortion on demand," and helped form the National Association for the Repeal of Abortion Laws (NARAL). Ehrlich made pleas for liberalizing abortion laws as early as 1965, as well as in *The Population Bomb*. Speaking for many environmental Malthusians in spring 1970, Nobel Prize–winning medical researcher George Wald wrote: "We must as rapidly as possible make convenient, safe, altogether legal, and cheap—again I would rather say free—means of abortion universally available."[20]

Many environmental Malthusians also supported calls for rethinking woman's social roles. They wanted womanhood separated from motherhood—that is, they wanted to find alternate ways for women to contribute to society besides as mothers. Raymond Dasmann had made this argument in *The Last Horizon* in 1963. In *The Population Bomb*, Ehrlich called for finding substitutes for the "satisfaction which many women derive from childbearing." In a 1970 article in *McCall's*, he elaborated, "Women should move to a position of greater equality with men and have a greater choice of life-styles. . . . A wide variety of professions should be thrown open to women immediately." Society, Ehrlich stressed elsewhere, had to "give women something else to do but have babies."[21]

Some in the women's movement, even the more radical "women's liberation" strand that emerged in 1968 and 1969, also worried about population growth. Part of the reason Lucinda Cisler disliked the focus on family within the family planning movement, she said, was that it "left little place for the interests of the individual woman or of the larger society." This included reining in population growth. Achieving a "zero growth-rate," she wrote, "must happen soon, regardless of the more equitable economic systems we may and should devise." She criticized Planned Parenthood for realizing only recently that "America, like India and Peru, should slow down." In fact, Cisler served on the national board of Zero Population Growth (ZPG), the grassroots organization that Ehrlich founded to reform American society.[22]

Like Ehrlich himself, ZPG was more complicated than is often remembered. It was founded in late 1968 by Richard Bowers, a lawyer, and Charles Remington, an entomologist and World War II veteran, in consultation with Lincoln Day, the chief of the U.N. demographic and social statistics branch, and Ehrlich, who became the first president. More than anything, ZPG was concerned with protecting the environment by reducing population growth. A mixture of science-minded people, mostly male, and women like Cisler, it thought globally, but acted locally. "The concern of this organization," Bowers wrote in a proposal outlining ZPG in early 1969, "will be worldwide, but in the beginning nearly all attention will be on the United States as a whole and on its individual fifty states." Highly decentralized, ZPG had local chapters around the country. In the 1970s, some ZPG members would take up immigration restriction, but that was not the concern in 1969 and 1970.[23]

To get Americans to "stop at two," ZPG tried many measures, but perhaps most consistently focused on expanding access to birth control and abortion. "The basic human right to limit one's own reproduction," the ZPG board of directors announced in 1969, "includes the right to all forms of birth control: to contraception, including sterilization, and to abortion. We therefore oppose all legislation and practices that restrict access to any of these means of birth control." Much of their rhetoric resembled Cisler's. "It is immoral and inhuman," another early statement of purposes and goals pronounced, "that in 1969 compulsory pregnancy should be part of the legal structure in America."[24]

As journalist Wade Green noted a few years later, ZPG was "anything but monolithic," especially on the issue of coercion. In 1969, cofounder Richard Bowers declared, "Voluntarism is a farce." Garrett Hardin, another major influence within ZPG, was also in the compulsionist camp. But voluntarists were equally vocal. "One of our major goals," ZPG Executive Director Shirley Radl testified before the California legislature in 1970, "is to provide the means for all members of society to voluntarily limit the size of their families. This means making all forms of birth control, including abortion, readily available." She added, "The uterus should be the concern of the owner and not of the State." That a single organization could bring together such seemingly disparate ideas shows that, at the time, the priority of the activism was on ending compulsory pregnancy through expanded access to birth control and abortion rights, not reproductive rights stressing women's autonomy, as it would be a few years later.[25]

Baby Boomers and Environmentalism

In 1968 and 1969, the birth-control strand of the women's movement joined forces with young people bursting with concern about population growth and equally frustrated by the Catholic Church's hidebound traditionalism. "It is among young people, particularly on college campuses," *US News and World Report* reported in early 1970, "that the idea of limiting sharply the size of families appears to be taking strongest root." *Life* reported much the same. In January 1970, Hugh Moore wrote of the "mounting student interest in environment, conservation and over-population" that has been "catching hold on college campuses throughout the country." In an April 1970 article in *Natural History* called "A Student Manifesto on the Environment," Pennefield Jensen placed population concerns at the center of her generation's sense of urgency.[26]

Many of these baby boomers had felt the squeeze of crowding from their cradles to college. They lived through a housing crunch in the late 1940s, a school and park crunch in the 1950s, and a college and employment crunch during the early 1960s. Growth was fastest on the coasts, and particularly acute in Sunbelt states like California and Arizona. California opened a school every week during the entire 1950s. Moreover, because baby boomers had been hearing from their parents and the press about the size of their generation for all of their lives, they had an unusual self-consciousness as a generation.[27]

Baby boomers had also grown up with a sense of crisis. In 1969, when the writer Joyce Maynard, then a teenager, read *The Population Bomb*, historian Adam Rome reports, she felt a rush of fear akin to what she had experienced

FIGURE 10 A student volunteer for Zero Population Growth (ZPG) trying to save the local marshland passes out population control leaflets to Connecticut commuters.
Life, January 1, 1970. Photo by Art Rickerby/Time Life Pictures/Getty Images.

during the Cuban missile crisis: "Not personal, individual fear but end-of-the-world fear, that by the time we were our parents' age we would be sardine-packed and tethered to our gas masks in a skyless cloud of smog." By the late 1960s, a host of crises, from stalemate in the struggle against communism to urban unrest to environmental problems paralyzed the nation, most of which Malthusians like Ehrlich connected to population growth. An Earth Day article at the University of Texas at El Paso captured this sense of systems collapse with a three-word headline: "Things Aren't Working."[28]

For many in this generation, Ehrlich had rock star appeal. One journalist reported on an Ehrlich appearance in Denver in the early 1970s: "The queues of people crowding in were nearly all young. . . . nearly all of them college and high school students." In the talk, Ehrlich ripped into middle-class American overconsumption. "He had the audience gripped and it struck me that there was more than a touch of the revivalist technique in his accusatory style, in the way he brought home to his eager listeners their own guilt and complicity." Part of Ehrlich's message was that overpopulation and overconsumption were problems young people in particular could do something about it. They could transform society through their own daily lives.[29]

One of the most prominent Ehrlich devotees among the younger generation was Stephanie Mills, the college student who, having recently read *The Population Bomb*, proclaimed at her Mills College graduation in June 1969 that the best thing she could do for humanity was to have no children. Mills had grown up in Phoenix, Arizona—a Sunbelt city that, like the Bay Area, was fast-growing and nature-defying. A person who liked to wear IUDs as earrings, Mills mixed together countercultural and feminist ideas in a toned-down version of population limitation. "Traditionally, commencement exercises are the occasion for fatuous comments on the future," she opened her commencement speech. Instead, she would tell it like it is. The promise of a rosy future was "a hoax." Her generation had been lied to. All that was supposedly good was hurtling them toward disaster. And yet "virtually nothing is being done by anyone with enough power." The situation demanded unconventional, even radical, action. We must, she said, "circumvent the so-called political realities."[30]

Mills, like others, decried the Catholic Church's resistance to birth control. "At the turn of the century," she pointed out, "people were arrested in New York for distributing birth control information, and only last year, Pope Paul the Sixth issues an encyclical which forbade the members of his flock to use contraceptives." Mills also echoed the themes of the women's movement. "Our identities as men and women," she complained, "are so conditioned by our reproductive functions." In an essay from March 1970, just a month before Earth Day, she called for "a re-orientation of child-bearing attitudes." Women should be encouraged to be something besides mothers: "Certainly women's roles must be expanded to encompass much, much more than the production of children."[31]

Mills also softened many of the most controversial aspects of environmental Malthusianism. By attacking modern health care for causing burgeoning populations, Ehrlich had made it seem as if he favored denying sick people medicine. Mills stressed that the real problem with modern medicine was that it had withheld birth control and abortion rights from women and their families for over a half century. In her 1970 essay, instead of identifying population growth as the root cause behind war, hunger, disease, and alienation, she stressed it only "intensifies" these problems. She also argued that population control would not create authoritarianism; it would prevent the high populations that encouraged authoritarianism: "More people are less individual; more people are less free. . . . A more populous nation becomes necessarily more authoritarian." Indeed, unless populations are reduced, more repressive governments might emerge down the road. "A danger inherent in the population problem is that the state may finally assume control of reproduction if the individual doesn't. Consider a state so powerful that it controls the reproduction of its citizens." That she could make such a claim just a year after Garrett Hardin's "Tragedy of the Commons" shows to what extent the coercion issue was overlooked at the time; the focus was more on fighting the powers that be and securing the right to birth control and to abortion.[32]

Republicans, Population, and the Environment

As the political was remaking the personal, the personal was also reshaping the political. In 1969, population planning continued to attract substantial support from both major parties. "Congressional interest and support in population problems was remarkably bipartisan," former congressman George H. W. Bush later recalled. This moment would not last: a decade later this broad appeal would be out of the question.[33]

Among Democrats, in addition to Lyndon Johnson and Ernest Greuning, prominent advocates for population planning included Stewart and Morris Udall, William Fulbright, Robert Packwood, Joe Clark, Joseph Tydings, Alan Cranston, and Gaylord Nelson. They were supported by many on the left-progressive side of the political spectrum. The editors of the *Progressive* magazine, for instance, called for "urgently needed legislation to speed population control."[34]

As George Bush's statement suggests, Republicans also supported population planning. "In the exuberance of our productive glut," leading conservative intellectual William F. Buckley wrote in the *National Review* in 1965, "we tended to assume that the procreative of the world could not possibly overtake our productive ingenuity. Well they have." He added, "That old dog Malthus turned out to be very substantially correct in his dire predictions, and there seems to be no point in waiting until the United States is like India before moving in on the problem." According to the 1968 Republican Platform, "The worldwide population explosion in particular with its attendant grave problems looms as a menace to all mankind and will have our priority attention." By this time, former president Dwight Eisenhower had become an outspoken supporter of population planning. In Congress, key supporters besides Bush included Representatives Robert Taft Jr. (Ohio) and Paul "Pete" McCloskey (California). Even conservative standard bearer Barry Goldwater supported population programs.[35]

Despite support from Buckley and Goldwater, it was the moderate Bush who best exemplified Republican interest in population matters. An oil executive elected to Congress from a Houston district in 1966, Bush was the son of a former U.S. senator from Connecticut—Prescott Bush—who had once lost an election in the late 1940s for supporting Planned Parenthood. In 1967, George Bush served on the Ways and Means committee that added birth control programs to the Social Security Act. In 1969, he led a group called the "Young Republicans for the Environment" and headed a Republican Task Force on Earth Resources and Population. Through population planning, Bush combined his family's patrician concern about poverty with an oilman's sense of the world's limited resources.[36]

Bush's approach to population, resources, and birth control might surprise Republicans of a later generation. Holding hearings from mid-1969 to early 1970, his task force investigated "the most critical problem facing the world in the remainder of this century": the interrelated issues of population growth, dwindling natural resources, and environmental degradation. The task force report, which Bush wrote, stressed global resource limits and called for limiting population. One of the "basic facts" about the earth is that it is composed of "finite resources." As a threat to the nation's resource base, population growth threatened the nation's security. Bush also linked population growth to social problems. Citing sociological studies, he warned that "density of population has been a prime cause for increased automobile traffic deaths, drug addiction, broken marriages, alcoholism, crime, homosexuality, suicides, venereal disease and heart attacks." As a solution, in addition to more liberal family planning and abortion laws, Bush called for a noncoercive population limitation program. Population growth problems, he wrote, are "of far greater magnitude and complexity" than could ever be solved through even "total availability of family planning services." The United States, he concluded, needed to set "a national goal to stabilize our population."[37]

The Ecologist in Chief and Systems Collapse

Perhaps the most surprising person to jump on the ecology bandwagon during the late 1960s was President Richard Nixon. Nixon had mentioned environmental problems in his inaugural address in January 1969, but within a few months his claim to leadership on the issue was in jeopardy. In early 1969, he had not only fumbled the response to the made-for-TV oil spill off the coast of Santa

Barbara but also dismissed a handful of prominent environmental initiatives by Democrats. "If Nixon enjoyed a 'honeymoon'" with his environmental policies, historian J. Brooks Flippen has written, "by summer it had long since ended." Nixon's environmental rhetoric seemed empty at best, cynical at worst. Republicans such as Bush called for action.[38]

Daniel Patrick Moynihan, the former Johnson aide who had crossed party lines to become Nixon's top advisor on urban affairs, steered Nixon toward the population issue. Moynihan had long linked poverty with population growth, and recent demographic and family planning studies had backed the relevant parts of his 1965 report, showing that poor and black populations had disproportionately high fertility rates and that birth-control programs were not reaching these communities. Moreover, a series of reports from the President's Office of Science and Technology, and lobbying from John D. Rockefeller III, a prominent Republican, supported his recommendation to Nixon. The issue also scored points politically, appealing to both race progressives and racist whites. In the campaign, Nixon had repeatedly said he would fix the "welfare mess."[39]

Although often overlooked, several international factors also pushed Nixon to take up the issue. Secretary of State William P. Rogers and AID director John A. Hannah had written to Nixon that population growth contributed to "crime, banditry and civil unrest in many developing countries." Additionally, as with domestic American poverty, Nixon wanted a new approach to poverty overseas. In August 1969, Nixon told Rudolph Peterson, the chairman of his Task Force on International Development, to rethink foreign aid and give particular emphasis to population planning. According to the minutes of the meeting, Nixon "stressed that population control is a must . . . a top priority national policy."[40]

An even more significant international concern also drove interest in population limitation. During much of 1969, as part of the détente initiative, Nixon and Moynihan had been searching for an apolitical activity that could redirect the North American Treaty Organization (NATO) away from confrontation with Soviets. Population limitation appeared just the right tool. It was worldwide, transnational, and seemingly something that most countries could agree to. To push the idea, Moynihan gathered together ambassadors from NATO member nations to form "The Committee on the Challenges of Modern Society." Science advisor Lee DuBridge also stressed the issue on his fall 1969 European trip.[41]

In July 1969, Nixon sent a special message to Congress about population growth. The "dramatically increasing rate of population growth," he wrote, ranked among "the most serious challenges to human destiny in the last third of this century." Nixon echoed the college group who said that nothing worked: "Many of our institutions are . . . under tremendous strain as they try to respond to the demands of 1969."[42]

Internationally, the problem was the third world. Because population was outstripping even expanded levels of production, underdeveloped countries faced obstacles that industrialized nations had never encountered. Imbalances threatened to impair individual rights, jeopardize national goals, and erode international stability. Whether motivated by the "narrowest perception of national self-interest or the widest vision of a common humanity," Nixon warned, "population growth is a world problem which no country can ignore."[43]

At home, Americans faced not a want of food but a shortage of "social supplies"—"the capacity to educate youth, to provide privacy and living space, to maintain the processes of open, democratic government." Rapid growth had created problems in housing, urbanization, health care, employment, and transportation. Unwanted or untimely childbearing also pushed people into poverty.[44]

Refuting Eisenhower's dismissal only a decade earlier, Nixon affirmed that the federal government had "a special responsibility" for addressing population issues. Indeed, he called for more federal planning, both expanded birth control programs as well as a national commission to study population policy.[45]

Nixon's message lent presidential weight to the cresting population planning movement. He reinforced the international themes that many in the movement had trumpeted since before the Draper Committee in 1959, and emphasized domestic quality-of-life concerns. Nixon did not shift the debate away from the food crisis in India and toward social problems at home, as some have argued, as much as add a new domestic emphasis to continuing and expanding international concerns.[46]

Just nine months before Earth Day, Nixon's message also lent momentum to the swelling environmental movement. Notably, Nixon voiced opinions that could have come off the tongue of Paul Ehrlich: "Perhaps the most dangerous element in the present situation is the fact that so few people are examining these questions from the viewpoint of the whole society." And: "What of our natural resources and the quality of our environment? . . . The ecological system upon which we now depend may seriously deteriorate if our efforts to conserve and enhance the environment do not match the growth of the population." Nixon also added momentum by spurring ecologists and environmentalists, such as Ehrlich, who thought he did not go far enough.[47]

The National Environmental Policy Act

Population concerns also informed national environmental legislation such as the National Environmental Policy Act (NEPA), one of the defining laws of the postwar environmental movement. Signed into law by Nixon on January 1, 1970, NEPA established the Council on Environmental Quality, the Environmental Protection Agency, and environmental impact statements. Title I of the act mentions "the profound impact of man's activity on the interrelations of all components of the environment, particularly the profound influences of population growth."[48]

NEPA's chief architect was Lynton Caldwell, a political scientist deeply worried about population growth. During the early 1960s, Caldwell more or less invented the field of environmental management out of whole cloth. Curiously, though, Caldwell had not been trained in conservation or demography. Like John Kenneth Galbraith and others, thinking about poverty and prosperity in the global context of the 1950s and 1960s had led him to see new problems and to forge novel ways of thinking, including about the role of the government vis-à-vis the environment.

After teaching American politics for almost a decade at Syracuse University, in 1954 Caldwell took an assignment coordinating the U.N. Technical Assistance program for Turkey and the Middle East. Later in the 1950s, Caldwell oversaw Indiana University's training program for public administration in Southeast Asia. One day in 1962, while in Hong Kong, Caldwell had an epiphany. While surveying the crowded city from Victoria Peak, it dawned on him that many of the developing societies he had worked in during the previous decade, places like Hong Kong, suffered from severe problems—especially environmental problems—that governments considered outside of their concerns. Caldwell committed himself to charting a legitimate constitutional path by which governments could address environmental matters. The result was a 1963 article in *Public Policy Review*, "Environment: A New Focus for Public Policy?," in which Caldwell called for a new policy focus, one designed to obtain "integrated planning and action, and to get coordination among the agencies and polices affecting [the] environment." This award-winning paper launched the field of environmental policy.[49]

Caldwell was clearly an environmental Malthusian. "An accelerating demand on all resources," he wrote in his seminal 1963 article, "is resulting from our burgeoning technology and from increasing populations, their needs, expectations, and changing ways of life." In a 1966 essay co-written with William Vogt, he pointed out that Americans misunderstood the population issue in fundamental ways: "Questions of population growth are seldom considered in relation to the total human environment." People think of the environment as a storehouse of resources whose inventories run low because of population growth. Instead, they should think of the environment as a "complex interdependent bio-physical system." People also overlooked how overpopulation would erode not just economic stability but also quality of life, freedom, and political and social stability. Human societies, he insisted, needed to be conceptualized—and managed—as "complex ecosystems within larger and more complex ecosystems."[50]

This environmental approach distinguished Caldwell's views from those of his contemporaries. Years later, political scientists Robert Bartlett and James

Gladden wrote that Caldwell "was alone in focusing on the distinctive, integrative character of the concept 'environment' and its implications for politics, public policy, and public administration." His experience overseas, thinking about population growth and natural resource questions, helped him see the world holistically. "My involvement in international technical assistance work during the 1950s and early 1960s was a major factor in my turn to environmental policy," he wrote later. "On-the-ground experience in Latin American, the Middle East, and Southeast Asia left me deeply skeptical of the prevailing theories and practices of U.S. and U.N. development economists and planners."[51]

By the late 1960s, Caldwell had become, according to biologist and conservationist F. Fraser Darling in 1967, the "leading thinker in biopolitics." In 1968, as a consultant to the Senate Committee on Interior and Insular Affairs, Caldwell authored "A Draft Resolution on a National Policy for the Environment," the document that became the foundation for NEPA, which was passed in late 1969 and signed into law soon afterward.[52]

Population Concerns and Earth Day

Population concerns also shaped the grassroots environmental politics of Earth Day. Among the largest demonstrations the United States has ever seen, Earth Day consisted of a day of speeches and marches spotlighting environmental problems. It was the brainchild of Senator Gaylord Nelson, a Democrat from Wisconsin and environmental Malthusian. Born in a town of seven hundred in the northwest part of the state, Nelson had fought in the army during World War II, including a stint in Okinawa. After training as a lawyer, he served as governor of Wisconsin before being elected to the U.S. Senate in 1962, where he supported Lyndon Johnson's Great Society programs but also, earlier than most, opposed the Vietnam War. Both in Wisconsin and the Senate, Nelson worked to bring conservation issues to the attention of the public. In 1963, he convinced President Kennedy to undertake an awareness-raising conservation tour. By the late 1960s, Nelson later recalled, the public was "was far ahead of the political establishment" in its environmental concern. He decided to do something about it. In September 1969, at a conference in Seattle, Nelson announced a nationwide demonstration akin to the anti–Vietnam War teach-ins, to be held the following spring.[53]

Nelson had wanted to try something similar for years, at least since the early 1960s conservation tour. What stirred action in late 1969? January's Santa Barbara oil spill and June's Cuyahoga River fire disturbed him greatly. But so did rapid population growth. Nelson credited Fairfield Osborn's *Our Plundered Planet* with first awakening his environmental sensibilities and spurring his interest in public service years earlier. As governor and senator, he had often stressed overpopulation. "To the best of my recollection," he wrote a constituent in 1970, "I have not given a speech on the environment in the last half dozen years in which I have not emphasized the disastrous consequences of the increasing population." Nelson's concern spiked in the late 1960s. Warning of humanity's "rampaging breeding," he announced that Americans no longer enjoyed an "endless frontier." The same month he called for an environmental teach-in, he cited a Paul Ehrlich article on the floor of the U.S. Senate: "Man is not only running out of food, he is also destroying the life support systems of the Spaceship Earth."[54]

Scores of prominent environmental leaders had also expressed concern in various forms about population growth.[55] National environmental organizations also took up the cause. Under its new director Phil Berry, the Sierra Club board passed a resolution urging the federal government and all states to cut programs that promoted growth, to promote education programs aimed at reducing population growth, and to limit aid to countries that did not have population programs. The National Wildlife Federation and the National Audubon Society also took strong stances on population limitation.[56]

The committee Nelson established to organize Earth Day included many environmental Malthusians, including Ehrlich, Sydney Howe of the Conservation Foundation, Congressman Paul McCloskey, and regional planner Harold Jordahl. Denis Hayes, the Stanford graduate hired to coordinate the planning efforts, also worried deeply about population growth. The committee received donations from people such as Lawrence Rockefeller, and when money got very tight, Howe came to the rescue with $20,000. The committee made population one of its top priorities. The purpose of Earth Day, according to a press release to newspapers and college campuses, was to raise awareness about "the environmental problems being created by our advancing technology and expanding world population." Later it reiterated concern about "dramatically increasing population problem."[57]

The issue was all over the press in late 1969 and early 1970. Among others, *US News and World Report*, *Redbook*, *Seventeen*, *Parents Magazine*, *Look*, and *Foreign Affairs* had articles. *American Heritage* had a story called "Catastrophe by the Numbers," and *Reader's Digest* reprinted Morris Udall's article, "Standing Room Only on Spaceship Earth." *Life* devoted the lead article of its special issue on the 1970s in January to population. Population growth, it warned, will most likely lead to "widespread starvations, violence and constant, bristling annoyance with our fellow space-usurpers."[58]

Population growth was a staple topic on Earth Day; speaker after speaker mentioned the problem. The focus was international, national, and local. Local groups worked population concerns into their activism. A group in Davis, California, organized a "Cork-the-Stork" parade. Another at Indiana University threw birth control pills into the crowd. As a warning of impending famine caused by the world's rising population, students at San Fernando State College in Los Angeles prepared a meal of tea and rice to give people a taste of a "hunger diet." Common bumper stickers and buttons at the time included Let's Not be Sardines; Overpopulation Begins at Home; Crowded Right Now??—Baby You Just Wait!; For Our Country, Limit Population; Smaller Families or Bigger Headaches; and The Population Bomb is Everyone's Baby. Pete Seeger sang about how "We'll all be doublin' in 32 years."[59]

College students and faculty, especially within biology departments, showed particular enthusiasm. Sample campus events included a "Population and Survival" day at Oberlin; a talk by a physician on "Population Control and Contraceptive Techniques" at SUNY-Buffalo; a panel called "Black, Brown and White on Population" at San Jose State College; a population seminar at Brigham Young University; a panel discussion, debate, and a talk called "Spaceship Earth: People, Poverty and Pollution" at UCLA. Events on population were common in high schools, too. At Atherton High School in Louisville, Dr. Daniel Webler of the Louisville Presbyterian Theological Seminary conducted an "experiment in overpopulation" by crowding together students on the school concourse. Dissent was not common, but did foreshadow disputes to come. In Wisconsin a group called Concerned Demographers questioned local ZPG members about coercion and possible abuses.[60]

By Earth Day, ZPG had grown tremendously. According to one account, as late at September 1969, it had just 700 members. But an appearance by Ehrlich on Johnny Carson's *Tonight Show* in early 1970 sent their numbers skyward. By later that year, ZPG had over 30,000 members in over 300 chapters around the country. Ehrlich liked to point out that ZPG is "about the only thing growing faster than the population." *Life* featured the organization in a cover story the week of Earth Day. On Earth Day itself, local ZPG chapters offered speakers, handed out literature at rallies, made presentations at high schools, and showed films such as "Population Ecology." They had particular strength near college campuses, bolstered by student members. The Yale chapter had 200 members; the University of Wisconsin–Madison branch had 500.[61]

The influence of biological models of population growth on ZPG groups was conspicuous, perhaps nowhere more so than in an article in the Madison chapter's Earth Day newsletter called "The Kaibab Herd and the Human Horde—Deadly Parallels." By the late 1960s, the Kaibab story, as interpreted by Aldo Leopold, had become a standard cautionary tale among conservationists and many biologists. One scholar, writing in 1973, noted that of twenty-eight general biology textbooks that had appeared since 1965, seventeen included accounts of the Kaibab episode. "The incident," he wrote, "has exercised an irresistibly attractive force." It also had been worked into exhibits at museums such as the Field Museum in Chicago and the Milwaukee Public Museum. The Madison chapter of ZPG applied the logic of the Kaibab to modern human society. What happened on the Kaibab in the 1920s, it said, "appears to be entirely analogous to the imminent human catastrophe." Recounting the oft-cited story of predator controls, population explosion, and population crash, the article pointed out that "our earth is large, but in many respects it is but an isolated plateau." The story was particularly tragic "not only because it happened, but primarily because many people foresaw the crash but in the face of public opinion were powerless to stop it."[62]

Driven by such biological models, the rhetoric about population growth at times grew strident. In 1968, a Sierra Club flier proclaimed that the "continuing population growth is the root cause of all conservation problems." "It is obvious to me," Congressman Morris Udall of Arizona wrote in late 1969, "that the destruction of wilderness and natural beauty and the pollution or poisoning of soil, air and water are caused by man's numbers overwhelming, at the very time he needs it most, the delicate base of nature that sustains him." Two weeks before Earth Day, Senator Nelson called population growth the world's biggest problem. If population growth continued unhindered, he said, "we might as well forget finding solutions to any of our social and environmental problems." A panel at the University of Wisconsin–Lacrosse was entitled, "Population: The Real Pollution."[63]

Like Ehrlich, Nelson connected everything from litter and open space to pollution and pesticides to population growth, saying overpopulation would lead to a world of "nothing more than an area of poisonous waters and choking air surrounded by mountains of garbage and debris." Reflecting two decades of novel thinking about nature protection, Nelson's idea of "environment" incorporated a broad social vision, including the problems of American inner cities, the Vietnam War, and even the Cold War. "Our goal," he declared the night before Earth Day, "is an environment of decency, quality, and mutual respect for all other human creatures and all other living creatures—an environment without ugliness, without ghettoes, without discrimination, without hunger, poverty, or war. Our goal is a decent environment in the deepest and broadest sense." Global environmental problems, he said, even overshadowed East-West tensions.[64]

The New Ehrlich

In the months leading up to Earth Day, few environmentalists were more visible than Paul Ehrlich. Helped by three appearances in as many months on the *Tonight Show*, *The Population Bomb* sold almost a million copies between January and April and another 700,000 by the end of the year. (It continued to sell well over the course of the decade.) Ehrlich gave scores of speeches around the country. During a five-day stretch in February, he spoke at five college campuses in four states, and appeared on the *Phil Donahue* show. In 1970, he gave a hundred lectures, published scores of articles, and appeared on two hundred radio and TV shows. The week of Earth Day, he was on the *Tonight Show* on April 16, the *Today* show April 17, and the *Mike Douglas Show* April 20. On Earth Day, Ehrlich spoke at Villanova University, as part of a whole day devoted to "The Population Crisis: Where We Stand Now."[65]

In ways most have overlooked, the Ehrlich of Earth Day differed from the Ehrlich of *The Population Bomb*. Partly in response to President's Nixon's population policies, which he found disappointing, and partly in response to criticisms from African Americans, which upset him, by early 1970 Ehrlich was speaking less about overpopulation in urban areas and more about white-middle-class overconsumption. Many environmentalists and supporters followed his lead.

In 1968 and 1969, several black critics in the San Francisco Area began to criticize Ehrlich, especially for talk of coercion. In a January 1969, *Palo Alto Times* article, comedian and activist Dick Gregory, a Bay Area resident, claimed that birth control was a plot of the "power structure" to control "the discontent and rebellious youth." He singled out for criticism those who blamed environmental problems on population growth. "Man's sprawling, undisciplined urban complexes, his concentrated and polluted misuse of natural environment, and his refusal to realistically use the resources nature has provided," he contended, "has done more to create a population problem than the natural results of human reproduction." Responding in a letter, Ehrlich evoked the universalistic logic of racial liberalism, saying that the threat of population growth was "colorblind." Later that year, Walter Thompson of Endeavor to Raise Our Size (EROS), an Oakland group that hoped to raise African American power by elevating the number of African Americans to sixty million by 1980, also assailed Ehrlich.[66]

Ehrlich deserved rebuke. Drawing from John Calhoun's studies connecting density and violence, *The Population Bomb* appeared to blame urban riots on overpopulation. Although only quick references, they were enough for readers in the supercharged aftermath of Watts, Newark, and Detroit. "Ehrlich," a Wichita reviewer wrote, "discusses many of the issues concerning modern man, such as environmental pollution, mounting violence, poverty, etc." A Wisconsin couple, thanking their local paper for printing excerpts from Ehrlich's book, were more explicit: "The increased welfare rolls . . . [and] the riots brought on by increasing social stresses in our crowded cities . . . are mere symptoms of a little recognized disease—the population explosion."[67]

Ehrlich's book also encouraged others to see urban poverty as an *environmental* problem. "This Nation, and indeed large areas of the entire planet," Representative Melvin Laird (R-Michigan) warned in 1968, "are destroying the balance of nature through unwise use of our natural resources, through constant and continuing pollution of the air and water, by overcrowding in our cities, and by a host of other abuses." Others were more direct. "In America," Morris Udall wrote in July 1969, "the problems of poverty, racial strife, transportation, the rotting of our central cities . . . can be traced directly to the problem of overpopulation." In a widely republished 1969 article in *BioScience*, biologist Walter Howard echoed Udall: "It is incongruous that student unrest is so great and race problems so much in the front, yet almost everyone seems unaware that the basic cause of most of these socio-economic stresses is overpopulation." "Surplus individuals," he stressed, "do not quietly fade away."[68]

Black concerns about population control had a real basis in history. White slave owners had manipulated sex and reproduction for their own benefit, and during the twentieth century some white elites pushed birth control and sterilization, often with the specific intent of reducing black populations. African Americans knew this history far better than most white environmentalists, who like Ehrlich, tended to stress their color-blind universalism and sympathy for the civil rights movement. But by 1969, racial colorblindness no longer held the same moral sway, especially as talk of coercion increased. By this point, many African American activists had begun to stress cultural and historical difference, including the history of mistreatment by public health officials. For this reason, many saw Malthusianism—one of the driving forces of environmentalism—as insensitive at best, and perhaps a disguised form of racial control.

In 1969 Ehrlich began altering his message. He started to routinely acknowledge that population planning was, to some extent, "a white racist plot." About one-third of all those people who advocate population control, he explained in one article, "actually mean control [of] Blacks, or the poor, not the white or the affluent." He continued to critique this position into the 1970s: "All too often," he wrote in 1972, "population control is viewed as a plot by rich white people to suppress the poverty-stricken and colored people of the world. And unhappily, in the minds of some members of our white racist society, that is precisely what population control means." Ehrlich also stopped saying birth control was "color-blind."[69]

Additionally, in 1969 Ehrlich began—far more often and far earlier than most critics realize—to refocus the issue of population growth as a problem of American middle-class overconsumption. Aldo Leopold, Fairfield Osborn, William Vogt, and Robert Rienow had stressed consumption and growth, as had Ehrlich himself in his attacks on "superconsumers" in *The Population Bomb*. Now Ehrlich refined the argument by targeting mostly white, middle-class American consumption. In an April 1969 letter to the editor of the *Los Angeles Times*, he noted that American population growth, not third-world population, was "the most serious threat to human survival." Americans, he explained, were "consumers and polluters par excellence, and the average American baby puts more strain on the life support systems of our planet than two dozen Indian or Latin American children." By early 1970, he had sharpened these points. "Each American child," he told *Time* in February, "is 50 times more of a burden on the environment than each Indian child." He urged *U.S. News and World Report* in March, 1970 to "look at the way in which we Americans destroy the environment and consume natural resources."[70]

Ehrlich also shifted his view about the Vietnam War. In *The Population Bomb*, he had generically blamed wars on overpopulation and resource scarcity, and other environmentalists had echoed his ideas. "War is the ultimate manifestation," biologist Richard A. Watson wrote, "of the lemming-like madness that grips large populations of men in great need." In July 1969, Morris Udall wrote that "the foundation of war can be laid in the Asian mud of prospective famine." By early 1970, though, Ehrlich had begun to see American efforts in Vietnam as a war to defend access to resources in Southeast Asia.[71]

A talk he gave in Wisconsin in mid-March 1970, just a month before Earth Day, gives a sense of Ehrlich's new emphasis on American consumption and imperialism. "I would like to point out that the population problem first of all," Ehrlich opened the address, "is primarily a problem of the affluent whites of the world.... In our country for instance, our minority groups—the Blacks, the Chicanos and so on—are generally sufferers from white pollution, not creators of it. I would like to point out that the Vietnam War ... is part and parcel of the whole thing. Our legions are marching over there and elsewhere in the world because we—a very small portion of the world's population—consume what is now estimated to be thirty-three and a third percent of the natural resources.... Fundamentally the rich of the world are still stealing from the poor." Americans, Ehrlich pointed out in *National Wildlife* in April 1970, "are really looting the world to maintain our level of affluence." In *Mademoiselle* that same month, he called Americans the "champion looters and polluters of the globe."[72]

In no small part because of Ehrlich, many other environmentalists had begun before Earth Day to attack middle-class American consumption. "Those who can 'afford' luxurious living," Walter Howard wrote in September 1969, "are largely utilizing many times their share of the limited food and other resources, and also they are contributing much more pollution to the environment than are the have-nots." Dismissing the "widely prevalent view that the poor are over-reproducing," George Wald wrote in the Spring 1970, "it is precisely the well-to-do and their children who make the most trouble—who are at once the biggest consumers and the biggest polluters." Ehrlich's new thinking was clearly visible in David Brower's approach to population planning in 1970: "start controlling population in affluent white America, where a child born to a white American will use about fifty times the resources of a child born in the black ghetto." For Ehrlich and ZPG, *Newsweek* reported in January 1970, the real villains were not greedy industrialists, but "consumers who demand ... new, more faster, bigger, cheaper playthings without counting the cost in a dirtier, smellier, sicklier world." Although the population movement brought together people with a wide variety of interests, those environmentalists who joined Ehrlich in expressing concern about population growth on Earth Day did so in large part out of concern about American middle-class overconsumption.[73]

With arguments about disparities in consumption and American imperialism, Ehrlich-style environmental Malthusians were able to win over some on the left. But some skeptics still worried that environmentalism would distract from other important matters. Writing in April 1970, journalist and former Democratic speech writer Ben Wattenberg believed it was "wrong, and dangerous, and foolhardy" to make population a crisis. "When Explosionists say, as they do, that crime, riots, and urban problems are caused by 'the population explosion,' it is just too easy for politicians to agree and say sure, let's stop having so many babies, instead of saying let's get to work on the real urban problems of this nation." Politicians, he explained, could too easily replace the billion-dollar price tag for the cleanup of Lake Erie with the 25-million-dollar price tag for birth control. President Nixon could earn easy political points for little real leadership. "I imagine there have been luckier Presidents," Wattenberg remarked, "but I can't think of any."[74]

Other critics worried about coercive programs overseas. "Paul Ehrlich is a nice man. He doesn't hate blacks, advocate genocide or defend the empire. . . . But his appeal, while barely denting the great waste-production economy, will only create the self-righteousness to impose America's middle-class will on the world," wrote journalist Steve Weissman in May 1970. "Long before even the least of the predicted ecological catastrophes comes to pass, such fears [of overpopulation] might well turn race on race, young on old, rich on poor."[75]

A few years after Earth Day, in the midst of the growing criticism of environmental Malthusians, a journalist asked Ehrlich why he had stressed population so single-mindedly between 1968 and the early 1970s. Ehrlich responded by stressing the intransigence of his opponents, "It looked as if we were up against religion and tradition and the very glue that holds society together." Much as when he was a graduate student fighting to convince biologists to think about taxonomic classification in new ways, Ehrlich saw himself in a battle with a powerful, reactionary force—as did many of his numerous allies, especially among young people, feminists, and environmentalists. For Ehrlich, environmental Malthusianism was not just a single-issue campaign but a critique of an entire way of organizing society. When environmental Malthusians of the late 1960s and early 1970s took aim at population growth, they also targeted "the very glue" that held society together—traditional attitudes toward sex, women, the family, and especially the consumption-based economy.[76]

To Ehrlich and his supporters, the struggle engaged the enlightened forces of reason with the benighted forces of tradition, superstition, and status quo. Indeed, perhaps because of Ehrlich's battle with conservative elements of the Church—a struggle that, in his mind, recapitulated the Enlightenment struggle of science and reason against medieval ideas of religious tradition—he lost sight of how Enlightenment values of universal science were just then coming under attack. Although in the crisis atmosphere of the late 1960s, the science-based Malthusianism of critics such as Ehrlich seemed new, penetrating, and potentially unifying, within just a few years it would be strongly attacked by both sides of the political spectrum.

8

We're All in the Same Boat?!

The Disuniting of Spaceship Earth

"Above all, *humanity as a whole* must take action first to stop population growth."

—Paul Ehrlich, 1970

"After the political exhaustion of 1968, and the heavy drain of our war in Vietnam, the nation badly needs to feel a common sense of mission and unity. But ecology can hardly make our differences disappear."

—John Lindsay, "The Plight of Our Cities," 1970

Ten days before the first Earth Day celebration, on April 13, 1970, an explosion aboard *Apollo 13* put its three-person crew at risk and forced the American spacecraft to abort a mission to the moon. Four days from reentering the Earth's atmosphere but with only two days of electricity and water, *Apollo 13* hung above the earth, and the fate of its passengers hung in the balance.

To a Texas doctor writing to President Richard Nixon, the incident provided a compelling analogy for the planet's environmental crisis. "Apollo 13, following an environmental disaster, is in peril with its finite supplies of oxygen, water and fuel," he wrote. "Earth with finite resources of air, water and land faces a similar peril." And time was running out: "Apollo 13 has until Friday to resolve its dilemma, Earth a very few years."[1]

Earth Day represented, in many ways, the apotheosis of the view of Spaceship Earth in jeopardy. Despite their other numerous divisions, many Americans could agree with the metaphor's core concept: that the earth's resources were limited and running short. The Spaceship Earth metaphor also suggested a sense of unity, a sense that "we are all in the same boat together" and that Americans and indeed the whole human race had better work together, or else they faced catastrophe—a sinking ship. "Survival on the frail spaceship Earth," the Washington State ZPG chapter proclaimed in their Earth Day newsletter, "depends on each of us—we sink or swim together." To many observers, the environment could bring together the country at a difficult time. "Ecology is one major political issue," *Life* reported in early 1970, "on which the country may be united."[2]

And yet, if the idea of global environmental crisis and concern about population growth seemed to bring people together in 1970, over the next decade and a half it began to divide them, sometimes bitterly. During the 1970s and 1980s, Americans across the widening political spectrum became increasingly critical of Malthusian environmentalism—not just the idea of unity but also the idea of crisis. "While population trends encouraged people to think of the world as a whole," historian Matthew Connelly has written, "it also provided new reasons and new ways to divide it." In one attack after another, critics on the right and the left targeted both Malthusian diagnoses and remedies. Even environmentalists split over population.[3]

The consequences were dramatic. In 1970, the population movement had reached a place that few could imagine just a half decade earlier. President Nixon had established a high-level commission to create a national population policy, and a national environmental movement had coalesced, at least partially in response to concerns about overpopulation. By the mid 1980s, though, despite isolated pockets, population had fallen off the national agenda almost entirely.

Several factors drove this change, including demographic trends. Internationally, although total population continued to grow, the rate of growth of the world's population began to level off and even decline beginning in the late 1960s—ironically, at just the moment that Ehrlich's *The Population Bomb* appeared. Domestically, the American population growth rate reached the replacement level in 1972. The food supply situation also improved dramatically because of the green revolution and other changes. In 1976, the Food and Agriculture Organization reported that food production had increased 0.4 percent per year since 1952, and that only seven countries, most in Africa, still had major food problems. Finally, policy changes played a role. By the early 1970s, the population movement had achieved much of what it had aimed for, including expanded legal access to birth control and abortion, as well as much greater funding for education and family planning services, especially for the poor.[4]

But significant cultural and political changes were afoot as well. Concern about population had peaked in the 1960s because of worry about poverty overseas and at home, and quality-of-life concerns. During the 1970s, though, new environmental measures began to remedy many quality-of-life problems, while the decade's economic downturn pushed aside those that remained. At the same time, the crises of poverty—in the inner city and the third world— dissipated. They also got more complicated. As new arguments arose on both the right and the left about the origins of poverty, Cold War strategy, and the proper role of government, it was no longer so easy to blame population growth. Added to these arguments were powerful cultural arguments against the ideas of sex, birth control, and the family that the population movement pushed for. At the same time, environmental Malthusians increasingly focused their "limits to growth" arguments on middle-class consumption, a widely unpopular position.[5]

Criticism from African Americans and Limits to Growth

One of the first signs of serious problems for Malthusian environmentalists came just a few months after Earth Day in April 1970. In June, at a national conference meant to bring together the various activists working on population and environment issues—the First National Congress on Optimum Population and Environment (COPE)—a group of African American participants stormed out. Consisting of twelve hundred delegates from over two hundred organizations, COPE, according to one journalist, "brought together about as liberal a bunch of white adults and young people as you can find.... Most of the delegates were people long committed to dealing with America's social, racial and economic injustices." The protesters, however, claimed that the conference was blind to the race and class biases shaping environmental priorities, including population. They had a point. The conference agenda spoke mostly to the concerns of the white middle class, and when the conference did address urban America, the conversation focused on overpopulation. The conference program suggested that participants could see overpopulation firsthand by taking a walk through "a teeming urban area." Walking through these areas, "it will be apparent that a dangerous crisis of ecological imbalance already exists." In a statement, the African American protesters complained that "the established priorities of the Congress demonstrate the failure to include ... input which reflects the interests of Black people." Taking aim at Spaceship Earth universalism, they rejected "the assumption of a common national life style." They particularly disliked the coercionist sentiments of Garrett Hardin and ZPG's Edgar Chasteen, although many speakers, including Ehrlich, emphasized voluntary measures. According to a second statement, "The purpose of this conference is to use those delegates invited to legitimize a preconceived vicious plan of extermination. This plan is one of systematic reduction of a specific population, namely blacks, other non-whites, the American poor and certain non-white and ethnic immigrants."[6]

Similar criticisms became common in the next several years, especially against coercion. Perhaps no one matched comedian-activist Dick Gregory for his mix of political critique and wit. In a 1971 *Ebony* article, Gregory criticized the "white folks ... who are interested in ecology and overpopulation" who ask him frequently why he rejects population control. "Personally, I've never trusted anything white folks tried to give us with the word 'control' in it," he wrote, adding, "I guess it is just that 'slave master' complex white folks have. For years they told us where to sit.... Now he wants me to sleep under the bed." Concerns about genocide help to explain why many African Americans perceived mainstream environmentalism to be racist and elitist. "Without the proper perspective," Julian Bond noted, "*The Population Bomb* becomes a theoretical hammer in the hands of angry, frightened, and powerful racists, as well as over the heads of black people, as the ultimate justification for genocide."[7]

African American political leaders expressed similar concerns. At a meeting of the National Commission on Population Growth and the American Future (NCPGAF), Jesse Jackson argued that "our community is suspect of any programs that would have the effect of either reducing or leveling off our population growth. Virtually all the security we have is in the number of children we produce." Other minority representatives concurred. Speaking in Spanish to the NCPGAF, Manuel Aragon argued, "What we must do is to encourage large Mexican American families so that we will eventually be so numerous that the system will either respond or it will be overwhelmed."[8]

Some African American critics attacked Malthusians for blaming others while overlooking their own tremendous resource consumption. The richest nations in the world, Howard University political scientist Ron Walters pointed out in 1974, "aimed their population control rhetoric at poor nations and poor residents of the United States, all of whom consumed fewer resources than well-to-do Americans." That same year, the national director of the Congress of Racial Equality (CORE), Roy Innis, criticized those who based their theories on "alleged world problems of space, natural resources, food production and health care." Echoing Walters, he wrote, "Overpopulation is a white man's problem. In his limited space, he squanders an extremely disproportionate share of the world's resources."[9]

Some blacks did express concern about population growth, and supported environmentalists. "The simple fact," the journalist and former U.S. ambassador to Finland Carl Rowan wrote in 1971, "is that even if American families average only two children over the next few decades, and growth through immigration were held at a net 400,000 a year, the U.S. population would still reach 300 million in 50 years. Whether it is 300 million or 400 million, there are people who will want to take a simultaneous look at the impact of this larger population on the economy, the environment, [and] new difficulties of governing."[10]

In response to criticisms, many environmentalists, led by Ehrlich, began to refocus their emphasis. By the COPE meetings in 1970, Ehrlich had already shifted his arguments about overpopulated cities. At COPE, he continued this shift, making a speech that might surprise many who tend to equate his population thinking with white bias. He opened with a characteristically blunt style: "Is the ghetto part of the environment?" Ehrlich then sided with the protesters. While environmental deterioration is "often viewed largely as an aesthetic problem," blacks faced much more dangerous environmental risks. "Pollution," he pointed out, "is something that kills him [the inner-city resident] stone cold dead." Cities brought environmental health risks. "If you raise your children in Los Angeles, California or Chicago, Illinois," he pointed out, "statistically, you are killing them off early. If you raise your kids there, they are likely to die young . . . because the air that they breathe contains something beyond that mixture of oxygen and nitrogen that we evolved to breathe." He also noted that African Americans typically carry higher loads of DDT than whites.[11]

Ehrlich encouraged whites to think twice before suggesting that blacks needed to control their populations. White Americans, he said, "are the ones who are doing the looting and polluting of the globe, not the blacks." "I see no reason," he continued, "why any black should listen to advice from any white until they have in our society precisely the same education opportunities, social opportunities, political opportunities, and so on." Instead, Ehrlich called for controlling the overpopulation of the affluent. He also called for the "de-development" of the "overdeveloped" countries. He emphasized these points with a retrofitted version of Spaceship Earth, one that highlighted race and class differences: "There is absolutely no way to live on a little spaceship with limited resources, with some people in steerage or third class, with the people in the first class cabins stealing food from the people in the third class cabins, waving large bombs at them, and expect them to sit still for it." Ehrlich reiterated these messages in numerous writings, including a 1971 revised edition of *The Population Bomb* that spoke not of "Developed Countries (DCs)" but of "Overdeveloped Countries (ODCs)." Concern with overdevelopment also drove Zero Population Growth.[12]

This attack on middle-class consumption—which focused the ideas of carrying capacity that came from East, Pearl, Leopold, and Vogt—helped define a central concern in American life during the 1970s: the "limits to growth." Worries about growth, historian James Patterson has observed, became a central "lament" of the decade. "The uncertainty and ambivalence about growth that had earlier appeared as an undercurrent," historian Robert Collins explained, "took on a new scope and stridency in the 1970s, causing many to doubt whether future growth was either possible or desirable." Many Americans came to agree with E. F. Schumacher's 1973 bestseller *Small Is Beautiful*, which called for a "maximum of well-being with the minimum of consumption." Following Kenneth Boulding's lead, several economists advanced limits-to-growth arguments, including "steady state" theorists like Herman Daly but also mainstream economists. Few books, though, were more influential than *The Limits to Growth*. A computer-aided version of Malthusian environmental arguments put out by the Club of Rome in 1972, *Limits to Growth* used elaborate calculations about population growth and resource use to make predictions. Human civilizations, it warned, would collide with upper limits within a century. In part because the computer gave it legitimacy, the book became a cultural phenomenon, ultimately selling millions of copies. Even magazines such as *Business Week*, after critiquing parts of its analysis, conceded the book's central point: "For all the criticism, practically everyone agrees that on a finite planet, growth must end sooner or later." According to Collins, the Club of Rome won the "war of public perceptions."[13]

In the hands of Ehrlich, ZPG, and many others, 1970s environmental Malthusianism began to focus more exclusively on American overconsumption. Ironically, this view may have helped answer some of the criticism that African American and other critics began to make, but it ultimately undermined environmentalism's appeal. Attacking middle-class consumption not only threatened American creature comforts but also set up Ehrlich and like-minded environmentalists for a direct clash with many working-class Americans, especially those Keynesians within the Democratic Party who held tight the dream of upward mobility through an expanded economic pie. The seeds of this conflict, which would play out politically in the end of the decade, were visible in the acrimonious public duel between the two best-known environmentalists of the early 1970s, Ehrlich and the biologist Barry Commoner.

The Dueling Critiques of Barry Commoner and Paul Ehrlich

Like Ehrlich, Commoner was among the new breed of "politico-scientists" who lobbied public officials to work the lessons of science into public policy. But Commoner had a more urban, working-class background than Ehrlich, and was more influenced by socialist ideas. Born in Brooklyn in 1917 to a Russian immigrant tailor and his wife, Commoner worked at a series of odd jobs to pay his way through Columbia, where he studied zoology. He went on to receive a Ph.D. in biology from Harvard in 1941. During World War II, as part of a navy team investigating the first tests of DDT for mosquito control, Commoner gained an early understanding of the dangers of pesticides. After the war, working for a Senate subcommittee, he helped pass the McMahon Act of 1946, which mandated civilian control of nuclear weapons. A group he founded in 1958 to spread knowledge about the dangers of nuclear fallout, the St. Louis Committee for Nuclear Information, played a major role in pressuring the American government to approve the Nuclear Test Ban (1963), which halted above-ground nuclear testing. During the mid-1960s, Commoner's St. Louis group researched the damaging effects of other new technologies, including detergents, insecticides, and plastics. In 1970, when *Time* put Commoner on its cover and labeled him the "Paul Revere of Ecology," the analogy suited him well: Commoner was a man of the people who used ecology to warn citizens of a danger they could not yet see.[14]

Commoner's first full-scale public attack on Ehrlich came in his 1971 book, *The Closing Circle: Nature, Man, and Technology*. This environmental classic detailed how pollution destroyed the intricate natural cycles that life on Earth depended upon, and pulled no punches in criticizing societies that ignored ecological patterns. It also pulled few punches in attacking environmentalists—meaning Malthusians—who paid too much attention to "a single facet of what in nature is a complex whole." Commoner equated population control with "political repression."[15]

So many rumors exist about the "Ehrlich-Commoner debate" that it is useful to detail what Commoner did not say in the book. He never said that population growth and affluence played no role in pollution and environmental problems. He also never said that population growth in the third world was not a problem. And he never said that he favored all-out economic growth. Nonetheless, Commoner missed few opportunities to attack Malthusian approaches. Blaming population growth for only 10 to 20 percent of environmental damage—far less than he ascribed to other factors—he rejected the issue as a core problem.[16]

Instead, Commoner's understanding of environmental history focused on the technological changes that accompanied postwar capitalism—the "huge array of new substances" resulting from a "sweeping revolution in science." "While production for most basic needs—food, clothing, housing—has just about kept up with the 40 to 50 per cent or so increase in population," he noted, "the kinds of goods produced to meet these needs have changed drastically." He singled out synthetic detergents, synthetic fabrics, aluminum, plastics, concrete, truck freight, nonreturnable bottles, and synthetic fertilizers. Commoner implied that, as long as technological problems were eliminated, the increased economic growth that Malthusian environmentalists had been railing against since the late 1940s was acceptable.[17]

Historians have suggested that Ehrlich and Commoner represent two fundamentally different ways of approaching environmental problems—with Ehrlich emphasizing models drawn from nature, and Commoner placing his view of nature and the environment within a larger social and often socialist analysis. This characterization is accurate and useful, but only to an extent. Commoner did call for restructuring the economy to make corporate producers more accountable, and Ehrlich did stress the inevitable crash predicted by population biology. But by the time of their debates, Ehrlich also had a strong political-economic vision—one that, although flawed in important ways, was itself strongly critical of capitalism, just differently so. Whereas Commoner, the man who had taken on the nation's nuclear testing program and its detergent companies, believed that the environmental problems were due to "changes in productive technologies," Ehrlich focused on consumption, the "superconsumers" he mentioned in *The Population Bomb*.[18]

Part of the debate boiled down to feasibility. What pillar of modern American society would be easier to recast—capitalist production or capitalist consumption? Put simply, was changing technology more realistic or getting people to use birth control and have smaller families? Ehrlich saw the practical and moral difficulty of denying people material goods and thus sought to keep populations low. But, Commoner countered, if the main problem was productive technologies such as new plastics and detergents, couldn't pollution problems be eliminated by simply eliminating those technologies? Doing so, moreover, would not require coercion or undoing the growth that had lifted so many working-class Americans into the middle class in the previous three decades.

As the 1970s unfolded, this difference would have large political implications. Commoner's views appeared much easier to align with the old left politics of Keynesian growth, a policy dear to many Americans, especially liberals and the working class. Ehrlich's anti-growth views, on the other hand, reached a political dead end. By trying to be more aware of race and class concerns with his critiques of middle-class consumption, Ehrlich was attacking a dream of material comfort to which many people aspired. With these attacks on middle-class consumption, Ehrlich may have raised more doubts about Malthusian environmentalists among the poor and the nonwhite, both at home and abroad.

The Ehrlich-Commoner debates played out on another level, as well. Commoner attacked Malthusians not just for their understanding of environmental history but also for their unqualified faith in scientific experts. By 1971, Commoner had been working tirelessly for over two decades to make governmental decisions about nuclear weapons and other environmental risks more open and accountable to the public. This experience made him nervous about the Malthusian call to put decisions about sex, families, and consumption in the hands of authorities. Ehrlich, on the other hand, was nervous about influential conservative religious groups and was unsure that ordinary people could be trusted with decisions about reproduction and consumption, which were influenced by strong biological forces. He had often argued that discussions about population growth demanded an input of rationality and science, especially from ecologists.

Here, however, was another scientist disagreeing with him. And Commoner had company. In June 1972, the British journal *Nature* attacked Ehrlich's tendency to argue that his critics simply did not understand the science involved. Arguments against Ehrlich, it wrote, were not "an obscurantist unwillingness to recognize problems that exist but rather a protest at the way in which he and others have made interesting and important problems appear too simple."[19]

Some went so far as to question science altogether. In an influential 1974 article in *Economic Geography* about population, Johns Hopkins geographer and early postmodern theorist David Harvey noted the dominance of human values, not objective rationality, in the Ehrlich-Commoner debates. "The lack of ethical neutrality in science," Harvey wrote, "affects each and every attempt at 'rational' scientific discussion of the population-resources relationship." Although largely sympathetic to Commoner's position, Harvey went beyond critiquing Ehrlich's position to claim that science itself—*all* science—was inherently biased. "The adoption of certain kinds of scientific methods," he wrote, "inevitably leads to certain kinds of substantive conclusions which, in turn, can have profound political implications." He concluded, "We are . . . forced to concede that 'scientific' enquiry takes place in a social setting, expresses social ideas, and conveys social meanings."[20]

Environmentalists were forced to align with either Ehrlich or Commoner, although many refused to pick sides. Singer and environmental activist Pete Seeger, for instance, decided that both population and technology deserved urgent attention. "Commoner has convinced me," he wrote, "that technology and our private profit politics and society must be radically changed and quickly. But I'm still working hard for Zero Population Growth (ZPG), because even if population growth is only 5 percent or 10 percent of the U.S. environmental crisis, it's a big world problem, and one can't expect others to work on it if we don't. The world is the concern of everyone." Seeger agreed with Commoner about coercion, though, adding, "I'm in favor of ZPG as soon as it can be achieved by persuasion, not force."[21]

Nixon's About-Face

Black nationalists and environmentalists like Commoner were not the only critics of Malthusian environmentalism. By 1972, Richard Nixon had also begun to move away from his earlier concern about population growth, rejecting most of the findings of the population commission that he set up, the National Commission on Population Growth and the American Future (NCPGAF). This was part of a larger ideological shift in the Republican Party toward more conservative positions.

Meeting routinely during 1971 and 1972 to discuss domestic population problems, the NCPGAF was chaired by Population Council founder John D. Rockefeller III, a Malthusian with a proclivity for universalism. According to historian Donald Critchlow, Rockefeller thought it "self-evident" that population growth causes social, economic, and political instability, and held great faith in "the Olympian objectivity of the experts." The divisions within the commission, however, showed that nothing was self-evident. The majority held that population growth was the most important problem facing the nation, but a smaller group composed of minority representatives and white liberals argued that power differentials, particularly race and class inequalities, were more important. Two other issues, both new to the political landscape, split the commission even further: abortion and immigration. Abortion was just then becoming a national issue; immigration would be a hot-button issue by the end of the decade.[22]

In addition, five of the twenty-three commission members formed an ecological faction. Supported by a letter-writing campaign encouraged by Paul

Ehrlich, they made anti-technology, anti-economic growth arguments. Writing Rockefeller directly in 1971, Ehrlich stressed that the combination of high population and high per capita consumption created unsustainable aggregate resource consumption levels. He did not see a need for coercion. Population control, he wrote, "can probably be achieved without coercion if we act with alacrity now." Instead, "strong governmental support for women's liberation might alone turn the tide." He also wanted to see presidential "jawboning" in support of small families.[23]

The commission, which issued its final report in June 1972, found that "in the long run, no substantial benefits will result from further growth of the Nation's population, rather that the gradual stabilization of our population would contribute significantly to the Nation's ability to solve its problems." It could not find "any convincing economic argument for continued population growth." Ehrlich was thrilled. The Nixon administration was not.[24]

Nixon disliked the leanings of the commission in part for social reasons. On May 5, 1972, he criticized the commission's work, especially its stance on legalized abortion: "I consider abortion an unacceptable form of population control." With such statements, Critchlow argues, Nixon "politicized" the population issue; he took what many had assumed to be a bipartisan, mostly scientific issue and used it to score political points. Abortion, though, was not the only social issue conservatives had on their minds. Nixon also frowned upon providing birth control to minors: "I do not support the unrestricted distribution of family planning services and devices to minors. Such a measure would do nothing to preserve and strengthen close family relationships." Behind the scenes, Nixon speechwriter Pat Buchanan was making what just a few years later would be called "pro-family" arguments.[25]

As Derek Hoff has emphasized, the Nixon administration turned against the NCPGAF report for economic reasons as well, drawing from conservative economists who began to challenge Malthusian assumptions in the 1960s. Examining the history of resource scarcity in *Scarcity and Growth* (1963), Harold Barnett and Chandler Morse's noted that, because substitutes for scarce resources could often be discovered or invented, resources tended to get less rather than more expensive in the long run. In *The Conditions of Economic Growth* (1965), Esther Boserup argued that population pressure often spurred technological innovations and a higher standard of living. The University of Chicago's Theodore Schultz helped create "human capital theory," which showed how population growth led to public investments, which in turn fostered economic growth. The conservative media also began to doubt Malthusian concerns. In 1970, the *Wall Street Journal* noted that, far from being a drag on the world's predicament, the world's growing population "might also be of help in finding solutions to the problem."[26]

The Nixon administration argued for shifting the scale of analysis for population problems. The problem was not absolute size, but poor distribution.

To make such arguments, they drew from demographers such as Conrad F. Taeuber, the associate director of the Census Bureau, who pointed out that 3,000 counties were losing population while 70 percent of the population squeezed onto 2 percent of the land. Ehrlich and other environmental Malthusians resisted this logic. Writing to Rockefeller, Ehrlich insisted the problem was not land but resource use, thus the size of the total population had to be reduced.[27]

Nixon's about-face on population issues, which came during the 1972 presidential election year, reflected part of a deliberate strategy to construct a permanent Republican majority. The president's new approach, especially his views on abortion, were aimed at conservative Catholics and anyone else whom he might draw away from the Democratic coalition. He hoped to win the votes of people like Mrs. T. W. Hodges of rural New Jersey, who found the "environmental" values of NCPGAF deeply offensive. In 1972, she wrote to John D. Rockefeller III, "God will not be mocked. He will strike not only the little people, but the big people (like you) as well, who consider the environment more important than human life. You talk about the dignity of quality existence, but are willing to commit murder, motivated by situation ethics."[28]

In the election, Nixon crushed Senator George McGovern, winning the Electoral College by a margin of 520 to 17. His victory included a sizeable number of independents and Democrats, people who would become crucial to Ronald Reagan's election in 1980. Nixon won 66 percent of independents and even 42 percent of Democratic voters. He won 60 percent of Catholics, 59 percent of the working class, and 57 percent of union households. Much of this victory was due to other issues, such as foreign policy, law and order, and race relations, but Nixon's rejection of Rockefeller's NCPGAF also played a supporting role.[29]

This rejection reflected only part of the criticism Nixon and his aides had, behind closed doors, for the commission's arguments. Nixon demurred from stronger language because of concern for Nelson Rockefeller, the brother of the NCPGAF's chairman John D. Rockefeller. In the early 1970s, moderate Republicans were still a crucial part of the party's coalition. As the decade progressed, however, conservative Republicans would grow less concerned about upsetting moderates.[30]

The World Food Crisis, Lifeboat Ethics, and Food First

After the NCPGAF, much of the concern about overpopulation was focused internationally. Problems in 1973 and 1974 seemed to bear out this concern. Appearing not long after *Limits to Growth*, the oil crisis of 1973 and 1974 gave a strong boost to Malthusian arguments, although it was caused by politics, not resource shortages. "For several years now environmentalists have been warning of an impending Armageddon," wrote John R. Quarles, deputy administrator of the Environmental Protection Agency. "It now appears they are correct." During these same years, an acute food crisis also hit, the result of bad weather and major crop failures in the Soviet Union, South Asia, and North America. The price of grain multiplied fourfold. Wheat climbed from $60 per ton in 1972 to $200 two years later; rice jumped from $130 to more than $500. In Bangladesh, Southeast Asia, and the Sahel region of Africa, hundreds of thousands faced dire problems and many died.[31]

The Nixon administration, despite its shift on population matters at home, continued to sound the alarms about the international situation. A 1974 National Security Council report under Henry Kissinger (NSSM 200) warned that the "political consequences of current population factors" created "political or even national security problems" for the United States. In a broader sense, NSSM 200 continued, "there is a major risk of severe damage to world economic, political, and ecological systems and, as these systems begin to fail, to our humanitarian values." The report even stressed environmental degradation: "In some overpopulated regions, rapid population growth presses on a fragile environment in ways that threaten longer-term food production." For remedies, the report emphasized that population limitation should be incorporated into all foreign aid, and recommended a significant expansion in AID funds "for population/family planning."[32]

Ehrlich and other Malthusians saw the oil and food crises as signs that their predictions were coming true. Some went even farther. In late 1974, Garrett Hardin published his most provocative article yet, "Living on a Lifeboat." Rejecting the Spaceship Earth model, which he called "suicidal," Hardin argued that humans were more like passengers on several lifeboats, where some boats, though full, were still taking on new members. In such a situation, altruism and charity were counterproductive. Food aid allowed countries—especially their incompetent or corrupt leaders—to ignore their population problems, which guaranteed that even more severe problems would reappear in the future. "Every life saved this year in a poor country," he wrote, "diminishes the quality of life for subsequent generations." Hardin even argued that famines, disease, and war were helpful: "The third-century theologian Tertullian expressed what must have been the recognition of many wise men when he wrote: 'The scourges of pestilence, famine, wars, and earth-quakes have come to be regarded as a blessing to overcrowded nations, since they serve to prune away the luxuriant growth of the human race.'"[33]

A series of critics took aim at Malthusian concerns about the third world. Barry Commoner had laid the foundation for these critiques in *The Closing Circle*. European powers had encouraged population growth in their colonies for assistance with material extraction, he said, then abandoned colonies when they found synthetic substitutes. Now these same people were warning of third-world overpopulation. Economic growth in these areas was not the problem, he added, but part of the solution. Third-world population growth would level off by the improvement of living standards, reduced infant mortality, old age insurance, and voluntary contraceptive practice. Believe in the demographic transition model, Commoner appeared to be saying.[34]

Ehrlich countered that Commoner had unfounded faith in "the self-regulation of human populations." Even in the unlikely event that a demographic transition began overnight, it would be a century before third-world growth rates decreased to current first-world rates. Ehrlich also attacked Commoner's "misplaced faith on man's ability to industrialize instantly without environmental damage." But he agreed that people of the underdeveloped world deserved better. Whereas Commoner imagined an expanding economic pie, Ehrlich focused on redistributing the existing pie. "Mankind's only chance for improving the lot of the poor significantly," he and Richard Holdren wrote, "lies in diverting energy and other resources from extravagant affluence in the DCs to necessity-orientated uses in the UDCs."[35]

Strong criticisms of Malthusian positions emerged at the World Population Conference in Bucharest in August 1974, where the U.S. delegation, led by Secretary of Health, Education, and Welfare Caspar Weinberger, pushed for the United Nations' "World Population Plan of Action," a plan calling for specific targets with specific deadlines. Denouncing the plan as a new form of Western control with hints of racism, China and other developing nations called for reordering the world's economy instead. "The best contraceptive," Indian Minister of Health Karan Singh proclaimed, "is development." Not all the criticism of the U.S. plan came from third-world advocates and Chinese Maoists. The biggest surprise of the conference came when John D. Rockefeller III, the force behind the Population Council for two decades, called for a "reappraisal" of population policy involving a broader approach, one that paid more attention to socioeconomic contexts.[36]

Perhaps the most influential critique of the population movement's view of third-world overpopulation came from Frances Moore Lappé and Joseph Collins in their 1977 bestseller *Food First: Beyond the Myth of Scarcity*. Although sympathetic, Lappé and Collins rejected the "scarcity scare" that they believed drove Malthusian environmentalism. Too often, they stated, the concern about population growth boiled down to a fear of the poor. The hungry "are not our enemies." The poor and their reproductive habits were not even the cause of hunger; instead, Lappé and Collins blamed hunger and environmental damage on power disparities in societies. To overlook these power disparities was to blame the victim.[37]

Unlike Barry Commoner and other Old Left Democrats, whose political roots were in the Depression and World War II, Frances Moore Lappé came of age in the late 1960s. Graduating from Earlham College in 1966, she shared the strong anti-imperial sentiments of the New Left. In 1971, she published *Diet for a*

Small Planet, a collection of recipes and analysis that sold more than three million copies, and that, in the words of one historian, "typified radicals' faith in the ability to combine personal therapy with political activism." In the book, Lappé argued that food shortages arise when pasture and grains are devoted to raising meat instead of feeding people directly. "A grain-fed North American steer," she noted, "ate 21 pounds of vegetable protein for every pound of protein it delivered to the steak eater." Her solution: vegetarianism.[38]

In *Food First*, Lappé and Collins disputed pretty much everything Malthusian environmentalists claimed. Poor peasant families had large families not from ignorance of their own self-interest but because children provided much-needed labor as well as old-age security. Poverty *created* rapid population growth, not vice versa. Moreover, overpopulation did not exist everywhere. Bristling at the tendency to lump together the "third world," Lappé and Collins noted that many underdeveloped nations actually had rapidly declining birth rates. Moreover, food production was not falling behind population growth everywhere. Between 1952 and 1972, 86 percent of those living in underdeveloped countries lived where food production had kept pace or better.[39]

Overpopulation was not the main cause of hunger, they said. Even regions with large food surpluses contained hungry people. Stanislaus County, California, had some of the world's most productive farmland, yet some of its residents went hungry. Conversely, even supposedly overpopulated countries—the real "basket cases" such as Bangladesh—had enough to feed themselves. Indeed, the world was not running short of food. The planet produced about two pounds of grain, or about 3,000 calories, per person, more than enough for every single person. "There is no such thing today as absolute scarcity," Lappé and Collins stressed.[40]

Lappé and Collins did not reject population growth as a problem, however. At the current rate, continued population growth "will certainly undercut the future well-being of all of us. That is self-evident." But other problems dwarfed population growth.[41]

The real cause of hunger was not natural, but social. Because of long-standing power differences exacerbated by colonial and postcolonial economic patterns, some people had access to the world's production of calories, while others simply could not afford to buy what was available. Bangladesh, for instance, could feed itself—but the nation's rich ate much more grain than the poor, as much as 30 percent more calories and twice the protein. "An elite few," Lappé and Collins wrote, "prevent the majority from having access to the country's resources." Worldwide, "the concentration of wealth and power" undermined food production and distribution. Consequently, even doubling the world's food supplies would not eliminate hunger.[42]

As with hunger, so with environmental degradation: the real causes lay not with population growth but with social inequalities. Ecological destruction, especially soil erosion, existed in much of the world, and populations had indeed grown. But no causal link connected the two. Instead, like Commoner, Lappé and Collins refocused attention on the forces of production, pointing to an exploitative economic system started under colonialism and continued under national elites. The powerful expropriated the best land for themselves, and pushed local farmers onto marginal, hilly land ill-suited for intensive farming, thus creating the appearance of overpopulation. Land degradation "turns out to be the result, not of the size of a country's population, but of other forces: land monopolizers who export nonfood and luxury crops that force the majority of farmers to overuse marginal lands; colonial patterns of taxation and cash cropping that continue today; well-meant but unenlightened 'aid' and other forms of outside intervention in traditionally well-adapted systems; and irresponsible profit-seeking by both local and foreign elites." Cutting the world's population in half overnight would not curtail degradation.[43]

Lappé and Collins also focused on consumption. But here, too, they pointed to vast disparities: "The four billion human beings on earth ... aren't four billion equal units at all. One person can represent a burden on agricultural resources many times greater than another." The impact of a plant-food diet on the cultivated farmland is relatively light. On the other hand, much greater pressure comes from an average American diet of animal foods produced by "shrinking annually 1800 pounds of grain into 250 pounds of meat."[44]

By overlooking these differences, Lappé and Collins argued, the "we're all the same" logic of Spaceship Earth in effect blamed the poor for problems caused by the rich. Doing so not only misallocated blame but also avoided redistributive remedies: "Often the very people who prefer to blame the poor themselves—and their breeding—for deplorable social conditions are those who stand to lose by the redistribution of power over productive resources."[45]

Such arguments further undermined the support for environmental Malthusians.

"Reproductive-Rights" Feminists and Malthusian Environmentalism

As these critiques about the Malthusian *diagnosis* of overpopulation emerged, a powerful critique of the *remedies* proposed by Ehrlich and others also gained in strength. Feminists in particular began to take issue with Malthusian environmentalists, especially as reproductive autonomy increasingly became a rallying cry within the women's movement in the 1970s. The key issue was coercion. Would controlling population mean controlling people, especially women? Here Barry Commoner raised a crucial question: "Who decides?"

In the late 1960s, many—not all—Malthusian environmentalists had joined Ehrlich and Hardin in calling for coercion, even as they sought expanded birth control and abortion rights. "Anything other than government control of conception may be self-defeating," biologist Walter Howard wrote in *BioScience* in 1969. "Perhaps what is needed is a system of permits for the privilege of conceiving, or compulsory vasectomies of all men and sterilization of all women ... responsible for two births." Paul McCloskey, a Republican congressman and member of Gaylord Nelson's committee organizing Earth Day, also walked the line between voluntary and coercive population planning: "It seems possible, therefore, that, should we one day exhaust our resources and efforts in obtaining voluntary use of universally-available family planning techniques, we will have to confront head-on the question of legal sanctions against individuals who intentionally or negligently sire or bear children they cannot or will not support." Howard summed up the issue, "No longer can we consider procreation an individual and private matter. Intercourse, yes, but not unregulated numbers of conceptions."[46]

The early 1970s witnessed intensified calls for coercion. In a 1970 article in *Science*, Garrett Hardin asked if parenthood was a right or a privilege. Because raising children had become a task that society at large was increasingly paying for, society as a whole should have to ability to control population size. Practically, this meant special constraints on women. "The Women's Liberation Movement may not like it," he wrote, "but control must be exerted through females." Divorce and remarriage made it impossible to assign responsibility to couples or to men.[47]

Although in the 1960s Malthusian environmentalists had aligned with some women's groups, during the 1970s, motivated not just by concerns about contraceptive access but also by larger questions of reproductive autonomy, feminists began directly attacking people like Hardin for believing that women's control of their own bodies should yield before the "greater good" of protecting the environment. Accordingly, the emphasis of the movement shifted from universalism to social difference. "Instead of radical egalitarianism," which believed "that women's liberation required the abolition of pregnancy and motherhood," historian Bruce Schulman has written, "feminists increasingly stressed the positive virtues of female biology and women's culture."[48]

Nothing showed the push toward "separate but different" more, Schulman notes, than the women's health movement, which encouraged women to push for more control over their own health care, especially reproductive health. In this powerful strand of the evolving women's movement, as several historians have noted, the struggle for abortion rights were crucial. In the very late 1960s, feminist groups began stressing that individual women, not doctors, lawyers, or legislatures, should have power to make decisions about abortions. Ironically, the phrase devised by Garrett Hardin—"abortion on demand"—became their rallying cry. This struggle intensified after the Supreme Court's *Roe v. Wade* decision in 1973.[49]

FIGURE 11 This photo accompanied an opinion piece by Edgar Berman in the *New York Times* called "We Must Limit Families by Law."

New York Times, December 15, 1970. The photo was taken by Beverly Hall sometime before July 1968.

Yet reaction to Malthusian calls for coercion also played an important role. Looking back over her own transformation, Barbara Seaman, the author of a crucial text in the health movement, *The Doctors' Case against the Pill* (1969), pointed to a disturbing talk she heard by a Malthusian doctor at a late 1960s conference. "The dangers of overpopulation are so great," Dr. Frederick Robbins, a Nobel Prize laureate and medical school dean, told the Association of American Medical Colleges, "that we may have to use certain techniques of conception control that may entail considerable risk to the individual woman." Shocked by this willingness to risk the health of individual women, Seaman began to argue for reproductive rights. "It is a most basic violation of civil rights," she wrote, "for the group that is not at any risk from reproduction (male) to control the group that is at risk (female)."[50]

Startling 1973 revelations about forced sterilizations within the United States also added to the idea of reproductive autonomy. The first well-publicized case involved Minnie Lee Relf, one of two African American sisters aged twelve and fourteen who were forcibly sterilized during a visit to federally funded family planning clinic in Alabama. In the late 1960s and early 1970s, one government official estimated, thousands of women a year were sterilized without their approval, using federal monies. Many of these were African American, Hispanic, and Native American. In 1973, fourteen states were considering legislation to require women on welfare to undergo sterilization. After the Relf case and others like it, African American feminists and the larger women's movement began to champion the right *to have* children.[51]

This reproductive rights strand of the movement attacked Malthusian environmentalists such as Ehrlich, who had argued that environmental protections would benefit both men and women. But Ehrlich's Spaceship Earth model had failed to recognize important biological and historical differences that women wanted recognized. A one-page handout distributed by a group called Women Against Genocide at one of Ehrlich's appearances in the early 1970s suggests the opposition among some women's groups. Entitled "Bomb Ehrlich," the handout questions giving power over reproductive decisions to the same government that ran the Vietnam War. It warned of the mass sterilization of women and racial genocide. "When it comes down to cutting the population to the 'ideal level' of 1 billion people," Women Against Genocide argued, "it will not be the rich and powerful who will 'go,' but the poor, Black and Brown people, and women who are manipulated, sterilized, chemically poisoned, murdered, etc."[52]

Feminist arguments against coercion gained steam throughout the decade. At the United Nations Conference in Bucharest in 1974, prominent women's advocates such as Germaine Greer, Betty Friedan, and Margaret Mead called for more awareness of the reproductive rights of women. Indeed, in making arguments for rethinking population problems, Population Council founder John D. Rockefeller III, pushed by advisor Joan Dunlop, also called attention to the gender bias of many programs.[53]

One of the earliest and most articulate reproductive-rights feminists to speak out against the likes of Ehrlich and Hardin was the historian Linda Gordon, first in articles and later in a 1976 book. Gordon came out of Boston-area socialist and feminist activist circles during the early 1970s. "Overpopulation is a problem in some parts of the world," she wrote in *Woman's Body, Woman's Right*. "The important point, however, is that overpopulation is not the only or even the main cause of poverty." Overpopulation was more often the result of poverty than the cause. By blaming the poor for their own poverty instead of political economic systems, the population planning movement had served capitalist interests. Most important, Gordon stressed that the population control movement worked against the interests of women. Organizations like Planned Parenthood promoted birth control as a way to stabilize marriages, not to promote women's autonomy. Moreover, population controllers "increasingly advocate various kinds of coercion" and often claim to support female emancipation "only on terms that they would define for women." The feminist view, by contrast, imagined women's emancipation "as a process designed and created by women themselves."[54]

In 1976, a population limitation program in India of horrifying proportions appeared to confirm feminists' worst fears. "We must act decisively and bring down the birth rate," Indian Prime Minister Indira Gandhi declared not long after establishing a state of emergency. "We should not hesitate to take steps which might be described as drastic. Some personal rights have to be kept in abeyance for the human rights of the nation." Launched in April 1976, the population program was novel in two regards: it enlisted all governmental agencies to help with the campaign, and it involved unprecedented levels of coercion, especially through sterilizations. In addition to offering cash incentives for sterilizations, officials often withheld basic governmental services—licenses, health care, electricity, and pay raises—until beneficiaries, often poor and low caste or of ethnic minority, found people to meet the quotas. Gandhi's politically ambitious younger son, Sanjay, pursued recklessly aggressive actions, including rounding up people from slums. In one year, eight million Indians, mostly men, received sterilizations, a vast increase over previous years. It is impossible to know how many were coerced, but there were widespread protests and Gandhi was eventually voted out of office at least in part because of these abuses.[55]

Americans could follow these events in occasional reports in the *New York Times* and several magazines. "The curtailment of civil liberties in India," Kai Bird wrote in the *Nation* in June 1976, "has taken a startling new form—compulsory sterilization." Bird described men being arbitrarily dragged off the streets, hauled to family planning clinics in garbage trucks, forcibly held down and sterilized. "In what is coming to be regarded as 'Indira's folly,'" the Catholic magazine *America* told its readers in November 1976, "millions of Indians are being coerced into sterilization." In that same article, a Catholic priest blamed Western organizations for encouraging India's reckless population programs and urged Americans, particularly American Catholics, to mount campaigns to reform them. The problems in India, he said, were "the direct result of forces operating in the world and not only in India." Historians such as Matthew Connelly have made similar arguments.[56]

As the Spaceship Earth idea of a single species all sharing the same problems increasingly came under attack, the coalition that had come together in the late 1960s in support of population programs—liberal Democrats, Republicans, and women's groups—fell apart with it.

Ehrlich Mellows, Slightly

A 1977 edition of Ehrlich's textbook, revised with Anne Ehrlich and Richard Holdren (a Stanford professor who would become science advisor to President Obama in 2009), showed that by the late 1970s—the years when, according to *The Population Bomb*, catastrophe was to hit—Ehrlich seemed to have mellowed somewhat, although not completely.

Ehrlich, Ehrlich, and Holdren opened the book defiantly, arguing that many of the claims of *The Population Bomb* had in fact come true. Not only had the world grown by a half billion people in the last seven years, but "famine has stalked the nomads of sub Saharan region of Africa and the struggling peasants in India." They pointed to other signs that "civilization has entered a period of grave crisis": the energy crisis, the world's economic downturn, and the greater accessibility of nuclear weapons.[57]

In the race between population and food supply, Ehrlich, Ehrlich, and Holdren acknowledged progress, both domestically and internationally, but stressed continuing problems, although in less dire tones than in *The Population Bomb*. Domestically, they noted that American fertility rates had unexpectedly come down, crediting the women's movement and their own population movement, but also the decade's economic uncertainty. However, the U.S. population was still growing because of demographic momentum, and high fertility rates could return. Moreover, the United States still lacked a national population policy, and 3.6 million women, many of them poor, were not getting the birth control services they desired. Internationally, their main concern was with absolute numbers and the declining quantity and quality of natural resources. They admitted the successes of the green revolution but saw a "slackening" in its spread. They warned of serious troubles ahead but without the tone of certain devastation. The outlook, they said with uncharacteristic understatement, is "not particularly bright."[58]

Ehrlich and his colleagues clarified or softened a number of earlier positions. They stressed that, because of its high consumption rates, the United States, not the third world, was the biggest environmental threat to the world. They insisted that critics had miscategorized them as "population hawks." In fact, they believed that population was not the only cause of poverty or environmental problems, and that the world needed not just population programs but also development programs. Regarding coercion, some of their general thinking remained unchanged. They continued to criticize the family planning movement for being oriented "to the needs of individuals and families, not of societies," and they seemed open to a number of coercive population control measures, which they described in some detail, seemingly not ready to dismiss them entirely. But their tone was different. Rather than involuntary measures, they said, they wanted "milder methods of influencing family size preferences" and greater accessibility of birth control, abortion, and sterilization. Milder methods meant propaganda for small families and efforts to redefine women's roles. "Anything that can be done," they wrote, "to diminish the emphasis upon these traditional roles and provide women with equal opportunities in education, employment, and other areas is likely to reduce the birth rate." In addition to expanded birth control programs around the world, they also called for land reform, basic health care, education (especially of women), improved child care, and better nutrition.[59]

Some daylight also became visible between the Ehrlichs and Garrett Hardin. The Ehrlichs discussed triage at length—suggesting that Bangladesh, the case the Lappé and Collins wrote about, best qualified for the "no hope" category—but remained noncommittal. More important, they disputed Hardin's lifeboat metaphor, ultimately arguing for "changing behavior in the rich lifeboats before adoption of a strict lifeboat ethic should be considered." This was because rich lifeboats wasted supplies and "regularly took more than they gave" to poor lifeboats. They also attacked Hardin's ideas that saving a life in poor countries today meant more death down the road. That statement, they said, was better applied to the rich. Every life preserved in the United States threatened the planet many times more than a life preserved in, say, Bangladesh. Finally, they pointed out, Hardin overlooked the fact that the poor lifeboats were arming themselves with nuclear and biological weapons to present a credible threat to the rich. "The fundamental interdependence of nations," they said, "is unlikely to disappear."[60]

Despite this attention to class differences, Ehrlich elsewhere disputed the conclusions of Lappé and Collins in *Food First*. Although admitting that there are "very grave problems of maldistribution and poverty that *must* be attacked" and that "considerable potential exists in most LDCs for expansion of food production," he countered that, ultimately, there are "environmental and biological ceilings beyond which yields cannot be pushed, especially on a sustained basis." Even if poverty was somehow eliminated and fertility drops as a result, which was by no means certain, global population levels would still reach at least ten to twelve billion, in which case "problems of *absolute* scarcity would far overshadow problems of maldistribution." Moreover, he insisted, reforming capitalism and eliminating inequalities was far easier said than done.[61]

Population, Immigration, and Borders

The issue of immigration illustrates another way the Spaceship Earth idea of species unity was fracturing during the 1970s, as it further undermined support for the environmental Malthusian position. In the 1950s and 1960s, because so few migrants came to the United States, Malthusians had spoken of immigration only as a theoretical threat. But the Immigration Act of 1965, which grew out of President Kennedy's family history of immigration and Cold War pressures on President Johnson to show the United States was not a racist nation, opened the doors to new migrants. By the mid 1970s, the new levels of immigration had become very visible, especially because roughly 80 percent of new migrants came from Asia and Latin America, not the traditional sources of migrants to the United States. The media and concerned officials offered many reports, often tinged with race and class bias, especially about those who came without official permission. In 1974, the chief of the Immigration and Naturalization Service described illegal migration to the United States as a "critical problem" that was "completely out of control." "We're facing a vast army that's carrying out a silent invasion of the United States," he said. In the late 1970s, the arrival of refugees from Vietnam, Haiti, and Cuba garnered additional attention. By the early 1980s, historian David Reimers was reporting, "Scarcely a week goes by without the national media featuring stories about the subject."[62]

Increased immigration coincided with the decline in the U.S. fertility rate. Most Malthusian environmentalists saw the upsurge as a frustrating threat to hard-fought progress in reining in the U.S. population. "We environmentalists felt, with relief, that the U.S. had just barely in the nick of time, begun to limit its population, and government aid to family planning clinics has been of enormous help," one ZPG leader wrote a few years later. "But other cultures and countries have not yet done this, creating an additional impact on our population as their people immigrate here." "It will be a sad situation," wrote Anthony Smith in *National Parks Magazine* in 1975, "if the American people achieve the seemingly impossible, the stability and reduction of their own numbers, but meet with defeat because hordes of invaders pour across our borders from nations which have not established population control." Smith wanted immigration reduced but not eliminated. ZPG took a similar route. In 1974, it called for an end to illegal immigration and a reduction in legal immigration to 10 percent of current levels.[63]

Concern about immigration remade organizations such as ZPG. While the overall goal of reducing environmental and social problems by cutting the U.S. population remained the same, many in the population movement shifted their priority from fighting for birth control and abortion rights toward restricting immigration. "During the next few decades," one ZPG leader predicted at the time, "we believe the world will see a new population issue emerge. Instead of family planning, it will be immigration." Not all environmental Malthusians, however, liked the new focus. Environmentalists more broadly were even more skeptical. Two decades hence, the Sierra Club would almost split over immigration and related questions about consumption, race, class, gender, and foreign policy. The seeds of these divisions were apparent in the 1970s.[64]

Among the most outspoken of the immigration restriction strand of the population movement of the 1970s was Garrett Hardin. Hardin's much-disputed "lifeboat" analogy in 1974 was as much about immigration as about the world food crisis. In the essay, Hardin noted that the poor occasionally "fall out of their lifeboats" and hope "to be admitted to a rich lifeboat, or in some other way to benefit from the 'goodies' on board." For members of a well-off boat, Hardin saw three options: allow new people on board even if doing so would exceed the boat's capacity of the boat and likely capsize it, admit just enough people for capacity, but without any safety margin, thereby again running the risk of sinking, or admit as many as possible while still maintaining a safety factor under capacity. Hardin favored the third option. He insisted that the issue was quantity of people, not quality. The origins of each new passenger mattered little, he said, compared to the fixed capacity of each boat. For a number of years Hardin had been moving away from Spaceship Earth universalism and toward a model that emphasized national borders. "Rights based on territory," he insisted in a 1971 essay, "must be defended." If regulating the global commons in reproduction was not sufficient, then protecting private interests was the only answer.[65]

At the time and for decades to come, the leader of the immigration restriction forces among environmentalists was ophthalmologist John Tanton, president of ZPG from 1975 to 1979. Tanton was running the Northern Michigan Planned Parenthood Federation, which he and his wife had founded, when he read *The Population Bomb*. He became one of ZPG's first members in 1969 and served on its board in the early 1970s. His early position on immigration, which he explained in a 1975 essay, grew from general concerns about population growth but also ideas drawn from Ehrlich and ZPG about needing, in Tanton's words, "to slow the economic growth of the developed nations, rather than stimulate it, and in turn to promote the economic growth of the less developed countries." Migration might be good for individuals, he argued, but clearly not for either of the societies involved or for the world as a whole. Source countries lost talented people, including doctors and health-care professionals crucial to family-planning programs. Migration hurt the United States because it increased economic activity and population growth. (Tanton worried that migrants would "bring their traditionally high fertility patterns with them.") The world overall suffered from migration because countries lost the incentive to address their own population problems, and because it "moves people from less consumptive lifestyles to more consumptive ones."[66]

Above all, for Tanton, the issue was fairness in a finite world. What happens when some countries limit their population growth before others? "As certain portions of the globe deal with their problems more effectively than others," he wrote, "they will stabilize more quickly. This will doubtless increase their attractiveness, especially if other regions are not making progress." The resulting pressures for international migration, he believed, would destabilize regions approaching stability. "Can *homo contraceptivus* compete with *homo progenitiva* if borders aren't controlled?" he wrote a decade later. "Or is advice to limit one's family [in the United States] simply advice to move over and let someone else with greater reproductive powers occupy the space?"[67]

Tanton may have originally been motivated by concerns about the absolute numbers of Americans, not race or class fears, but some ZPG members worried about the "quality" of those coming to the United States in the 1970s, fretting that Hispanic culture would debase American society. Migrants from Mexico, one ZPG member wrote in 1974, would "dominate a large part of the U.S. and influence the political structure." Another, noting in 1978 that "WASPs" were "now in the minority," spoke of social anarchy. "There is bad trouble brewing," she wrote, "and it's not just the using up of our resources by overpopulation, for I see chaos, anarchy and revolution first." In later years, Tanton would make similar cultural arguments.[68]

Ehrlich originally lent his name to the campaign to curb immigration. But as with the larger population issue, he quickly complicated his view, in this case with a 1979 book on migration from Mexico, *The Golden Door*. A well-documented tome co-written with his wife Anne and historian Loy Bilderback, *The Golden Door* made about as informed and left-leaning an argument about immigration as possible without completely rejecting the issue. It included interviews with migrants themselves, debunked exaggerated numbers and claims of lost jobs, and showed awareness of the "the historic ethno-racial prejudice against Hispanics." It repeatedly stressed how U.S. power, capitalism, and inequalities had shaped migration. Indeed, current migration patterns were part of a "pattern of exploitation of the Mexican labor pool by American interests." Americans, the authors wrote, wanted a free flow of goods and capital and even people across the border when it served their economic interests, but complained when they thought such an open border wasn't helping them.[69]

Recognizing that it was poverty in Mexico, where the richest 20 percent controlled 60 percent of the country's wealth, that sent migrants to the United States, the book called for U.S. foreign policy "designed to create jobs and improve living conditions" through targeted, smart development, as well as family-planning programs. For the United States, Ehrlich, Ehrlich, and Bilderback sought a plan for a constant overall population, and only then a decision about how to reach that number, whether by natural increase or immigration or some combination. The main issue was total numbers, not ethnic composition. Without an overall population limit for the United States, they believed that, as one reviewer put it, laissez-faire immigration policies were better than racist xenophobia. Although it's possible that this was just rhetoric designed to deflect criticism, by the late 1970s the Ehrlichs had been making similar points for almost a decade. They also did not support the immigration restriction movements within the Sierra Club, either in 1998 or 2004.[70]

In 1979, frustrated with ZPG's unwillingness to push harder for immigration restriction, John Tanton and allies formed FAIR, the Federation for American Immigration Reform. Although originally concerned about the environmental problems that grew from immigration and population growth, FAIR in the 1980s began to make economic arguments, as well. Some observers have claimed that FAIR also made culturally nativist arguments, while others say that FAIR members resisted such arguments, forcing Tanton and allies to form yet other organizations, such as U.S. English, an organization that sought to establish English as the official language.

Since the late 1970s, immigration has continued to divide environmentalists, even those concerned about population growth.[71]

Conclusions

During the heady days surrounding the first Earth Day in 1970, Paul Ehrlich and other Malthusian environmentalists believed that population growth created not just environmental problems such as litter, pollution, and a shortage of green space but also social and political problems such as inner-city unrest and third-world poverty. In making this argument, they often appealed to the species universalism and sense of interconnection that swept through biology and larger American society in the wake of World War II. A reaction to Nazi Germany, this view stressed understanding human beings as a single species, not a collection of smaller racially or nationally defined species. In the late 1960s, as divisions within American society appeared to be growing wider, this biologic universalism carried great appeal. In fighting for the common survival of humanity, Ehrlich and others suggested, human beings needed to overlook or downplay what divided them.

Within months of the first Earth Day, however, cracks began to appear within the consensus for tackling the population problem. Within a few years, people from all points of the political spectrum began attacking Malthusianism. A common target for these critics was Ehrlich and the "we're-all-in-the-same-sinking-boat" logic of *The Population Bomb*, which stressed the essential unity of the human species and the unity of its home, the planet Earth, in a time of crisis. Criticisms came from African American critics, fellow environmentalists, conservative Republicans, New Left anti-imperialists, and reproductive-rights feminists—all of whom stressed disaggregating analysis and emphasized some form of cultural and historical difference. Some stressed ethnicity and race, others national borders, yet others gender differences—and some even attacked the possibility of scientific knowledge itself. Whereas once population limitation appeared the middle ground that could bring many sides together, by the end of the 1970s it had became the common landmark against which a host of different political groups, at least in part, defined themselves.

9

Ronald Reagan, the New Right, and Population Growth

"For the first time, they [politicians] may even have to come out against motherhood."
—Walter Cronkite, reporting on Earth Day, 1970

The controversies about population growth in the 1970s show the complicated concerns, interests, and divisions that surrounded the environmental movement in the wake of Earth Day. These debates had fascinating political fallout in a decade that, although often overlooked, continues to shape the contemporary political landscape. Overall, the population and limits-to-growth issue hurt Democrats and helped Republicans. Nothing shows this better than the varied fates of Jimmy Carter and Ronald Reagan in the late 1970s. Carter came to the Oval Office determined to find a way to help the United States live within its limits. "Dealing with limits," he recalled shortly after leaving office, had been "the subliminal theme" of his presidency, and concern about population growth was part of this. By this time, most environmentalists had gravitated toward the Democratic Party. Within the party, though, both the Old Left and the New Left found aspects of the Malthusian views of environmentalists like Paul Ehrlich troubling. Ultimately, Carter's emphasis on limits would cost him dearly: it would divide the Democratic Party and help lose the 1980 election.[1]

As the Democratic coalition was ripping at the seams, a new political movement was sweeping over the American landscape. "Most establishment experts," historian Bruce Schulman writes about the surprising 1980 election, "had simply not heard the thunder on the right." This thunder was the "New Right"—the conservative wave that remade the Republican Party during the 1970s and early 1980s. More conservative on economic, social, and international issues than other Republicans, the New Right not only helped put Ronald Reagan in the White House in 1980 but also became one of the defining features of American politics for decades to come.[2]

Population politics helped shape the new Republican Party in surprising ways. President Reagan molded his political personality in the late 1960s and 1970s in opposition to the population limitation movement, especially environmental Malthusianism. Whereas Malthusians generally saw new government action—programs designed to spread birth control and abortion rights and even

FIGURE 12 Many conservatives continue to remain deeply skeptical of environmental problems because of the perceived "population dud."

Illustration by Taylor Jones for *Hoover Digest*, Fall 2001.

to reinvent family structure—as necessary to prevent oncoming disaster, Reagan believed that market-based solutions, combined with policies to promote economic growth, would yield a rosy future. Reagan was the optimistic flipside to the doom and gloom of the environmental Malthusians, a sunny "anti-Ehrlich" who reveled in flouting environmental limits.[3]

Jimmy Carter, Malthusians, and Malaise

Before the ascendance of Reagan and the New Right, though, there was Jimmy Carter. Carter came to office at a time when concern about population growth and limits was still strong. In the mid-1970s, driven by the oil and food crises, and buoyed by the success of the Club of Rome's *Limits to Growth*, environmentalists saw great urgency and little need to compromise. "Environmentalists Surge Onward, Strengthened by Oil Cutoff," read a 1974 headline. "What is called for is no less than a restructuring of society," argued Frank Tysen, a professor of environmental management. Some, though, raised an eyebrow at such prescriptions. "Environmentalists are approaching an issue," wrote Larry Prior of the *Denver Post*, "that until a few months ago was considered political suicide to even discuss: Will the public be willing to change its life-style? Can people be appealed to above the belly?" He warned environmentalists to proceed cautiously: "Nobody knows yet whether environmental concerns can survive during a period of economic hardship, since the movement so far has been a product of affluence, a luxury of the post-industrial society rather than a necessity." Carter would be the test, focusing on "limits to growth" programs at home, and population limitation and environmental protection overseas.[4]

As governor of Georgia in the early 1970s, Carter was known for racial liberalism and environmental concern. He became a hero of environmentalists across the country for blocking a dam already under construction by the U.S. Corps of Engineers on a small but scenic Georgia river. In his 1976 presidential campaign, he tapped into a growing network of environmental activists. While Republican opponent Gerald Ford said he would always chose jobs and growth over environmental protection, Carter rejected the tradeoff, but ultimately declared, "if there is ever a conflict, I will go for beauty, clean air, water, and landscape." As president, he opposed several dam projects and attempted comprehensive energy reform.[5]

Often derided for lack of leadership, Carter aggressively took on the problem of environmental recklessness and overconsumption. From the beginning of his administration, he sent signals that he prioritized reconciling the nation to new limits. "More is not necessarily better," he announced in his inaugural address, "even our great nation has its recognized limits, and . . . we can neither answer all questions nor solve all problems. We cannot afford to do everything."

On the environment he was pointed: "the quality of our natural surroundings is threatened because of avarice, selfishness, procrastination and neglect." Not long afterward, in a now famous episode, he signaled his intentions to take seriously the new age of limits by wearing a cardigan sweater in a televised fireside chat to the nation.[6]

Carter's commitment to limits lasted throughout his presidency. In the infamous "malaise" speech on energy policy of July 1979 (in which he never actually used that word), Carter criticized the belief that "our nation's resources were limitless," as well as American habits of "self-indulgence and consumption." "Human identity," he lamented, "is no longer defined by what one does, but by what one owns." Echoing these concerns later in the year, he suggested a broad overlap of foreign and environmental policy: "We have a keener appreciation of limits now—the limits of government, limits on the use of military power abroad; the limits on manipulating, without harm to ourselves, a delicate and a balanced natural environment." In his farewell address in January 1981, he declared the environment one of three big issues facing the United States, specifically mentioning "the demands of increasing billions of people." Even while attempting to be positive, he sounded pessimistic. "But there's no reason for despair. Acknowledging the physical realities of our planet does not mean a dismal future of endless sacrifice."[7]

Carter's views signaled a break with not only Richard Nixon and most Republicans but also previous Democratic leaders. They were a far cry from John F. Kennedy's "New Frontier" and Lyndon Johnson's guns-and-butter "Great Society." "Among the qualities that distinguished Carter" from earlier Democratic presidents, historian Robert Collins has written, "was his appreciation that the United States had indeed left behind the exuberant growth of the immediate postwar years and entered a new season of limits." Indeed, Carter was departing from the idea of an expanded economic pie that had held together the Democratic coalition since the 1930s and 1940s.[8]

Carter's sense of limits no doubt owed to the intertwined domestic and international frustrations that the United States faced: the apparent failures of the Great Society and the War on Poverty, the debacle in Vietnam, the rise of the Soviet Union to nuclear parity and the hardening of Cold War borders, the oil crisis and loss of American economic independence, and the economic slowdown that ended the great postwar boom. But Carter's deep religious beliefs also informed his embrace of limits. He was the nation's first self-described "born-again" Christian president. "The Baptist moralist in Carter," Bruce Schulman has written, "found the irresponsible waste of precious resources morally repugnant." Carter's "visceral Puritanism," according to another historian, caused him to stress the signs and meaning of decline. Whereas other politicians saw obstacles as a chance to show American determination and achievement despite adversity, Carter saw insuperable barriers.[9]

A White House conference on limits to growth organized by Carter's Commerce Department in 1978 showed the political risks of such an approach. The White House Conference on Balanced National Growth brought over 500 delegates to Washington for five days in late January 1978. They faced tremendous challenges. The oil crisis had sent the price of staples skyrocketing, and concerns about economic fairness and environmental quality were voiced that no previous Democratic presidents had to take on. Moreover, the conference planners had few tools to work with, as galloping inflation had disarmed the standard Keynesian pump-priming techniques. Not surprisingly, the meetings descended quickly into intractable debates. "Within hours of the opening of the conference's first sessions," reported journalist Thomas Oliphant, "it was apparent that one person's balance is another's cause for grief." The conference's chairman, Governor Jay Rockefeller of West Virginia, noticed "the incredible array of tensions involved with growth." Divides emerged everywhere, especially between economic interests, between regions, and between growthists and environmentalists.[10]

According to historian Collins, the Carter administration found itself in the "unenviable political position" of having to arbitrate "among the various interests competing for portions of a stable or even shrinking pie." It responded by trying what Collins calls "evasive decentralization": downplaying what the federal government could do and emphasizing the state and local nature of the problem.[11]

In the international arena, too, Carter also had trouble addressing environmental limits. In 1977, he put together a high-level commission to study overseas environmental degradation and resource depletion. In 1980, the commission issued its report, *The Global 2000 Report to the President: Entering the Twenty-First Century*. Consolidating reams of scientific data, the report concluded that "if present trends continue, the world in 2000 will be more crowded, more polluted, less stable ecologically, and more vulnerable to disruption than the world we live in now." The report described many significant problems—even drawing early attention to the problem of climate change. And yet by 1980 it seemed that the American public had wearied of the negative tone.[12]

One sign came from the *New York Times*. Since the 1950s, the *Times* had regularly trumpeted environmental and population problems. But by the late 1970s, while it continued to press on environmental concerns, the *Times* suggested more moderation in dealing with population. In a 1978 editorial called "The Population Bomb, Reconsidered," it called for bringing "balance" to "a field often weighed down by dire predictions." A population catastrophe was "not inevitable." Malthus was "mistaken in his conclusions two centuries ago, and so, it appears, are his successors. The demographic sky may be overcast, but it is not falling." Two years later, in a 1980 editorial called "The Chicken Little Syndrome," the *Times* found the tone of President Carter's Global 2000 report overly glum. Noting that the report actually suggested a modest improvement in global human welfare by 2000, it asked, "Why are soothsayers so relentlessly pessimistic these days?" It then spoke of crisis fatigue: "Doomsday cries have already debased public discourse. Instead of goading people to action, they increasingly produce indifference or despair." The stage was set for another kind of politician.[13]

Rockefeller Republicans and the Emergence of the New Right

As Carter was struggling with limits, the Republican Party was reinventing itself. For decades, moderates in the tradition of Theodore Roosevelt and Dwight Eisenhower had held significant power. In foreign policy, this group rejected the isolationism of Republicans like Ohio's Robert Taft in favor of a robust international presence. Although some were more hawkish than others, they tended to agree with most Democrats on the ends and means of fighting the Cold War, and even, led by Nelson Rockefeller, to support international development programs. In the economic arena, they grudgingly accepted a modicum of government intervention, pushed by a sense of noblesse oblige developed during the Great Depression and a Cold War concern for strong central government. On social issues, they pursued middle-of-the road or even liberal policies growing from the turn-of-the-century Social Gospel and the Progressive Movement. The Rockefeller family, in particular, was known for giving money to various moderate and liberal social causes.[14]

Conservatives had often criticized moderate Republicans during the 1950s and 1960s, but rarely did they seem capable of making a lasting impact. Barry Goldwater was slaughtered by Lyndon Johnson in 1964, and Richard Nixon, who won the 1968 election at least in part because of the unpopular war in Vietnam, embraced a mix of conservative, moderate, and liberal policies that have confounded historians ever since. On top of this, conservatives in the postwar decades were, according the historian James Patterson, "famously disputatious." Economic conservatives, those who favored a free market, and social conservatives, those who wanted to protect and reinforce social values they viewed as traditional, routinely butted heads. It seemed unlikely that they could put aside their differences long enough to build a lasting majority.[15]

But in the 1970s, American conservatives found a way to do so. The trick was to organize in opposition to the "sixties," an imagined decade of excess and decline that contrasted with the ostensible national glory of the 1950s. Three groups provided the main energy of the New Right: a bring-back-America's-strength group who argued that timidity had lost the Vietnam war and that Nixon's détente policies were tantamount to walking with the devil; an anti–big government, pro-market group who fumed at the economic interventions of the

Kennedy and Johnson administrations; and the "cultural right," a collection of mostly evangelical Christians and conservative Catholics who believed that the 1960s had legitimized a permissive culture that was undermining the nation's traditions, especially "family values."

For these groups, the leader of the moderates—Nelson Rockefeller—became a villain, especially in 1976 when he seemed the default choice to be Gerald Ford's running mate. The governor of New York in the 1960s and a frequent candidate for president, Rockefeller seemed too soft on international, economic, and social issues. Conservatives rallied to remove him from the ticket, and in a rare rejection of their presidential candidate's original choice, succeeded in replacing him with the more conservative Robert Dole.

Many early supporters of population planning, such as General William Draper and George H. W. Bush, had been Rockefeller Republicans. In a book called *Our Environment Can Be Saved* (1970), Nelson Rockefeller, himself, spoke of "the significant issue of overpopulation." He added, "With the growing pressures of population and industrialization on our natural resources all of us living this century must be environmentalists." Perhaps no one, however, represented this group better than John D. Rockefeller III, Nelson's brother and founder of the Population Council. Before Earth Day, population limitation as a political issue had belonged as much to Republicans as to Democrats.[16]

Conservative reaction to environmental Malthusianism first emerged during the presidency of Richard Nixon, gained steam in the mid-1970s in reaction to the limits-to-growth movement and the oil crises, and eventually became the strong economic and social critiques seen in the Reagan years. Opposition to Malthusian ideas, especially environmental Malthusianism, helped unite the various New Right groups: foreign policy hawks who believed in a strong military, not "nation-building" economic development programs; libertarian conservatives who frowned upon government interventions in the economy; and cultural conservatives who abhorred what they perceived as an attack on the family. Historians should add the reaction to concern about population growth, especially environmental Malthusianism, to the list of the origins of the New Right.[17]

"People Are Ecology, Too": Malthusianism, the New Right, and Ronald Reagan

As governor of California, home to the nation's largest and fastest growing population, as well as Paul Ehrlich, Garrett Hardin, and many other strident Malthusians, Ronald Reagan was forced to address population policy earlier than most politicians. His evolution on the topic is revealing. Although in 1969 and 1970 he took the middle ground on population and environment policies, by 1971 and 1972, he was showing hints of the anti-Malthusian stance he would use to define his core political identity. This identity, framed against the backdrop of

1970s environmentalism, was on full display by the time of the energy crisis of 1973–1974.

In the late 1960s, Reagan accepted and supported many environmental premises. "Americans, at last, are beginning to realize man can no longer ignore his own damaging impact on his overall environment. We must begin to weigh that impact in every step that affects the quality of the air we breathe, the water we drink and the living space we inhabit," he wrote in a business magazine in 1970. He added, "We have permitted air and water pollution to become a national disgrace—a peril that threatens permanently to alter the delicate balance of ecology that preserves a livable natural environment." As governor, he took measures against pollution and DDT, and often supported his conservation unit against development projects.[18]

In population policy, Reagan showed moderation, occasionally speaking out about fast growth. At the Governor's Conference on California's Changing Environment, which he organized in November 1969, he made population one of his chief talking points, highlighting that by 2000 California would have fifty million residents, each producing five pounds of waste a day. Reagan's comments surprised the *San Francisco Chronicle*'s Harold Gilliam, who commented, "There were times when we wondered whether we were listening to Ronald Reagan or David Brower." Moreover, as governor, Reagan did not block calls for birth control programs and abortion rights.[19]

Not long afterward, the seeds of Reagan's later views about population growth began to emerge. In a 1970 press conference, Reagan argued that the market would solve California's population problem. The state was "filling up," he said, but "you see nature take its course"—nature, in this case, being the market's feedback mechanism. "Right now the projections of ten and twenty years ago for population are no longer valid . . . the curve of increase is beginning to level off." Limits did not loom: "I think we also still have room."[20]

In a speech to a YMCA group later the same month, after reiterating his support of government distribution of birth control, Reagan directly addressed Malthusians who called for "population control." "If, by this," he said, "they mean a wide-scale effort on the part of the public and private organizations to inform the people on the need for birth control—based on individual and family decision—I would be in accord." He continued, "If, however, they are advocating some form of compulsory sterility on a mass basis, then I would be strongly opposed." Reagan was not alone in taking this stance, but his two reasons are significant. The first was moral: "Who would play God?" he asked. "Is it less barbaric to stop a life from starting than it is to stop a life from continuing?" The second regarded the role of government: "Give government the power to preempt what should be a matter of individual choice and conscience and you will have given government the power to control every other aspect of personal life."[21]

Addressing the American Petroleum Institute in November 1971, Reagan articulated his most fully formed critique of environmental Malthusianism to that point. In this important speech, he gave voice to a number signature values—pro-Americanism, concerns about national security and freedom, and faith in markets and technology—all wrapped together in the rhetoric of the frontier. It was a moral vision of what made America great. Reagan had refined these views as General Electric spokesman during the 1950s, and now the environmental Malthusians had given him a spotlight and a perfect foil.

Reagan opened by attacking the crisis mentality that gripped the "doomsday crowd": "We seem to live in an age of simplistic overstatement and false propaganda. We used to have problems. Today, we have crises." For proof, he pointed to "worry about over-population" and the "threat of imminent mass starvation." What particularly irked Reagan was the gloomy view of America. "Their pervasive pessimism is anti-technology, anti-industry and includes opposition to the defense program we must have to maintain the very freedom that allows them to speak their minds and stage their demonstrations. From all this has come a downgrading and even a reviling of the most prosperous and advanced society in the world." The doomsaysers, he believed, ignored all progress in American society since the 1930s.[22]

Reagan specifically attacked population control and the idea of zero population growth. "Population control is one of their popular causes. Zero population growth is the rallying cry. The specter of mass starvation, of people standing elbow-to-elbow . . . is raised as the frightening prospect if we do not take drastic steps to curb the birth rate. Some of the steps proposed involve a kind of regimentation Americans have always found unacceptable." The truth about population, he said, makes the doomsayers sound "a little melodramatic and downright silly." America's birth rate was dropping and the nation still held vast amounts of space. "If you put America's total population in the land area of only two states—California and Texas—you would have a lower population density than any country in Western Europe. And 48 empty states left over." He ignored Ehrlich's point that the planet lacked resources, not space.[23]

Reagan countered with characteristic optimism. Market feedback would take care of population problems: "Things like excessive population growth and decline have a way of balancing themselves out to avoid the Doomsday predictions." He showed faith in technology, even pesticides: "A faster way to achieve the mass starvation the Doomsday prophets worry about might be to do one of the things they advocate—abandon the use of agricultural chemicals and pesticides." Reagan made clear, though, that this did not mean ignoring the problems of pesticides. "As a matter of fact, California has been steadily phasing out the use of DDT. The amount being used today is only 2 percent of what it was just ten years ago."[24]

For Reagan, worries about population growth, resource shortages, and technology paled before his Cold War faith in the United States. The Malthusian view clashed with his trust in freedom and the market, and the basic goodness of America, especially when compared to the Soviet Union. "This is the most dynamic, humane, forward-looking society in the world. We do care about the oppressed, the disadvantaged, the minorities. Freedom and individual dignity are as important to us as the technology that made them possible. Whatever the doomsday myth-makers say, this is the brightest hope of men who seek a brighter tomorrow."[25]

In other venues, Reagan celebrated economic growth. Well before the oil crisis in 1973, he began to angle for working-class voters disgruntled with the Democrats' slow- and no-growth policies. In a speech to the American Federation of Labor in March 1973, he called for meeting the nation's environmental crisis "without blindly disrupting the jobs and economic life of thousands of California workers and their families." Both extremes of the argument, he explained, are wrong. Those who want to "cover the whole state with concrete and punch holes in it for the houses" are wrong. But "equally wrong is that extreme group on the opposite side who would tell us that we cannot build a home for ourselves unless it looks like a bird's nest or a rabbit hole." Parks, he said, should not be "just something that [people] must stand on the edge and look at it because someone wants to go in there and study the bushes that are growing." He singled out limits-to-growth environmentalists, attacking their "panic about overbuilding" California and "hysteria" about population growth in the United States. "People," he stressed, "are ecology too."[26]

While many Americans viewed the oil crisis of the fall of 1973 as a sign that the environmental Malthusians were on to something, Reagan, like many Republicans, was moving in the other direction. In a speech given—fittingly, considering his frontier ideology—to the Cattleman's Association in December 1973, he sketched a vision of the American people's relation to resource scarcity that, in stressing a lack of unity, will, and optimism, seemed to sum up his critique of 1960s liberal culture.[27]

After acknowledging environmental problems, Reagan stressed that "saving the environment" cannot mean calling "a halt to all development, all construction, all drilling and mining." Instead he called for renewed American unity, determination, and self-reliance. "Our people have always been willing to work together to meet a great national challenge. I believe they will meet the energy crisis with the same spirit of cooperation, compromise, and determination. . . . What has been lacking is a sense of national purpose, a determination to get on with the job of making America self-sufficient in fuel and energy." In particular, he called for optimism. "In the past few months, we have heard quite a bit about what we do not have in the way of energy resources and prospects. I think it is time we start remembering what we do have, in natural resources and in the technical capacity to find new sources of energy."

Indeed, overcoming the energy crisis took on a kind of moral meaning—it could help remedy the deeper problems dogging the nation. The energy problem "just might be an historic opportunity for our country to again demonstrate the inherent strength of our system, by accelerating the technology we must have to meet our energy needs." Doing so "will not be easy or cheap and like other problems we face as a country and as a people, the answer can be summed up within a four letter word: work." Finally, he added that Americans "will not find the answers by looking always to government. We must start looking to ourselves."[28]

The population issue and energy crisis of the early 1970s provided Reagan with a chance to articulate the ideas—and the optimistic tone—that would make him a formidable presidential candidate later in the decade. His views aligned well with specific concerns of the two groups then gaining strength within the Republican Party: social conservatives and economic conservatives.

The Cultural Right and Malthusian Environmentalism

The cultural right, which included the evangelical Christians known as the "religious right," first emerged as a social and political force during the 1970s. "In this decade," historian Paul Boyer has written, "the nation's evangelical subculture emerged from self-imposed isolation to become a powerful force in mainstream culture and politics." Although the environmental movement was gaining strength these same years, for the most part the two groups remained at arm's length and increasingly supported different political parties.[29]

This antagonism is somewhat surprising. The two groups shared many values, especially deep misgivings about modern life. Significant numbers of both groups abhorred consumerism, favored traditionalism, and rejected individualism in favor of community bonds. Moreover, strands of both movements rejected mechanistic, despiritualized views of the world. Indeed, in the early 2000s some groups on the cultural right began to show a strong interest in environmental causes, seeing stewardship of god's creation—the Earth—as a central part of their theology. Given this, why has it taken so long for social conservatives and environmentalists to recognize common interests? When asked this very question recently, one green evangelical offered a one-word answer: "Population." As both groups were taking shape during 1960s and 1970s, the politics of population growth created a gulf between them that was difficult to bridge, despite their other similarities.[30]

The most obvious issue separating the cultural right and population planners was abortion, the issue that Richard Nixon had used so skillfully in dismissing the findings of Rockefeller's National Commission of Population

Growth and the American Future in 1972. Well before the Supreme Court's legalization of abortion in 1973, this issue had begun to stir strong feelings. Because so many environmental Malthusians had played a key role in fighting for legalizing abortion—such as Garrett Hardin and Ehrlich—many "right to life" advocates conflated the environmental and abortion rights movements, even though not all population planners, and certainly not all environmentalists, had advocated abortion rights.[31]

Birth control also troubled cultural conservatives—not all but many. Conservative Christians, like the conservative Catholics influenced by the papal encyclical of 1968, stressed that humans should not interfere with God's will for human families. They disapproved of family planning programs in the United States and the spread of birth control by American foreign aid organizations. This was true from the early 1970s, when conservatives like John Steinbacher, an outspoken critic of sex education, attacked the U.S. Agency for International Development (USAID) in *The Child Seducers* (1970), and continued into the early 1980s, when conservative groups lobbied President Reagan to cancel or restrict family planning programs in China.[32]

Members of the emerging cultural right also frowned upon the acceptance of sex that had transformed American society in the previous decades—and that environmental Malthusians had encouraged. While most Americans had been comfortable with the sexual revolution, many conservatives were not, and they organized in the 1970s. New Right groups such as the Conservative Caucus headed by Howard Phillips, the Committee for the Survival of a Free Congress led by Paul Weyrich, and the National Conservative Political Action Committee chaired by Terry Dolan saw "the discontent spawned by sexual issues as a force that could propel their politics into power," according to historians John D'Emilio and Estelle Freedman.[33]

In particular, many conservative groups disliked the growth of sex education programs in public schools, which began in the United States in the mid-1960s. The leading sex ed advocacy group in the United States, the Sexuality Information and Education Council of the United States (SIECUS), was founded in 1964. Sex education became a potent issue. "Nothing strikes terror into the hearts of parents more," historian Gerald Grob wrote about the New Right, "than the suggestion that potent forces beyond their personal or local control—e.g., in the media, the educational bureaucracy, the medical establishment—compete with them for control of their children's upbringing." Indeed, the anti-sex ed movement helped spur the New Right into action. Battles over sex education, historian Janice Irvine has written, "mark an early moment when the nascent Christian Right recognized the mobilizing power of sexuality."[34]

Critics often conflated the sex ed movement with the population limitation movement. John Steinbacher helped organize some of the first anti-sex ed protests in Anaheim, California, in 1968. In *The Child Seducers*, his 1970 call to arms, John Steinbacher attacked what he called the "control-the-people-through-controlling-the-population crowd." Concern about sex ed appeared later in the decade as well. "The problem is," Constance Horner wrote about sex education in the *New York Times* in 1980, "that one person's belief in the danger of overpopulation conflicts with another's rejection of government's role in the business of reproductive decision-making."[35]

The connection between environmentalists and the sex education movement was real. Few people had called for education about sex and birth control as loudly as Malthusians such as Paul Ehrlich. Disputes over sex education broke out at the peak of concern about overpopulation. Ehrlich made his stance very clear in *The Population Bomb*, calling for a federal law requring sex education in schools that began in elementary school. Ehrlich pushed supporters to request that local media teach about sex education, and even to have their children take IUDs to school for "show and tell." It's hard to imagine that these suggestions went over well with cultural conservatives.[36]

Conservatives also loathed the evolutionary framework that many environmental Malthusians embraced, often as their core belief. Many environmentalists were militant atheists, part of an urban, coastal culture that had, since at least the Scopes trial in the 1920s, routinely disparaged and sometimes mocked creationists. Scientific training often reinforced these sentiments. Cultural conservatives, on the other hand, not only rejected evolution as antithetical to the Bible, but also increasingly attacked views of a despiritualized world in which God did not reign and people were just "machinelike things." They often attributed these views to modern biology. Writing in *Whatever Happened to the Human Race?* (1979), Francis Schaeffer and future Reagan administration surgeon general C. Everett Koop rejected the belief that "Man is a part of nature, in the same sense that a stone is, or a cactus, or a camel" because this idea posited that human beings are "by chance more complex, but not unique." Tim LaHaye also attacked secular humanism in another popular social conservative book, *The Battle for the Mind* (1980). Such antipathy to a despiritualized world often brought together conservative Protestants and conservative Catholics. While many population planners believed that man had to take responsibility for the planning of reproduction with birth control and other measures, many cultural conservatives preferred to leave such matters in the hands of God.[37]

A long op-ed in the *Kansas City Star* in 1972 by a member of the Missouri Citizens for Life specifically mentioned Paul Ehrlich. Dismissing the population projections of environmentalists as flawed, local resident Frances Frech rejected calls for a new family structure. "A crowded nation where children are accepted and loved is more desirable," she wrote, "than an uncrowded one where children are something to be prevented if possible, killed before they are born or battered after they get here." Her utter revulsion at the materialist rationality of Ehrlich's approach was particularly clear: "Enough. Enough. I've had enough.

Don't torture me any more with numbers. People are not numbers. Maybe when computers start having babies, the methods of projection used by the alarmists will work."[38]

Perhaps what disturbed cultural conservatives most was the state of the American family. "As the women's movement grew in strength and social conservatives mobilized in opposition," historian Marjorie J. Spruill writes about the 1970s, "distinctly different views on women and their social roles led feminists and antifeminists into a highly visible struggle for influence." Activists such as Phyllis Schafly and Beverly LaHaye frowned upon many of the changes that postwar life had brought the family, and often blamed the women's movement and the sexual revolution. They wanted society to support "traditional" families where the woman's main responsibility was to raise children and support the husband. They despised attempts, such as by Malthusians, to downplay children as the center of family life and women's identities. Cultural conservatives also hated what they saw as the tendency of environmental Malthusians to see babies as a problem. "Mothers don't have babies anymore," one conservative critic complained in the *Kenosha News* in 1972, "they have 'disasters.'" Ironically, just as these conservatives were attacking environmentalists for supporting feminists, many feminists were beginning to attack environmentalists for not being feminist enough.[39]

One of the most influential social conservatives angered by Malthusian politics was Paul M. Weyrich, a leader of the New Right who played a large role in the 1980 election, especially in rallying Catholic voters for Reagan. In August 1970, Weyrich wrote an article in a Catholic weekly illustrating his growing sense that a new political movement was urgently needed. He had just witnessed the passage of a 1970 population and birth-control bill in the U.S. Senate. What particularly appalled him was not just the content of the bill but also that it had passed with almost no debate, even from Republicans. A "billion dollar birth control and population 'research' authorization" bill had become law "without a single word of opposition." He concluded that conservative Catholics were almost powerless on Capitol Hill and needed "truly Catholic lobbyists" to fight for "their point of view in Washington." Over the next decade, he would become such a lobbyist.[40]

A second influential social conservative was Reverend Jerry Falwell, one of the founders of the "moral majority" and a crucial organizer of conservative Christians in the 1980 election. In *Listen, America!*, which he published in 1980, Falwell lashed out at much of what environmental Malthusians stood for. He attacked humanism because it "recognized evolution as a source of man's existence," and what he called naturalism, which "looks on man as a kind of biological machine." To naturalists, "the birth of a child is no different than the birth of an animal. Man lives a sort of meaningless existence in life, and it really doesn't matter what significance he thinks he has or what goals he is headed for." Decrying a "vicious assault upon the American family," Falwell defended the procreative family. "God gave authority and dominion over the creation and told him to multiply and replenish the earth," he wrote, at roughly the same time of Carter's *Global 2000* report. "The family is that basic unit that God established, not only to populate but also to control and contain the earth." A woman's call to be a wife and mother, he added, "is the highest calling in the world." Significantly, Falwell also lamented America's obsession with "materialistic wealth," a view that many environmentalists shared.[41]

Taking issue with the "species universalism" of the Malthusian environmentalists—the idea that all humans were the same and part of a single group, the cultural right, like many feminists, rejected the blindness to gender difference that Ehrlich and others aspired to. What they wanted recognized, however, was not women's history of oppression but women's special roles as mothers. Society, they believed, should do all it can to support motherhood.

Ironically, the cultural right's ethical position was not too distant from that of environmental Malthusians. Both rejected the emphasis on individual rights that started sweeping through society during the 1960s and continued into the 1970s—the so called "me" decade—and both highlighted the need for individuals to sacrifice for the greater good. Ehrlich believed that individual interests should yield before the environmental interests of the entire species; cultural conservatives argued that the individual concerns of women should give way to the needs of the family and country.

But in the late 1970s and early 1980s, their differences far outnumbered their similarities. Instead of coming together with environmentalists over shared concerns about consumption-based modernity, cultural conservatives joined force with market conservatives to become the backbone of the New Right of the Republican Party.

Economic Conservatives and Population Growth

Another key issue driving changes in the Republican Party during the 1970s was concern about the growing power of the federal government. This was a change from the 1950s and 1960s, when, just out of the Depression and World War II and fearing communism more than anything, many Republicans tolerated a large governmental presence. As the Cold War consensus unraveled during the late 1960s and early 1970s, however, libertarians within the Republican Party began to more strongly assert their views about small government. Beginning with Ronald Reagan's failed presidential campaign in 1976, this group slowly gained power within the party. They argued that market forces could solve whatever problems population growth presented.[42]

No one represents what the rise of free-market libertarianism meant for population policy better than Julian Simon, a marketing professor-turned-economist who became one of the most outspoken critics of Ehrlich and the environmental Malthusians during the late 1970s. Simon's chief contributions to the debates on population growth came with clear and persuasive writings and through a spirited decade-long public debate with Ehrlich that included a well-publicized wager. Drawing from a variety of conservative thinkers, Simon deployed a libertarian ideology not prominent within the Republican Party for decades, and helped lay the groundwork for Ronald Reagan's reversal on population planning during the early 1980s.

Ironically, for much of the 1960s, Simon supported Malthusian ideas. After studying at Harvard and the University of Chicago, he taught advertising and marketing for several years before growing deeply concerned about population growth during the mid-1960s. He conducted research on population issues and did consulting work for various population planning organizations. At one point in 1967, he even tried to convince demographer Frank Notestein of the Population Council that, without financial incentives, family planning programs in India would not work. Next to the nuclear arms race, Simon believed, nothing posed a greater threat to peace than uncontrolled population growth.[43]

In 1969, Simon began to change his mind. Reading the works work of Simon Kuznets, Richard Easterlin, Alfred Sauvy, Colin Clark, Esther Boserup, Harold Barnett, and Chandler Morse convinced him to revisit the assumptions of Malthusianism. He also experienced an epiphany. While waiting for a meeting at USAID's Washington office in the spring of 1969 to discuss a population limitation project in the developing world, he noticed a street sign pointing to the Iwo Jima memorial. The sign, as he later recounted, reminded him of the eulogy a Jewish chaplain had given over the dead on the Iwo Jima battlefield, something like, "How many who would have been a Mozart or a Michelangelo or an Einstein have we buried here?" Suddenly it dawned on Simon that he had no right to work so that fewer human beings were born, each one of whom "might be a Mozart or a Michelangelo or an Einstein?"[44]

Simon's chief intervention in the debates on population growth came with two books: *The Economics of Population Growth* (1977), a hefty tome full of statistics, and *The Ultimate Resource* (1981), a reader-friendly attack on Ehrlich and other Malthusians. In these works, Simon disputed Ansley Coale and Edgar Hoover's classic 1958 study about India's economy, arguing that Malthusians overlooked the feedback mechanisms that exist in human societies. If social observers looked over the long haul, they would find that as population growth created scarcities, societies did not just passively accept a lower standard of living. Instead, they responded with technological innovation, either inventing ways to circumvent scarcities or devising substitutions. Simon concluded that as populations grow, the innovation that accompanies that growth usually results in higher standards of living.[45]

For proof, Simon pointed to the price trajectories of various metals in the world market. Over time, he argued, even as humans consumed greater amounts of each metal, their price in real terms actually went down. This countered common sense. The key to Simon's argument was that, as a resource like copper became scarce, people discovered substitutes and found technological alternatives to whatever had grown scarce. After these substitutions, the original resource actually became easy to come by and thus less expensive. The more copper used, Simon pointed out, the cheaper it became.

Ironically, Simon's argument drew from an argument about resource interconnection. He never claimed that the supply of a certain commodity was infinite, but that, in a world full of possible substitutes, a shortage of one material would ultimately not matter. Resource use did not take place in a vacuum, but in a world replete with different resources, with no end to possible substitutes. To say this, though, was to deny a foundational pillar of postwar environmental thought: that humans lived in a world of limits.

Simon also made a point about relativity, disputing the Malthusian assumption of a stable, self-evident world. Environmentalists often discussed feedback loops, but only in terms of nature: production caused environmental problems that eventually decreased the planet's productive capacity. But he highlighted another feedback loop: the way human beings responded to resource scarcities. Natural resources, he stressed, were always mediated by social institutions, especially political systems and markets, which themselves changed as technology and human needs shifted. To Simon, scarcity was not a stable fact of the world but something human beings could respond to.[46]

In sum, Simon disliked Ehrlich's "we're all in a sinking ship" Spaceship Earth argument because it suggested that human beings were helpless. He countered that, sensing the trouble around them, people would respond to problems by devising solutions. They would, in other words, build better ships, better bailers, or better fishing lines. Population growth was not a bad thing. Indeed, the more people in the boat, the more minds to devote to devising solutions. Government need not intervene.

In addition to his arguments about substitution, resource interconnection, and feedback loops, Simon also offered two significant moral arguments. First, he disputed Ehrlich's view of human beings as mere animals dependent upon a web of food chains, which he said reduced them to consumers and nothing more. Humans were creative producers as well, making many of the resources they needed and in some cases a surplus. People also produced knowledge, technology, music, and culture. "Perhaps most significant of all," he wrote, "is the contribution that the average person makes to increasing the efficiency of production through new ideas and improved methods." Each additional person was not just another mouth to feed and degrader of the environment, but quite the opposite. Whereas Ehrlich distrusted man's new inventions, Simon celebrated technology: it was both a sign of what separated humans from animals and a guarantee of humanity's future.[47]

Simon's second moral objection to the Malthusians regarded the dignity of human life. According to Simon, Ehrlich viewed preventing the birth of a human as akin to preventing the birth of a chicken or a goat—nothing more. Simon himself did not object to birth control and abortion except when used in the name of preventing population growth. To him, the more people, the better. "Enabling a potential human being to come into life and to enjoy life is a good thing," he wrote, "just as enabling a living person's life not to be ended is a good thing."[48]

1980 Election

Malthusianism and the "limits-to-growth" movement helped shape the 1980 electoral campaign. By 1980, Jimmy Carter's concern about limits, reversing fifty years of a very successful Democratic Party political strategy, had created tremendous stress within the Democratic coalition, including clashes with organized labor and liberal intellectuals. "Again and again," historian James Patterson has written, Carter's "refusal to pursue a big-spending, Keynesian agenda touched off bitter rows with [Senator Edward] Kennedy and other liberals in his own party." Carter, himself, later noted that "even in . . . the spring of '77, I was already getting strong opposition from my Democratic leadership in dealing with economics. All they knew about it was stimulus and Great Society programs." Party factions clashed in both 1978 and 1980. Even many of Carter's advisors disagreed with him. By the election, Carter's refusal to push economic growth led to a rare primary challenge to a sitting president, in this case from Senator Ted Kennedy. The party was, according to one historian, "adrift without a compelling theme."[49]

In the general election between Carter and Reagan, the limits-to-growth issue formed an important background issue. At one point Carter's running mate, Vice President Walter Mondale, said the fundamental difference between the campaigns was reflected in their long-term views, such as Carter's concern about overpopulation, resource depletion, and environment degradation outlined in the Global 2000 report. When asked about this, Reagan countered that predictions are often wrong and that no one could predict technological breakthroughs. Predictions, he said, "cannot foresee what is going to happen, what is going to come in the line of technology."[50]

Using a pro-growth agenda borrowed from the Democrats' playbook (although minus its social agenda), Reagan tried to win over key Democratic constituencies, such as the white working class. Meeting with Teamsters in Ohio in October, he reiterated his "people are ecology, too" argument from earlier in the decade. "Just possibly, some . . . have gone beyond the point of just being concerned about the responsibility for seeing that the environment is protected in a responsible manner," he said. "In reality, what they believe in is no growth." In some ways, as historian Timothy Stanley has noted, Reagan resembled Ted Kennedy: "They both believed that consumption and growth could heal the economy."⁵¹

Reagan backed his rhetoric with policies prioritizing growth. "The basic aim of the policy we were trying to implement," Republican Party chairman William Brock later explained, "was to restore growth." The 1980 Republican Platform read, "We strongly affirm that environmental protection must not become a cover for a 'no-growth' policy and a shrinking economy. Our economy can continue to grow in an acceptable environment." This was a long way from the "limits-to-growth" movement that Ehrlich and others had pushed for, and that Jimmy Carter had mostly embraced.⁵²

An article from the *Wall Street Journal* on Earth Day, 1980, suggests how limits-to-growth Malthusianism had helped undermine the political appeal of environmentalism. The movement, the author notes, had started in the "mainstream," but then "a theme that was a little stranger than their laudable desire to clean up the water" made it controversial—"a thoroughgoing dislike for industrialism and capitalism." Some environmentalists spoke "in the most alarmed terms about . . . how necessary it was for us to retreat to a simpler life less encumbered by goods and services." This new theme made it possible for anti-environmental critics to garner allies that "they couldn't have corralled otherwise."⁵³

Reagan's optimism helped him win in 1980 and in 1984. Historian Patterson writes, "Americans seemed drawn to his supreme self-confidence and especially to his optimism, which led him to assert that the United States had by no means entered an 'Age of Limits.'" Reagan's "unfailingly upbeat message," he added, "contrasted with the atmosphere of 'malaise' around Carter." In his inaugural address, Reagan highlighted his differences with Carter: "We're not, as some would have us believe, doomed to an inevitable decline." In the campaign four years later, Reagan asked voters to choose "between two different visions of the future, two fundamentally different ways of governing—their government of pessimism, fear, and limits, or ours of hope, confidence, and growth."⁵⁴

As during other moments during the postwar period, during the 1970s and 1980s, too, ideas about population growth and the environment interwove with other aspects of American life: the patterns of poverty and affluence, the shape and size of families, the social roles of men and women, the relations between races and classes, the meaning of material things and increased consumption, the promise and perils of technology, and the possibilities and limits of science. Ideas about population growth and the environment were also central to the political negotiations—within families, nations, and the international community—surrounding all of these issues. In the 1970s and 1980s, thinking about population growth and the environment—and associated ideas of scarcity and abundance, progress and degradation, interconnection and isolation, individuality and communal responsibility, providence and human autonomy, central authority and freedom, innovation and self-destruction—were not just sideline stories, but a central part of important historical changes, changes that still shape American life in the early twenty-first century.

Conclusion

The Power and Pitfalls of Biology

Three questions have driven this study: What caused the wave of Malthusian concern about population growth and environmental problems that swept over the United States in the twentieth century, especially after World War II? How large a role did this wave play in postwar American society and especially the birth of the environmental movement? What impact did Malthusianism leave on the environmental movement and the way Americans today understand interactions between humans and their natural surroundings?

From a historical perspective, it's remarkable that so many Americans ever grew so concerned about population growth. Although Europeans had often blamed poverty and scarcity on overpopulation, Americans had typically seen the world as a place of abundance—at least until the end of the nineteenth century. Few early Americans would have predicted that the United States would see a wave of Malthusian concern sweep through the country. The same could be said looking back from the last two decades of the twentieth century. In 1990, after all the recent disputes about population growth, it would have been hard to imagine that only twenty years earlier many Americans, including top officials of both major parties, had made reducing population growth one of their main concerns. How did this happen?

My argument is that from the 1940s to the early 1970s, an unusual alignment of historical forces—international and domestic, material and cultural—made Malthusianism very attractive, and then in the 1970s these forces mostly dissipated. Among the most important factors were physical changes, both in the environment and populations. The global ecosystem changed dramatically from 1900 to 2000, and one of the most obvious changes was in the number of homo sapiens walking the earth. Never before had the number of humans on the planet been as high or grown as fast as between 1950 and 1990. Because of the baby boom and a drop in mortality, the American population during these same years also set records for both absolute numbers and growth rates.

Dramatic international events directed a powerful spotlight on these material changes. World War II showed the importance of interconnection, and as the country's diplomats and businessmen searched a shrinking world for resources, markets, and political allies during the Cold War decades, many people grew increasingly worried that poverty-induced political instability would draw the nation into another global war. War and fear of war lay behind the explosion of concern about population. Many saw population growth as a source of instability abroad and a potential threat to U.S. national security and abundance. Those who grew the most worried thought not just about resource imbalances but also about the declining capacity of the planet to provide high-quality resources. The first to make such arguments were Raymond Pearl and Edward Murray East, who in the wake of World War I recognized that not only increased demand but also limited and declining capacity posed worrisome problems. Concerns escalated after World War II, when the United States replaced Great Britain as the world's strongest power in an international system that was more interlinked economically and politically than ever before. Surveying the wreckage of the war, William Vogt and Fairfield Osborn pushed the architects of the new world order to not ignore the environmental imbalances that they believed had first ignited the recent wars. In subsequent decades, because of the Cold War, environmental issues and even reproductive patterns in places such as India and Indonesia became issues of U.S. national security.

Geopolitical competition spread to the far corners of the planet but also insinuated itself into each and every aspect of domestic life, as well. This period was marked by not just its globalness but also its totalness. In the name of the national security, the United States mobilized and monitored the material resources of much of the country and even much of the planet. It should not come as a surprise that some mid-twentieth century conservationists began speaking of our "total" environment.

During the 1950s and 1960s, new technologies greatly affected the balance of people and resources. Many people—especially top American policymakers—placed tremendous faith in spreading modern technologies, especially programs to increase food supply through green-revolution hybrid seeds. Big changes also occurred in the history of reproduction, including new birth-control technologies such as the birth-control pill and the intrauterine device, as well as new attitudes about sex and women's roles. These new technologies gave hope to many Malthusians. But most environmental Malthusians, the most militant Malthusians of the 1960s and 1970s, placed little faith in these technological solutions. For them, birth control could not contain biology and biologically driven cultural imperatives, and the green revolution could not dodge ecological realities.

Several developments in biology encouraged the militancy of the environmental Malthusians. Ecological models developed early in the twentieth century stressed the role of consumption and food chains in interconnected systems. New models of analyzing populations as a whole that emphasized limits came into being. Aldo Leopold's well-known model of population irruption and crash—although developed for deer and other wildlife in the 1930s and distilled into allegorical form during World War II—seemed to epitomize the apocalyptic tenor of the postwar period. In an interconnected, technologically dependent world in a headlong rush for development, it was not hard to imagine human society reaching systems collapse—and dying "of its own too much." Second, because of the Malthusian origins of some of Darwin's key concepts, Malthusian thinking was never far from the surface of modern biology. Adding to this was a mid-century revolution in biology that highlighted the importance of evolution within biology and, within evolution, the importance of overpopulation and scarcity. Finally, Malthusianism appealed to biologists with a proclivity for reductionism. In an era of seemingly intractable problems such as entrenched poverty both overseas and at home, Malthusianism seemed to cut straight to the underlying problem. It seemed able to explain things that other approaches, especially culturally premised and technology-dependent modernization theories, could not. In a world where the choice seemed to be between hunger and political instability on one hand and green revolution programs requiring environment-destroying pesticides on the other, reducing population growth seemed to be a biologically informed and transcendent solution.

Concern about third-world problems during the 1950s and 1960s overlapped with brewing crises at home. After World War II, great demographic shifts reconfigured American life, especially in and near cities. Millions of southern blacks moved to urban centers across the country, and millions of whites moved from cities and farms to the suburbs. The baby boom added massive numbers of children to these cities and suburbs, eventually creating deep worries about both places. The problems of the inner cities appeared to mirror those the United States faced overseas: what to do about the poverty that ignited instability? The problems within suburbs were different, of course, but also could be seen as a struggle between the high quality of life and the impoverishment associated with a high quantity of life. Traffic and sprawl, the loss of personal autonomy, the homogenization of culture, the monotony of the landscape—each problem could be blamed on population growth, although not without downplaying other important causes. As cities grew more squalid, a sense of creeping poverty—or, at least, a fear of impending decline—spread in middle-class places. In a crowded interconnected world threatened by homogeneity and violence, uncrowded wildernesses also gained in appeal, but seemed all the more threatened. Places without people became at once more meaningful, and more endangered. All of these domestic problems, mostly unknown when Vogt and Osborn wrote their

Malthusian bestsellers in the early postwar period, reached a point of crisis during the late 1960s. The "feel" of overpopulation included a sense of global resource scarcity but also national and local crowdedness. It was the flawed genius of Malthusians like Paul Ehrlich to bring together all of these problems, and propose solving them, that brought population growth such attention in the late 1960s.

An unusual political climate also contributed. The Cold War created a coming together of political forces in the United States in which long-standing ideological positions were softened in the name of fighting communism. In this "consensus," conservatives generally agreed to brook the greater role of the federal government in all aspects of life, including reproduction, and liberals generally became less critical of the ways that the nation's capitalist system disadvantaged certain classes of citizens. Economic growth became a uniting obsession. And once poverty was connected to national security and new birth control technologies became available, population planning became both a technically and politically feasible option for government officials, a seemingly logical extension of the national and global welfare state. The late 1960s was probably the last moment when population control arguments could have gained the wide appeal that they did.

Examining the history of postwar Malthusianism not only shows its origins in international, scientific, and social trends of the times; it also shows how, in important ways, concern about population growth shaped these aspects of postwar American society. By linking social stability, both at home and abroad, to resource imbalances, and resource imbalances in turn to poor population planning, fears of population growth shaped both local and global politics. Among other things, postwar Malthusianism influenced American Cold War strategies and international development policies, new reproductive technologies such as the "pill" and the IUD, the popular perception of places like the "third world," the "inner city," and the suburb, the shift from Cold War to détente, the women's liberation movement, the reproductive rights movement, and the emergence of the New Right. Although we should not exaggerate the impact of Malthusianism, we cannot fully understand any of these historical changes without it. Ideas about population and environmental imbalance were woven into much of the fabric of postwar American society.

Most important, without understanding postwar American Malthusianism and its origins, we cannot understand the explosion of environmentalism in the late 1960s and early 1970s. Population concerns, heightened by international and domestic worries that seemed to be cresting and converging in the years between 1965 and 1970, help explain the tone, timing, and priorties of the early environmental movement.

The standard story about environmentalism does a better job describing the differences between early twentieth-century conservation and postwar environmentalism than it does explaining the transition from one to the other. In general, early conservationists called for bringing order to the laissez-faire world of resource use in rural hinterlands by making productive industries less wasteful through more efficient planning and technology. In contrast, postwar environmentalists were generally driven by apocalyptic anxieties about modern industrial technology and an ecological sense of interdependence to protect and preserve places in both extremes of civilization's impact: in massive metropolitan areas as well as in remote wilderness areas. The best explanation of environmentalism, offered by Samuel Hays, holds that postwar prosperity drove a consumerist search for a higher "quality of life" that included all sorts of environmental amenities. But Hays says little about what caused the change, or as Adam Rome has pointed out, why environmental activism exploded when it did in the late 1960s. Why did the first Earth Day happen in 1970 and not earlier or later?[1]

While no doubt the changes that Hays describes were crucial to the development of environmentalism, they went hand in hand with other concerns—about international relations, national security, cities, and family patterns, both at home and abroad—that the history of postwar Malthusianism helps us see more clearly. In particular, reintegrating Malthusianism into the history of environmentalism allows us to see that the environmental movement was as much a story about poverty—a low "quality of life"—as about the high quality of life associated with prosperity. "If we talk of environment," U.S. senator and environmental pioneer Edmund Muskie wrote, "soon we must talk of poverty." Indeed, long before Rachel Carson's *Silent Spring*, it was fear about a return to economic depression and global war during the late 1940s that first prompted the American public to think seriously about the danger posed by industrial civilization's disregard for nature's limits. By connecting American national security to the poverty believed to result from resource and population imbalances, a line of thinkers from Fairfield Osborn and William Vogt to Hugh Moore and Gaylord Nelson convinced Americans to pay more attention to environmental issues. As poverty became crucial to foreign relations and thinking about war, so did concerns about natural resources.[2]

Seeing environmentalism as at least in part a discussion about poverty allows us to see that it was a story not only about humanity's relations with nature but also about America's relations with the rest of the world, race and class relations, and family structure. Although Malthusian environmentalists often embraced a universalist rhetoric about the human species, their ideas of population growth and reproductive behavior took shape and played out in a world where particular race, class, and gender ideologies held sway. During the 1950s and 1960s, the populations growing the fastest around the world were poor and nonwhite, and, of course, the people having these babies were women. Whether intentionally or not, when Malthusian environmentalists warned about international population growth, they were warning about the increase of impoverished people of color. In this context, when they suggested expanded birth control programs, Malthusians were often filling a long-held desire on the part of these women and their families to control their own family patterns. But when they suggested strong, and sometimes coercive, methods in order to restore ecological balance and preserve quality of life, they were suggesting policies that would, more often than not, restrict women's control over their own reproductive systems. This meant that, unavoidably, most Americans saw population growth through the lens of nation, race, class, and gender. The converse was also often true: Americans often conceived of racial, class, gender and national difference in terms of population issues. Through population matters, these categories became a crucial part of international relations. It is here that the global and the public most clearly linked to the local and the personal. "While intercourse remains an individual and private matter," one biologist wrote in 1969, "procreation must become of public concern."[3]

During the late 1960s, it seemed that crisis was sprouting everywhere—the third world was falling grip to poverty, famine, and communism as conventional development programs failed; the nation's inner cities were going up in flames; and the suburbs, the redoubt of a high quality of life, were being overrun by the masses. At this moment of multiple crises, Malthusians like Paul Ehrlich suggested they had the "master key"—the one thing that would solve all these problems. Malthusians presented population planning as a low-cost way to solve every problem from overseas famine and international communism to inner-city unrest, traffic jams, and crowded parks at home. This brought Malthusianism and environmentalism broad appeal.

During the 1970s, however, the configuration of international and domestic events changed in ways that eroded the appeal of population planning arguments. In particular, five things happened. First, globally the birth rate began to drop and the green revolution vastly increased food production, without—as yet—the dire ecological consequences environmentalists worried about. Second, birth rates dropped in the United States, and immigration became the main driver of U.S. population growth. Third, the federal government passed legislation establishing government support for birth control programs both at home and abroad, and it also passed a slew of strong environmental laws, including a National Environmental Policy Act incorporating Malthusian ideas about the "total" environment. Moving toward implementation also clarified who would be bearing the costs of population planning as well, and exactly how high the costs would be. Third, poverty became less of a national priority than during the 1960s. This was true at home, where the civil rights movement, the "war on poverty," and inner city unrest came to an end, but especially abroad where, because of détente and the end of the America's long ground war in Vietnam, Cold War competition between the Soviet Union and the United States grew less tense and less direct. Moreover, even where Cold War tensions remained, fighting poverty became a much less important strategy than during the 1960s. Critics of foreign aid and nation building dominated on both the right and the left.

And finally, as the early Cold War consensus split, some of the concerns that had been muted during the Cold War returned to the public agenda. New awareness about social class emerged, as did deep worries about technical experts and an overweening central government. During the 1950s and 1960s, the Cold War climate had fostered a number of essentially apolitical, technology-based remedies for poverty. Americans adopted approaches to fighting poverty that, despite surface differences, in effect ended up blaming poverty on individuals rather than on class barriers within a larger system. But as the Cold War consensus broke down in the late 1960s and 1970s, many Americans turned toward more systemic explanations. This made Malthusianism seem far less a progressive new approach and far more a reactionary diagnosis that blamed the poor for their own poverty and threatened a cherished right. That many of the poor were women of color added to the problem. At the same time, many Americans also began to think differently about the role of the government. During the 1950s and early 1960s, in large part because of the legacy of the Depression and World War II and the imperatives of the Cold War, Americans were much more willing to grant the government broad authorities than they had been previously or since. During the 1970s, Americans of all persuasions became far more skeptical about granting the government such power.

In addition to these five factors, environmental Malthusians also lost appeal because, responding to claims that their species-wide arguments overlooked deep inequalities, they increasingly joined their concerns about overpopulation with attacks on middle-class overconsumption. This had been an important strand of environmental Malthusianism since the 1930s, more important than the racial worries often believed to motivate them, but it became far more focused in the 1970s. Explaining that Spaceship Earth had different berths for the rich and the poor, environmental Malthusians took aim at the overdevelopment of the United States and ecological extravagance of its middle classes. Although in many ways accurate, anti-materialistic and anti-consumption arguments—even in the watered-down forms pushed by President Jimmy Carter in the late 1970s—proved to be exceedingly unpopular. "At heart," historian Steve Gillon writes of baby boom generation, "the Boomers were consumers, not revolutionaries."[4]

By the 1980s, the seeds of what have become common anti-environmental arguments had become conspicuous. Many historically marginalized groups such as African-Americans and women felt that, by not directly renouncing the use of coercion for population control, many Malthusian environmentalists showed an easy willingness to use repressive state power, while others saw strong Malthusian calls for immigration restriction as a similar kind of elitist social engineering. And many working-class and third-world advocates believed that the Malthusian call for slowing or even stopping economic growth unfairly hurt their interests. On the other end of the political spectrum, both social and economic conservatives strongly attacked environmental Malthusianism—the former because of its emphasis on birth control and its perceived attacks on motherhood and family life, the latter because of the Malthusians' lack of faith in markets and human ingenuity. The narrow focus on population planning cast a long-lasting pall on the environmental movement's relations with some potential allies and helped put an anti-regulation, pro-growth New Right Republican—Ronald Reagan—into the White House. To this day, exaggerated Malthusian claims of doom continue to inform Republican mistrust about climate science and other environmental warnings.

In retrospect, the debates about population growth that swept over the United States in the postwar decades showed both the power and pitfalls of applying biological models to human society. On the one hand, placing humans within nature allowed environmentalists to explain and emphasize the biological forces that shape our world—how we depend on raw materials, how energy and these materials cycle through our bodies and our societies, and how heredity and evolution shape our populations. In particular, far more effectively than most, Malthusian environmentalists such as Fairfield Osborn, William Vogt, and Paul Ehrlich showed that we are consumers as much as we are producers, and that, combined, our out-of-control appetites have a tremendous cumulative impact on the planet. For a nation that often acted as if the constraints of the physical world mattered little, this focus on sustainability was an invaluable contribution, and it helped to create a popular environmental movement in the 1960s.

On the other hand, even as the history of population thinking shows the need to see humans as part of nature, subject to all its laws, it also shows the importance of recognizing how humans differ from other animals. Because environmental Malthusians often rigidly overemphasized the biological context for human actions to the exclusion of historical and cultural contexts, they sometimes misdiagnosed social problems and called for unnecessarily drastic, even inhumane, remedies. In applying this biological model so single-mindedly, many environmental Malthusians often lost sight of the crucial ways that human beings were not "just another" species. Humans have culture and history and technology. They have historically conditioned feelings, thoughts, and memories that together add up to cultural identities as complicated and diverse as any ecosystem. These differences are just as real as the nutrients that flow through our bodies. Forgetting that means losing sight of important race, class, and gender differences, and also means losing sight of the complicated political atmosphere in which environmental activists must operate.

Too often modern Americans feel the need to place things either in the category "culture" or "nature" and overlook the ways that many things, including human beings themselves, are both cultural and natural at the same time. Few environmental issues show this better than population growth. Ultimately, if we are to solve the problems presented by global issues such as population growth, including climate change, we will have to understand the ways in which human beings are subject to nature's laws but are not just another species.

Epilogue

Since peaking in the late 1960s, the growth rate of the world's population has dropped steadily. In 1987, the number of people added each year to the population reached its greatest point, and each year since then the population has grown by a smaller and smaller number. In 2003, the median woman worldwide reached replacement fertility, another sign that population will eventually, perhaps not too long from now, begin decreasing. Two other demographic developments from the last decade began to steal the spotlight: globally, older people began to outnumber young people, and urban residents to outnumber rural residents. Both of these changes are remaking societies around the world, just as rapid population growth had previously.[1]

And yet, the world's population continues to climb each day. Because the base population is so large—the highest ever—even adding a smaller percentage still amounts to a tremendous number of people. Population growth won't peak for another twenty to forty years. The world may be adding fewer people each year, but it is still adding people.

Exactly how high the population will reach, no one can say. The largest generation in world history—a global baby boom of sorts—is now coming of age. Whether we land softly with a population of just over eight billion or reach between ten and eleven billion will depend on their reproductive decisions in the next decade or so. No one can say exactly what the social, political, and environmental consequences of this greater population will be. Will we reach the planet's ecological breaking point? Since the 1960s, despite spotty progress made by environmental groups around the world, environmental problems from biodiversity loss to water shortages to climate change continue to grow, especially as more and more countries reach the stage of "high mass consumption" that Walt Rostow hoped for in the 1960s. Reflecting on these trends, population scholar Laurie Mazur concludes, "We are living in a pivotal moment."[2]

In discussing the future, it's hard not to talk about the past, especially the history of predictions about the future. Malthusians in particular have to contend with a long history of predictions. Revisiting *The Population Bomb* in a 2009 article, Paul Ehrlich and his wife Anne have acknowledged some mistakes. The population, both globally but especially in the United States, has not grown as fast as anticipated in 1968. And since then, fewer people have died of hunger—a total of 300 million—than the billions they predicted. True, the number of malnourished was higher than in 1968, but not enough to account for the difference. In pointing to these mistakes, Ehrlich and Ehrlich said nothing about the calls for coercion that appear in *The Population Bomb* in several places. Instead, they emphasized what they say the book got right: its critique of overconsumption, its prediction about declining oceans, and the attention given to climate change. They were also right, they claim, about the serious ecological problems of the green revolution—the overreliance on chemicals and monocultures. Since 1968, additional concerns have emerged about hybrid seeds, especially about the overdraft of groundwater they require. The green revolution may have raised food production, but its medium-term success was "bought at a high price of environmental destruction." Indeed, the Ehrlichs argue that the book's fundamental point—that "the capacity of Earth to produce food and support people is finite"—is still "self-evidently correct." Its message is "even more important today." Signs of potential collapse—both environmental and political—are growing, they say.[3]

Not everyone, however, finds the Ehrlichs' argument self-evident, including some sympathetic to environmental concerns. Among the most outspoken recent critics is British journalist Fred Pearce. In a recent overview of population history called *The Coming Population Crash and Our Planet's Surprising Future*, Pearce chose to emphasize the positive, although he, too, warned of potential environmental problems. He pointed out that, because of the green revolution, humans over the last half century have brought only 10 percent more land in cultivation, yet more than doubled their food production. He acknowledged that the green revolution has stumbled in the last decade, with a slowing in the rate of increase in food production, and that water shortages are getting so severe that they could result in "billions of hungry people." But, he stresses, food shortages are not inevitable; the numbers are no worse than in the 1960s and great possibilities exist for technological improvements. Relatively simple technologies, such as drip methods of irrigation, can solve many shortages. Moreover, population growth actually can turn out to be an advantage. "As countries move from high to low fertility," Pearce notes, drawing from recent research, "they experience a period of a couple of decades when demographic conditions for rapid economic growth are nearly perfect." This "demographic window" exists because the bulge in the population has moved through the years when they require investment but don't create much, and reached a point where they produce and innovate at a very high level. The world is just entering that period. How productive this generation is, compared to how many children they produce, could make all the difference.[4]

Although clear differences separate Pearce's optimism from Ehrlich's caution, it would be easy—and wrong—to exaggerate the differences. Much of what separates them is rhetorical, especially differences about the sharp debates of the past, which still hang in the air. Even as one stresses the negative and the other the positive, both see many of the same dynamics—population growth, unsustainable technologies, and consumption—creating unprecedented environmental problems that need to be addressed.

NOTES

ABBREVIATIONS

ANB	*American National Biography*, ed. John Garraty and Mark Carnes (New York: Oxford University Press, 1999)
CFP	Conservation Foundation Papers, New York Zoological Society
CR	*Congressional Record*
DH	*Diplomatic History*
DKP	Donald King Papers, Denver Public Library
EH	*Environmental History*
EP	Paul Ehrlich Papers (SC0223), Dept. of Special Collections and University Archives, Stanford University Libraries, Stanford, California
GHP	Garrett Hardin Papers, University of California–Santa Barbara
HMF	Hugh Moore Fund Collection, Public Policy Papers, Department of Rare Books and Special Collections, Seeley G. Mudd Manuscript Library, Princeton University, Princeton, New Jersey
HSTL	Harry S. Truman Presidential Library, Independence, Missouri
IPPF	International Planned Parenthood Federation Collection, Sophia Smith Collection, Smith College, Northampton, Massachusetts
LBJL	Lyndon B. Johnson Presidential Library, Austin, Texas
LP	Leopold Papers, University of Wisconsin—Madison
JKP	Julius Krug Papers, Library of Congress
JDR	John D. Rockefeller III Papers, Rockefeller Family Archive, Sleepy Hollow, New York
MSP	Margaret Sanger Papers, Sophia Smith Collection, Smith College, Northampton, Massachusetts
NP	Gaylord Nelson Papers, State Historical Society, Madison, Wisconsin
NPM	*National Parks Magazine*
NYHT	*New York Herald Tribune*
NYT	*New York Times*
NYZS	New York Zoological Society, Bronx, New York
OPNYZ	Fairfield Osborn Papers, New York Zoological Society, Bronx, New York
OPLOC	Fairfield Osborn Papers, Library of Congress
PB	*Population Bulletin*

PCP	Population Council Papers, Rockefeller Family Archive, Sleepy Hollow, New York
PDR	*Population and Development Review*
PPFA	Planned Parenthood Federation of America Papers II, Sophia Smith Collection, Smith College, Northampton, Massachusetts
RFP	Rockefeller Foundation Papers, Rockefeller Family Archive, Sleepy Hollow, New York
RRL	Ronald Reagan Presidential Library
SCB	*Sierra Club Bulletin*
SCP	Sierra Club Papers, Bancroft Library, University of California-Berkeley
VP	Vogt Papers, Denver Public Library
WSJ	*Wall Street Journal*

PREFACE

1. "Fighting to Save the Earth from Man," *Time*, February 2, 1970, 56–63.
2. Not long before this pronouncement, Eisenhower had actually met with one of the main popularizers of the idea of biological interconnection in the postwar years, Fairfield Osborn, author of the 1948 book *Our Plundered Planet* and the subject of chapter 2. John O'Reilly, "Interview: The Zoo's Showman," *NYHT*, June 14, 1953.
3. Leopold, "On a Monument to a Pigeon" in Aldo Leopold, *A Sand County Almanac, and Sketches Here and There* (New York: Oxford University Press, 1949), 116–119. Also see Curt Meine, *Aldo Leopold: His Life and Work* (Madison: University of Wisconsin Press, 1988), 483. For more on Leopold's postwar concerns, see chapter 2.
4. In a recent article, Adam Rome identifies the three most common explanations of the origins of environmentalism: postwar affluence, technologies such as nuclear weapons and DDT, and ecological science. Rome adds information on the environmental activism of three previously ignored groups: post-materialism liberals such as John Kenneth Galbraith and Lyndon Johnson, middle-class women, and countercultural groups. Although Rome's reevaluation is convincing, the emphasis is still within the boundaries of the traditional nation-state. (Although, to be fair, Rome does point out that environmentalists borrowed organizing tools from antiwar protesters.) Adam Rome, "'Give Earth a Chance': The Environmental Movement and the Sixties," *JAH* 90, no. 2 (2003): 525–554. The emphasis on abundance comes from Samuel P. Hays, *Beauty, Health, and Permanence: Environmental Politics in the United States, 1955–1985* (New York: Cambridge University Press, 1987). A notable exception is Edmund Russell's excellent study of the chemical industry. See Edmund Russell, *War and Nature: Fighting Humans and Insects with Chemicals from World War I to Silent Spring* (New York: Cambridge University Press, 2001). Historians, however, have written about American environmental diplomacy and environmental impact abroad. See Kurkpatrick Dorsey, *The Dawn of Conservation Diplomacy: U.S.-Canadian Wildlife Protection Treaties in the Progressive Era* (Seattle: University of Washington Press, 1998); Richard P. Tucker, *Insatiable Appetite: The United States and the Ecological Degradation of the Tropical World*, concise rev. ed. (Lanham, Md.: Rowman & Littlefield, 2007); John Soluri, *Banana Cultures: Agriculture, Consumption, and Environmental Change in Honduras and the United States* (Austin: University of Texas Press, 2005). On the need to bring together environmental and diplomatic history, see Kurk Dorsey, "Dealing with the

Dinosaur (and Its Swamp): Putting the Environment in Diplomatic History," *DH* 29, no. 4 (2005): 573–587. Quotations from Russell, *War and Nature*, 2; and Richard Tucker, "War" in Shepard Krech et al., *Encyclopedia of World Environmental History* (New York: Routledge, 2003), 1284–1291; 1290.

5. Mary L. Dudziak, *Cold War Civil Rights: Race and the Image of American Democracy* (Princeton: Princeton University Press, 2000); Carol Anderson, *Eyes off the Prize: The United Nations and the African American Struggle for Human Rights, 1944–1955* (New York: Cambridge University Press, 2003); Thomas Borstelmann, *The Cold War and the Color Line: American Race Relations in the Global Arena* (Cambridge: Harvard University Press, 2001); Richard Grove, *Green Imperialism: Colonial Expansion, Tropical Island Edens and the Origins of Environmentalism, 1600–1860* (New York: Cambridge University Press, 1995); Peder Anker, *Imperial Ecology: Environmental Order in the British Empire, 1895–1945* (Cambridge: Harvard University Press, 2002).

6. Two excellent examples of population control as eugenics on a world scale are Matthew Connelly, *Fatal Misconception: The Struggle to Control World Population* (Cambridge: Harvard University Press, 2008); and Fred Pearce, *The Coming Population Crash and Our Planet's Surprising Future* (Boston: Beacon, 2010).

INTRODUCTION

1. Stephanie Mills, "Mills College Valedictory Address," in Bill McKibben, *American Earth: Environmental Writing since Thoreau* (New York: Library of America, 2008), 469.

2. For Johnson, see Phyllis Piotrow, *World Population Crisis: The U.S. Responds* (New York: Praeger, 1973), chapter 10. "Of Human Life," *NYT*, July 30, 1968. For Nixon, see Donald T. Critchlow, *Intended Consequences: Birth Control, Abortion, and the Federal Government in Modern America* (New York: Oxford University Press, 1999), chapter 5; and J. Brooks Flippen, *Nixon and the Environment* (Albuquerque: University of Mexico Press, 2000), 37–38.

3. Paul R. Ehrlich, *The Population Bomb* (New York: Ballantine Books, 1968), 46; *Newsweek*, September 2, 1974; Garrett Hardin, "The Tragedy of the Commons," *Science* 162 (December 1968): 1243–1248; and "Living on a Lifeboat," *Bioscience* 24, no. 10 (1974): 561–568. For more on Ehrlich and Hardin, see chapters 6 and 7, respectively.

4. Adam Rome, "'Give Earth a Chance': The Environmental Movement and the Sixties," *JAH* 90, no. 2 (2003): 525–554. Rome, of course, adds to the classic study of environmentalism by Samuel Hays: Samuel P. Hays, *Beauty, Health, and Permanence: Environmental Politics in the United States, 1955–1985* (New York: Cambridge University Press, 1987).

5. Gaylord Nelson, *CR*, September 12, 1969.

6. Paul Ehrlich, "Overcrowding and Us," *NPM*, April 1969, 10. Matthew Connelly's *Fatal Misconception* does an excellent job examining the global population limitation movement, but does not delve deeply into the environmental side of the population movement: Matthew Connelly, *Fatal Misconception: The Struggle to Control World Population* (Cambridge: Harvard University Press, 2008). Those who focus more on environmental Malthusianism include Björn-Ola Linnér, *The Return of Malthus: Environmentalism and Post-war Population-Resource Crises* (Isle of Harris, UK: White Horse Press, 2003); Ronald Walter Greene, *Malthusian Worlds: U.S. Leadership and the Governing of the Population Crisis*, (Boulder, Col.: Westview, 1999); and Fred Pearce, *The Coming Population Crash and Our Planet's Surprising Future* (Boston: Beacon, 2010). Also see

Hays, *Beauty, Health, and Permanence*, chapter 7; and Robert Gottlieb, *Forcing the Spring: The Transformation of the American Environmental Movement* (Washington, D.C.: Island Press, 1993), 254.

7. Patricia James, *Population Malthus: His Life and Times* (London: Routledge, 1979).
8. Quoted in Pearce, *The Coming Population Crash*, 3.
9. Malthus quoted in J.F.J. Toye, *Keynes on Population* (New York: Oxford University Press, 2000), 16; and in Pearce, *The Coming Population Crash*, 5. For Darwin and Russell, see Donald Worster, *Nature's Economy: A History of Ecological Ideas*, 2nd ed. (Cambridge: Cambridge University Press, 1994), 149–155.
10. David Potter, *People of Plenty: Economic Abundance and the American Character* (Chicago: University of Chicago Press, 1954). For exceptions, see David M. Wrobel, *The End of American Exceptionalism: Frontier Anxiety from the Old West to the New Deal* (Lawrence: University Press of Kansas, 1993); and Derek Hoff, *The State and the Stork: The Population Debate and Policymaking in United States History* (Chicago: University of Chicago Press, forthcoming), chapter 1.
11. Wrobel, *End of American Exceptionalism*, chapter 3. Donnelly quoted ibid., 40.
12. Samuel P. Hays, *Conservation and the Gospel of Efficiency: The Progressive Conservation Movement, 1890–1920* (Cambridge: Harvard University Press, 1959), 2; Gifford Pinchot, *The Fight for Conservation* [1910] (Seattle: University of Washington Press, 1967), 48; William Cronon, "Landscapes of Abundance and Scarcity," in Clyde A. Milner, Carol A. O'Connor, and Martha A. Sandweiss, *Oxford History of the American West* (New York: Oxford University Press, 1994), 633. For resources and overseas expansionism, see Wrobel, *End of American Exceptionalism*, chapter 5; Richard P. Tucker, *Insatiable Appetite: The United States and the Ecological Degradation of the Tropical World*, concise rev. ed. (Lanham, Md.: Rowman & Littlefield, 2007).
13. Turner quoted in Wrobel, *End of American Exceptionalism*, 113. Because Malthus had rejected birth control, some historians use the term "neo-Malthusians" to refer to those Malthusians who promoted contraception. However, because most, if not all, of those concerned with population growth after World War II believed in birth control, I will ignore this technical difference and use "Malthusian" and "neo-Malthusian" interchangeably.
14. Joseph Wood Krutch, "A Naturalist Looks at Overpopulation," in Fairfield Osborn, ed., *Our Crowded Planet: Essays on the Pressures of Population* (Garden City, N.Y.: Doubleday, 1962), 207–214; Letter to the Editor, *SCB* 50 (June 1965): 20.
15. Of course, before 1600, as J. L. Finkle has pointed out, a good century was one with any population growth at all. Finkle, "Politics of Population Policy," in Neil J. Smelser and Paul B. Baltes, *International Encyclopedia of the Social & Behavioral Sciences* (New York: Elsevier, 2001), 11793–11798. Also see U.S. Census Bureau, International Database; "Transitions in World Population," *PB* 59, no. 1, March 2004, 3–40; World Population Division, *World Population Prospects: The 2002 Revision* (New York: United Nations, 2003); and Lester Brown, *World without Borders* (New York: Random House, 1972), 5.
16. *Economic Report of the President, 1995* (Washington, D.C.: Government Printing Office, 1995), 311, table B-32. For U.S. demographic history, see Michael S. Teitelbam and Jay M. Winter, *Fear of Population Decline* (New York: Academic Press, 1985), esp. chapters 4 and 5; Richard A. Easterlin, "Twentieth-Century Population Growth" in Stanley L. Engerman and Robert E. Gallman, *The Cambridge Economic History of the United States*, vol. 3 (New York: Cambridge University Press, 2000), 505–548; and Michael French,

"The US Population since 1945," in *US Economic History since 1945* (New York: Manchester University Press, 1997), 3–20.

17. Stewart L. Udall, *The Quiet Crisis* (New York: Holt, 1963), 189.
18. For environmentalism, see note 4 in the preface.
19. John Kenneth Galbraith, *The Affluent Society* (Boston: Houghton Mifflin, 1958); Hays, *Beauty, Health, and Permanence*; Lincoln H. Day and Alice Taylor Day, *Too Many Americans* (Boston: Houghton Mifflin, 1964); on "Chinification," see William Paddock and Paul Paddock, "The Explosion in Humans," in *Topics in Animal Behavior, Ecology and Evolution*, ed. Vincent Dethier (New York: Harper & Row, 1971), 123.
20. "Planned Parenthood News," Planned Parenthood Association Chicago Area, June 1950, PPFA. For thinking about the relations between the intergroup or international relations and family ideas, I get inspiration from Elaine Tyler May, *Homeward Bound: American Families in the Cold War Era* (New York: Basic Books, 1988); and Linda Gordon, *The Great Arizona Orphan Abduction* (Cambridge: Harvard University Press, 1999).
21. See Connelly, *Fatal Misconception*; Allan Chase, *The Legacy of Malthus: The Social Costs of the New Scientific Racism* (New York: Knopf, 1977); Mark A. Largent, *Breeding Contempt: The History of Coerced Sterilization in the United States* (New Brunswick: Rutgers University Press, 2008); Ian Dowbiggin, *The Sterilization Movement and Global Fertility in the Twentieth Century* (New York: Oxford University Press, 2008).
22. Johanna Schoen, *Choice & Coercion: Birth Control, Sterilization, and Abortion in Public Health and Welfare* (Chapel Hill: University of North Carolina Press, 2005).
23. Paul A. Colinvaux, *Why Big Fierce Animals Are Rare: An Ecologist's Perspective* (Princeton: Princeton University Press, 1978), 222.
24. William Vogt, *Road to Survival* (New York: W. Sloane Associates, 1948), 284. Foreign relations historians have not focused much on consumption. Three important exceptions include Emily S. Rosenberg, *Spreading the American Dream: American Economic and Cultural Expansion, 1890–1945* (New York: Hill and Wang, 1982); Victoria De Grazia, *Irresistible Empire: America's Advance through Twentieth-Century Europe* (Cambridge: Harvard University Press, 2005); Tucker, *Insatiable Appetite: The United States and the Ecological Degradation of the Tropical World*. The literature on twentieth-century American consumption is too vast to list, but for consumer culture, see Lizabeth Cohen, *A Consumers' Republic: The Politics of Mass Consumption in Postwar America* (New York: Knopf, 2003); for consumption patterns, see Susan Strasser, *Waste and Want: A Social History of Trash* (New York: Metropolitan Books, 1999); for consumption and environmental sensibilities, see Gregory Summers, *Consuming Nature: Environmentalism in the Fox River Valley, 1850–1950* (Lawrence: University Press of Kansas, 2006); for studies of materialism and affluence, see Daniel Horowitz, *The Anxieties of Affluence: Critiques of American Consumer Culture, 1939–1979* (Amherst: University of Massachusetts Press, 2004).
25. For the world's demographic future, see Laurie Ann Mazur, *A Pivotal Moment: Population, Justice, and the Environmental Challenge* (Washington, D.C.: Island Press, 2009); Pearce, *The Coming Population Crash*.

CHAPTER 1 MALTHUSIANISM, EUGENICS, AND CARRYING CAPACITY IN THE INTERWAR PERIOD

1. Raymond Pearl, *The Biology of Population Growth* (New York: Knopf, 1925), 1.
2. Edward Murray East, *Mankind at the Crossroads* (New York: Scribner, 1923), 9.

3. Ibid., 8, 7. Matthew Connelly provides an excellent summary of efforts to monitor populations at this time in *Fatal Misconception: The Struggle to Control World Population* (Cambridge: Harvard University Press, 2008), chapter 1.

4. John Maynard Keynes, *The Economic Consequences of the Peace* (London: Macmillan, 1919); Lothrop Stoddard, *The Rising Tide of Color against White World-Supremacy* (New York: Scribner, 1920). In making these distinctions, I follow John H. Perkins, *Geopolitics and the Green Revolution: Wheat, Genes, and the Cold War* (New York: Oxford University Press, 1997); and Alison Bashford, "Nation, Empire, Globe: The Spaces of Population Debate in the Interwar Years," *Comparative Studies in Society and History* 49, no. 1 (2007): 170–201.

5. Along with Alison Bashford, but unlike other historians, I stress the emphasis on soil, carrying capacity, and conservation in understanding this strand of Malthusianism. Matthew Connelly, for instance, downplays the emphasis on carrying capacity before World War II and places the late 1940s environmental Malthusians, especially Vogt, mostly within a eugenics tradition. Garland Allen traced Paul Ehrlich directly to Pearl and eugenics. See Bashford, "Nation, Empire, Globe"; Connelly, *Fatal Misconception*, 40; Garland Allen, "Old Wine in New Bottles: From Eugenics to Population Control in the World of Raymond Pearl," in *The Expansion of American Biology*, ed. Keith Benson, Jane Maienschein, and Ronald Rainger (New Brunswick: Rutgers University Press, 1991), 232.

6. East, *Mankind*, 10. Historians who emphasize eugenics include Allen, "Old Wine in New Bottles," 231–256; Connelly, *Fatal Misconception*, 52, 59, 61. For exceptions, see Perkins, *Geopolitics and the Green Revolution*, 123; Bashford, "Nation, Empire, Globe," 171, 181.

7. Allen, "Old Wine in New Bottles," 251.

8. Raymond Pearl, "Karl Pearson, 1857–1936," *Journal of the American Statistical Association* 31, no. 196 (1936): 653–664; 654; Connelly, *Fatal Misconception*, 38.

9. Melissa Hendriks, "Raymond Pearl's 'Mingled Mess,'" *Johns Hopkins Magazine* 8, no. 2 (April 2006): 50–56.

10. Ibid.; Daniel J. Kevles, *In the Name of Eugenics: Genetics and the Uses of Human Heredity* (New York: Knopf, 1985), 69.

11. Raymond Pearl, *The Nation's Food: A Statistical Study of a Physiological and Social Problem* (Philadelphia: W. B. Saunders, 1920), 17. Also see Sharon E. Kingsland, *Modeling Nature: Episodes in the History of Population Ecology* (Chicago: University of Chicago Press, 1985), 64.

12. Pearl, *Biology of Population Growth*, 18; Kingsland, *Modeling Nature*, 97. Also see Sabine Hohler, "The Law of Growth: How Ecology Accounted for the World Population in the Twentieth Century," *Distinktion* 14 (2007): 45–64.

13. Pearl, "Some Eugenic Aspects of the Problem of Population," in *Eugenics in Race and State*, ed. Charles B. Davenport et al. (Baltimore: Williams and Wilkins, 1923), 214, as quoted in Connelly, *Fatal Misconception*, 59; Allen, "Old Wine in New Bottles," 245. Here Connelly uses Pearl to suggest that other Malthusians of the interwar period held similar racial beliefs.

14. Raymond Pearl, "Differential Fertility," *Quarterly Review of Biology* 2, no. 1 (1927): 117; Pearl, quoted in Kevles, *In the Name of Eugenics*, 122.

15. Pearl, "Differential Fertility," 116.

16. Raymond Pearl, "The Population Problem," *Geographical Review* 12, no. 4 (1922): 639–640; "closed universe" quote from Pearl, quoted in Allen, "Old Wine in New Bottles," 245. For consumption, see Charles McGovern, "Consumer Culture and Mass Culture," in Karen Halttunen, ed., *A Companion to American Cultural History* (Malden, Mass.: Blackwell, 2008), 183–197; and Gregory Summers, *Consuming Nature: Environmentalism in the Fox River Valley, 1850–1950* (Lawrence: University Press of Kansas, 2006), chapter 4. For carrying capacity, see Curt Meine, *Aldo Leopold: His Life and Work* (Madison: University of Wisconsin Press, 1988), 134; Christian C. Young, *In the Absence of Predators: Conservation and Controversy on the Kaibab Plateau* (Lincoln: University of Nebraska Press, 2002), 40–42, 172; Nathan Sayre, "The Genesis, History, and Limits of Carrying Capacity," *Annals of the Association of American Geographers* 98, no. 1 (2008): 120–134.
17. Pearl, *Biology of Population Growth*, 171.
18. Pearl, "Differential Fertility," 116. Emphasis in original.
19. Royal Chapman, "The Quantitative Analysis of Environmental Factors," *Ecology* 9, no. 2 (1928): 114, quoted in Sayre, "The Genesis, History, and Limits of Carrying Capacity," 128; Eugene Odum, *Fundamentals of Ecology* (Philadelphia: Saunders, 1953). For Pearl and Chapman, see Kingsland, *Modeling Nature*, 95.
20. East, *Mankind*, viii, 12.
21. Edward Murray East, "A Mendelian Interpretation of Variation That Is Apparently Continuous," *American Naturalist* 44 (1910): 65–82. Also see Diane Paul, "Edward Murray East," in John Garraty and Mark Carnes, eds., *American National Biography* (New York: Oxford University Press, 1999), 240–242; "Dr. Edward East, Noted Geneticist," *NYT*, November 10, 1938; Donald F. Jones, "Biographical Memoir of Edward Murray East," in *Biographical Memoirs* (Washington, D.C.: National Academy of Sciences, 1945); R. A. Emerson, "Edward Murray East," *Science* 89, no. 2299 (1939): 51; Karl Sax, "Edward East," *Records of Genetics Society of America* 8 (1939): 11–12; O. E. Nelson, "A Notable Triumvirate of Maize Geneticists," *Genetics* 135, no. 4 (1993): 937–941. Jones provides a complete list of East's publications.
22. Jones, "Biographical Memoir of Edward Murray East," 229.
23. East, *Mankind*, 187, 299, 181.
24. Ibid., 344, 75, 103.
25. Ibid., 179, 179, 194.
26. Ibid., 166–168, 160, 168, 56, 179. H. H. Bennett would later use the phrase "permanent agriculture." Hugh H. Bennett, *Soil Conservation* (New York: McGraw-Hill 1939), vi.
27. East, *Mankind*, 149, 10.
28. Ibid., 197, 191, 194, 235.
29. Ibid., 3, 39, 37. Also see vi.
30. Ibid., 303, 350.
31. Ibid., 113–114.
32. Ibid., 75.
33. Neither historians of Leopold nor historians of Malthusianism have focused on the influence of Leopold's ideas of carrying capacity on post–World War II environmental Malthusians. The literature on Leopold is otherwise excellent. See Susan L. Flader, *Thinking Like a Mountain: Aldo Leopold and the Evolution of an Ecological Attitude toward Deer, Wolves, and Forests* (Lincoln: University of Nebraska Press, 1974); Meine, *Aldo*

Leopold; Paul Sutter, *Driven Wild: How the Fight against Automobiles Launched the Modern Wilderness Movement* (Seattle: University of Washington Press, 2002); Julianne Lutz Newton, *Aldo Leopold's Odyssey* (Washington, D.C.: Island Press/Shearwater Books, 2006).

34. Newton, *Leopold's Odyssey*, 43.
35. Ibid., 57, 59.
36. Leopold, "Some Fundamentals of Conservation in the Southwest," unpublished manuscript, in Susan Flader and J. Baird Callicott, eds., *The River of the Mother of God and Other Essays* (Madison: University of Wisconsin Press, 1991), 95; Aldo Leopold, *Game Management* (New York: Scribner's, 1933), 26, 172.
37. Newton, *Leopold's Odyssey*, chapter 3.
38. Leopold and his coworkers identified two deer irruptions in the 1890s, two from 1900–1909, four from 1910–1929, seven from 1920 to 1929, fourteen from 1930 to 1939, and fifteen in the five years between 1940 and 1944. Thomas R. Dunlap, *Saving America's Wildlife* (Princeton: Princeton University Press, 1988), 69. For the Kaibab, see Dunlap, *America's Wildlife*, 65–66; Thomas Dunlap, "That Kaibab Myth," *Journal of Forest History* 32, no. 2 (1988): 60–68; Young, *Absence of Predators*. A recent study by Dan Binkley et al. using the age structure of aspens appears to have confirmed Leopold's main conclusions; D. Binkley, M. Moore, W. Romme, and P. Brown, "Was Aldo Leopold Right about the Kaibab Deer Herd?" *Ecosystems* 9 (2006): 227–241.
39. Dunlap, *America's Wildlife*, 71. For other researchers who studied wildlife populations during these years and for quantification and wildlife ecology, see ibid., 74.
40. Alexander Carr-Saunders, *The Population Problem: A Study in Human Evolution* (Oxford: Oxford University Press, 1922); Peder Anker, *Imperial Ecology: Environmental Order in the British Empire* (Cambridge: Cambridge University Press, 2002), 101–102.
41. Leopold, *Game Management*, 49, quoted in Frank Golley, "Human Population from an Ecological Perspective," in Michael S. Teitelbaum and Jay M. Winter, *Population and Resources in Western Intellectual Traditions* (Cambridge: Cambridge University Press, 1989), 200. Leopold, "Conservation Ethic," *Journal of Forestry* 31, no. 6 (October 1933): 634–643; 639, 643, 635.
42. Aldo Leopold, "A Biotic View of the Land," *Journal of Forestry* 37, no. 9 (1939): 727–730; 728. Newton emphasizes the importance of this essay in *Leopold's Odyssey*, chapter 6.
43. Leopold, "A Biotic View," 730.
44. Ibid., 729.
45. Leopold, "Ecology and Politics," undated, I, LP, 10–6, 16. Reprinted in Flader and Callicott, eds., *The River of the Mother of God and Other Essays*, 281–286.
46. Leopold, "Wildlife in American Culture," *Journal of Wildlife Management* 7, no. 1 (January 1943): 5–6.
47. Aldo Leopold, *A Sand County Almanac, and Sketches from Here and There* (New York: Oxford University Press, 1949), 129–133. Leopold wrote several memorable essays in 1943 and 1944. See Meine, *Aldo Leopold*, 701; Newton, *Leopold's Odyssey*, 313–315.
48. Dunlap, *America's Wildlife*, 65; also see 77. For New Deal conservation and Leopold's criticism, see Neil M. Maher, *Nature's New Deal: The Civilian Conservation Corps and the Roots of the American Environmental Movement* (New York: Oxford University Press, 2008), 167. For Leopold's influence on later environmentalists, see chapters 3, 6, and 7 of this book.

49. John Maynard Keynes, *The General Theory of Employment, Interest and Money* (New York: Harcourt, Brace, 1936). For Keynes, see D. E. Moggridge, *Maynard Keynes: An Economist's Biography* (New York: Routledge, 1992); and Robert Skidelsky, *John Maynard Keynes, 1883–1946: Economist, Philosopher, Statesman* (New York: Penguin Books, 2005).

50. John McNeill, *Something New under the Sun: An Environmental History of the Twentieth-Century World* (New York: Norton, 2000), 335–336. Also see Robert M. Collins, *More: The Politics of Economic Growth in Postwar America* (Oxford: Oxford University Press, 2000), 42.

51. Thomas Malthus, *In Principles of Political Economy Considered with a View to Their Practical Application* (1820) in *The Works of Thomas Robert Malthus*, ed. E. A. Wrigley and D. Souden (London: W. Pickering, 1986). Also see J.F.J. Toye, *Keynes on Population* (New York: Oxford University Press, 2000), 191–193.

52. John Maynard Keynes, "Some Economic Consequences of a Declining Population," *Eugenics Review* 29, no. 1 (1937): 13–17; 17. Keynes's chief U.S. ally, Alvin Hansen, made similar arguments about population. See Alvin H. Hansen, "Economic Progress and Declining Population Growth," *American Economic Review* 29, no. 1 (1939): 1–15. There were Keynesians who, following a line of reasoning from Keynes's essay, thought it possible to increase growth without population increases. For "stable population Keynesians," see Derek Hoff, *The State and the Stork: The Population Debate and Policymaking in United States History* (Chicago: University of Chicago Press, forthcoming).

53. For the adoption of Keynesian approaches, see Lizabeth Cohen, *A Consumers' Republic: The Politics of Mass Consumption in Postwar America* (New York: Knopf, 2003), 55; Collins, *More*, 10–11. For more on the shift to consumption during the 1930s, see Kathleen G. Donohue, *Freedom from Want: American Liberalism and the Idea of the Consumer* (Baltimore: Johns Hopkins University Press, 2003). For "gross national product war," see Russell Weigley, *The American Way of War* (New York: Macmillan, 1973), 146. Hansen and Colm quotation from Memo, "Postwar Employment," October 9, 1944, Box 1, Colm MSS, Truman Library, as quoted in Collins, *More*, 17; Robert R. Nathan, *Mobilizing for Abundance* (New York: McGraw-Hill, 1944), 98, as quoted in Cohen, *A Consumers' Republic*, 116 (emphasis added). Roosevelt, *Public Papers and Addresses of Franklin D. Roosevelt*, 12: 574–575, as quoted in Collins, *More*, 15.

54. U.S. Council of Economic Advisers, *Second Annual Report to the President* (Washington, D.C., 1947), as quoted in Collins, *More*, 20; see also Collins, *More*, 16.

55. On postwar Keynesianism, see Charles S. Maier, "The Politics of Productivity: Foundations of American International Economic Policy after World War II," *International Organization* 31 (Autumn 1977): 607–633; and G. John Ikenberry, "Creating Yesterday's New World Order: Keynesian 'New Thinking' and the Anglo-American Postwar Settlement," in *Ideas and Foreign Policy: Beliefs, Institutions, and Political Change*, ed. Judith Goldstein and Robert O. Keohane (Ithaca: Cornell University Press, 1993), 57–86.

56. Franklin Roosevelt, "State of the Union Address," January 6, 1941, as quoted in Cohen, *A Consumers' Republic*, 56; Robert Skidelsky, *Keynes: The Return of the Master* (New York: PublicAffairs, 2009), 101. For the Marshall Plan as a Keynesian program, see Michael J. Hogan, *The Marshall Plan: America, Britain, and the Reconstruction of Western Europe, 1947–1952* (New York: Cambridge University Press, 1987), 94, 429.

57. Quotes from Cohen, *A Consumers' Republic*, 119 and 9. For Keynesianism and growth, see Collins, *More*, chapter 1.
58. Aldo Leopold, *Game Management*, 20 and 45, italics added.
59. Leopold, "Postwar Prospects," *Audubon* 46, no. 1 (January–February 1944): 26–29; 27; Leopold, *A Sand County Almanac*, vii and ix.

CHAPTER 2   WAR AND NATURE

1. *Time*, November 8, 1948, 27. For the election, see "Before Breakdown," Hempstead, N.Y. *Newsday*, October 19, 1948, Osborn Scrapbook, Container 10, OPLOC; "Malthus Goes West," *Economist*, December 18, 1948.
2. *Time*, November 8, 1948, 27.
3. Unidentified editorial, September 25, 1948, 1948 Scrapbook, OPLOC; Fairfield Osborn, *Our Plundered Planet* (Boston: Little, Brown, 1948), viii; William Vogt, *Road to Survival* (New York: W. Sloane Associates, 1948), 142, 143. Before World War II, ecology had been the province of a small number of biologists themselves struggling to win acceptance within larger scientific circles. For the history of ecology, see Donald Worster, *Nature's Economy: A History of Ecological Ideas*, new ed. (New York: Cambridge University Press, 1985).
4. Osborn, *Our Plundered Planet*, 200; Vogt, *Road to Survival*, 17.
5. Richard Tucker, "War," in Shepard Krech, John Robert McNeill, and Carolyn Merchant, eds., *Encyclopedia of World Environmental History* (New York: Routledge, 2004), 1284–1291; 1288; Osborn, *Our Plundered Planet*, 136.
6. For Osborn, see "Osborn, Fairfield," *ANB*, 783–784; "Osborn, (Henry) Fairfield, Jr." in *Biographical Dictionary of American and Canadian Naturalists and Environmentalists*, Keir Sterling et al., eds. (Westport, Conn.: Greenwood, 1997), 596–597; "Osborn, Fairfield," in *Current Biography* (New York: H.W. Wilson Company, 1949), 463–465; "Osborn, Fairfield," in Anne Becher, *American Environmental Leaders: From Colonial Times to the Present* (Santa Barbara, Cal.: ABC-CLIO, 2000), 617–618; Andrew Jamison and Ron Eyerman, *Seeds of the Sixties* (Berkeley: University of California Press, 1994), chapter 3. For the New York Zoological Society, see Jonathan Peter Spiro, *Defending the Master Race: Conservation, Eugenics, and the Legacy of Madison Grant* (Burlington: University of Vermont Press, 2009), 31–52.
7. Spiro, *Defending the Master Race*, 88–91.
8. For the exhibit, see "Mayor Opens Zoo's New 'Veldt,'" *NYT*, May 2, 1941; and William Bridges, *A Gathering of Animals: An Unconventional History of the New York Zoological Society* (New York: Harper & Row, 1974). For Osborn's quotation, see Fairfield Osborn, *NYZS Forty-Seventh Annual Report* (for 1942) (New York, 1943), 4. For zoos, see Nigel Rothfels, "Paradise," in *Savages and Beasts: The Birth of the Modern Zoo* (Baltimore: Johns Hopkins University Press, 2002); and Elizabeth Hanson, *Animal Attractions: Nature on Display in American Zoos* (Princeton: Princeton University Press, 2002), 131, 152, 176.
9. William Bridges, *Gathering of Animals*, 460; and Fairfield Osborn, "On the Farm," *Animal Kingdom: Bulletin of the NYZS* 45, no. 4 (July–August 1942): 82. Also see "City Folk to See Cow at First Hand as Nation's Only Zoo Farm Opens," *NYT*, July 12, 1942.
10. Through his Audubon job, Vogt met birders from around the nation, including Roger Tory Peterson. Indeed, without Vogt's initial encouragement, Peterson might never have written his famous *A Field Guide to the Birds*, which was dedicated to Vogt. See

William Vogt, "Some Notes on WV for Mr. Best to Use as He Chooses in Connection with a Possible Article," no date (probably from mid-1950s), Box 5, FF21, VP. For Vogt, see entries in *ANB*, 1999; *Biographical Dictionary of American and Canadian Naturalists and Environmentalists*; "Vogt, William," in *Current Biography*, 1953, 638–640; Anne Becher, *American Environmental Leaders: From Colonial Times to the Present* (Santa Barbara, Cal.: ABC-CLIO, 2000), 827–828; as well as chapter 3 of Andrew Jamison and Ron Eyerman, *Seeds of the Sixties* (Berkeley: University of California Press, 1994); and Greg Cushman, "The Lords of Guano: Science and the Management of Peru's Marine Environment, 1800–1973," Ph.D. dissertation, University of Texas-Austin, 2003.

11. William Vogt, *Thirst on the Land: A Plea for Water Conservation for the Benefit of Man and Wild Life* (New York: National Audubon Society, 1937), 15. For Vogt's anti-drainage campaign, see Maureen McCormick, "Of Birds, Guano, and Man," Ph.D. dissertation, University of Oklahoma, 2005, 50–60.

12. Fairfield Osborn, *NYZS Forty-Sixth Annual Report* (for 1941) (New York, 1942): 4; *NYZS Forty-fifth Annual Report* (for 1940) (June 1, 1941): 3; Osborn, untitled editorial, *Animal Kingdom: Bulletin of the NYZS* 45, no. 2 (March–April, 1942): 33.

13. "The New York Zoological Park," *Science* 93 (May 16, 1941): 467–468; 467. Gregg Mitman makes a similar point in reference to Osborn, in his *Reel Nature: America's Romance with Wildlife on Films* (Cambridge: Harvard University Press, 1999), 86. Also see Mitman, *The State of Nature: Ecology, Community, and American Social Thought* (Chicago: University of Chicago University Press, 1992), 146–201; and Fairfield Osborn, "The Zoological Society in These Times," *Science* 93 (March 21, 1941), 269–270.

14. Fairfield Osborn, "The Lesson Not Yet Learned," *Animal Kingdom: Bulletin of the NYZS* 45, no. 5 (September–October 1942): 137; Osborn, *Our Plundered Planet*, 19; Osborn, "Conservation and War," *Animal Kingdom: Bulletin of the NYZS* 47, no. 3 (May–June 1944): 49. The turn-of-the-century conservationist John Muir also thought of humans as inherently destructive.

15. Fairfield Osborn, *The Pacific World: Its Vast Distances, Its Lands and the Life upon Them, and Its Peoples* (New York: Norton, 1944), xv. Also see Fairfield Osborn, "Genesis of a Book," *Animal Kingdom: Bulletin of the NYZS* 47, no. 2 (1944): 27–32. Quotations from Osborn, *Forty-ninth Annual Report of the NYZS* (for 1944) (New York, 1945), 4. For the origins of *Our Plundered Planet*, see Frank Rasky, "Vogt and Osborn: Our Fighting Conservationists," *Tomorrow* 9, no. 10 (June 1950): 5–9; 5.

16. Fairfield Osborn, "The Significant Future of the Zoological Society," *Animal Kingdom: Bulletin of the NYZS*, 48, no. 5 (October 8, 1945): 125. For conservation in earlier decades, see Osborn, *Fiftieth Annual Report of the NYZS* (for 1945) (New York, 1946): 4; Fairfield Osborn, *Fifty-second Annual Report of the NYZS* (for 1947), (New York, 1948), 3.

17. Vogt borrowed the phrase from his friend Frank Chapman. See William Vogt, "Best Remembered," manuscript, 1962, Box 4, FF2, VP.

18. Ibid.; "Hunger at the Peace Table," *Saturday Evening Post*, May 12, 1945, 17.

19. Rasky, "Vogt and Osborn: Our Fighting Conservationists," 7.

20. For the Vogt-Leopold relationship, see Box 3, FF12, VP. Also see the Aldo Leopold papers (LP).

21. William Vogt, "Aves Guaneras: Informe," *Boletin 18* (March 1942).

22. "Fellowship Application, Guggenheim Foundation, 1940," Box 5, FF19, VP.

23. Vogt published several reports from this research: "The Natural Resources of Mexico—Their Past, Present, and Future"; "The Population of Costa Rica and Its

Natural Resources"; "The Population of El Salvador and Its Natural Resources"; "The Population of Venezuela and Its Natural Resources" (Washington, D.C.: Pan American Union, 1946). Quotation from *New York World-Telegram*, January 18, 1946. For this stage of Vogt's career, see McCormick, "Of Birds, Guano, and Man," chapter 3.

24. Jay N. Darling, "It's Going to Be Universal Training for These, Anyway," 1917; "The Most Critical Race in History," 1917; "Time to Take an Inventory of Our Pantry," 1936; "Speaking of Labor Day," 1939; University of Iowa Special Collections (Iowa City, Iowa); Jay N. Darling, *Poverty or Conservation, Your National Problem* (Washington, D.C.: National Wildlife Federation, n.d.), 8.

25. Guy Irving Burch and Elmer Pendell, *Population Roads to Peace or War* (Washington, D.C.: Population Reference Bureau, 1945), 21; Vogt, *Road to Survival*, xvi. For more on Burch, see Eric B. Ross, *The Malthus Factor: Population, Poverty, and Politics in Capitalist Development* (New York: Zed Books, 1998), 71. Gifford Pinchot also wrote frequently to President Franklin Roosevelt about war and nature, including population pressures. See Gifford Pinchot to FDR, May 1945, Clark Clifford Papers, Presidential Speech File, Box 38, Point IV-Miscellaneous Folder, HSTL. Also see Char Miller, *Gifford Pinchot and the Making of Modern Environmentalism* (Washington, D.C.: Island Press, 2001), 372–376.

26. Fairfield Osborn, "The Urgency of Conservation Education," speech, annual convention of the National Association of Biology Teachers, March 30, 1946, "The Urgency of Conservation Education" folder, Box 1, RG 2, OPNYZ, 5; Vogt, "Hunger at the Peace Table."

27. Matthew Connelly, *Fatal Misconception: The Struggle to Control World Population* (Cambridge: Harvard University Press, 2008), 119; John McNeill, *Something New under the Sun: An Environmental History of the Twentieth-Century World* (New York: Norton, 2000), 17.

28. Vandenberg's Statement, President's Committee on Foreign Aid, 1947, Box 5, HSTL; Herbert Hoover, "Testimony before the Senate Appropriations Committee," *CR*, June 19, 1947, 7415; "Before Breakdown," Hempstead, N.Y. *Newsday*, October 19, 1948, Osborn Scrapbook, Container 10, OPLOC; Statement by the President on Receiving Secretary Krug's Report "National Resources and Foreign Aid," October 18, 1947, in John T. Woolley and Gerhard Peters, *The American Presidency Project* (online), Santa Barbara: University of California (host), Gerhard Peters (database), http://www.presidency.ucsb.edu/ws/?pid=12777. Also see Memo, July 9, 1947, President's Committee on Foreign Aid, 1947, Box 5, HSTL; Statement by the President, September 25, 1947, President's Committee on Foreign Aid: Scope of Recommendations folder, President's Committee on Foreign Aid, 1947, Box 1, HSTL; Statement of Secretary Krug before the Foreign Affairs Committee of the House on the European Recovery Program, January 22, 1948, House Foreign Affairs Committee folder, Box 66, JKP.

29. Osborn, *Our Plundered Planet*, 33–35; Vogt, *Road to Survival*, 15.

30. Osborn, *Our Plundered Planet*, 43, 41; Vogt, *Road to Survival*, 16.

31. Osborn, *Our Plundered Planet*, 34; Vogt, *Road to Survival*, 15, 207, and 15.

32. Vogt, *Road to Survival*, 299, 80.

33. Osborn, *Our Plundered Planet*, 50–53; Vogt, *Road to Survival*, 42, italics in original. Osborn thanked Bennett for reading drafts of *Our Plundered Planet*.

34. Osborn, *Our Plundered Planet*, 76, 161. Osborn and Leopold began corresponding in the early 1940s. In March 1947, Osborn invited Leopold to be on the advisory board for the

new conservation organization he was developing, the Conservation Foundation. See Christopher Lewis, "Progress and Apocalypse: Science and the End of the Modern World," Ph.D. dissertation, University of Minnesota, 1991, 70–71.
35. Vogt, *Road to Survival*, 90.
36. Osborn, *Our Plundered Planet*, ix; Vogt, *Road to Survival*, 72–73, 193.
37. Alexander Carr-Saunders, *The Population Problem: A Study in Human Evolution* (Oxford: Oxford University Press, 1922); and Peder Anker, *Imperial Ecology: Environmental Order in the British Empire* (Cambridge: Cambridge University Press, 2002), 101.
38. Osborn, *Our Plundered Planet*, 39, 41.
39. Osborn, *Our Plundered Planet*, viii, ix; Vogt, *Road to Survival*, 54 and 286.
40. Vogt, *Road to Survival*, 86, 53 and 273; ibid., chapter 2, "Energy from Earth to Man," 18–44; Osborn, *Our Plundered Planet*, epigraph, 48. It's not known whether Vogt had read the limnologist Ray Lindeman's 1942 article on energy flows. See Lindeman, "The Trophic-Dynamic Concept in Ecology," *Ecology* 23 (1942): 399–418.
41. Osborn, *Our Plundered Planet*, 198; George L. Peterson, "Conservation May Decide American-Russian Contest," unidentified newspaper and unknown date, 1948 Scrapbook, Container 9, OPLOC; Vogt, *Road to Survival*, 148, 41, 41, 44, 44, 165.
42. Richard Ely, "Conservation and Economic Theory," in *The Foundations of National Prosperity* (New York: Macmillan, 1917), 6; "National Academy of Sciences Conference on Population Problems: Morning Session, Friday, June 20, 1952" (Williamsburg Conference, 1952), Folder 720, Box 85, Subseries 5, Series 1, RG 5, JDR, 13.
43. Osborn, *Our Plundered Planet*, 43; Vogt, *Road to Survival*, 146.
44. Vogt, *Road to Survival*, 37, 34, 133, 285. Vogt told *Tomorrow*, "Even though you are on record as bitterly anti-communist, Washington is ready to misinterpret any criticism you might make." Frank Rasky, "Vogt and Osborn: Our Fighting Conservationists," *Tomorrow*, 7. Also see Lewis, "Progress and Apocalypse."
45. Osborn, *Our Plundered Planet*, 192–193; Vogt, *Road to Survival*, 265, 74.
46. Vogt, *Road to Survival*, 282, 280, 211, 280. Osborn called for birth control several years later in Fairfield Osborn, *The Limits of the Earth* (Boston: Little, Brown, 1953).
47. Critics include Allan Chase, *The Legacy of Malthus: The Social Costs of the New Scientific Racism* (New York: Knopf, 1977); Connelly, *Fatal Misconception*; Fred Pearce, *The Coming Population Crash and Our Planet's Surprising Future* (Boston: Beacon, 2010).
48. Vogt, *Road to Survival*, 13, 77, 13, 15.
49. Vogt, *Road to Survival*, 210, 211, 281. Emphasis in the original. Osborn also worried about U.S. assistance to Greece. See Osborn, *Our Plundered Planet*, 108.
50. Vogt, *Road to Survival*, 282, 224, 227, 197, 216, 212–213. Vogt's view of Asia, however, was not without ambivalence. At times, Vogt spoke of Asia as home to several of the world's great cultures; at other times he equated Asian civilizations, because of their simplicity of structure, to a sponge.
51. Vogt, *Road to Survival*, 282–283.
52. Ibid., 76, 152, 161.
53. Vogt to Leopold, December 15, 1940, Box 3, "Vogt" folder, LP; Vogt to Leopold, June 11, 1941, Box 3, "Vogt" folder, LP.
54. Vogt, *Road to Survival*, 162, 264.
55. Osborn, *Our Plundered Planet*, 34, 24, 25. For universalism, see David Hollinger, *Postethnic America: Beyond Multiculturalism* (New York: Basic Books, 1995), 51–57;

William Graebner, *The Age of Doubt: American Thought and Culture in the 1940s* (Boston: Twayne, 1990), 69–70.
56. Osborn, *Our Plundered Planet*, 19–21. It is possible that Osborn was building upon Aldo Leopold's views of violence in "A Biotic View of the Land."
57. Vogt, *Road to Survival*, 285.
58. Graebner, *Age of Doubt*; James Bernard Kelley, "The Motherless Earth," *New York City New Leader*, August 21, 1948, 1948 Scrapbook, OPLOC; Christopher Morley quoted in ad, "The Book That Can Change Our World," *NYT Book Review*, September 19, 1948.
59. Among others, *Look, Life, Vogue,* the *Baltimore Sun*, the *New Yorker,* and the *Saturday Review of Literature* gave *Our Plundered Planet* endorsements. For positive reviews of Vogt, see the *New Yorker*, August 21, 1948, 78–79; Vance Johnson, "Frightening Look at the Planet's Future," and San Francisco *Chronicle*, August 8, 1948; "Coming: A Hungry 25 Years," *Life*, April 26, 1948, 30; Russell Lord, "The Ground from under Your Feet," *Saturday Review of Literature*, August 7, 1948, 1948 Scrapbook, OPLOC; *Financial Times of London*, January 10, 1949.
60. *Atlantic Monthly*, April 1948, 71–76; William Vogt, "Road to Survival," *Readers Digest* 54 (January 1949), 139–160; "Look Applauds Fairfield Osborn," *Look*, July 20, 1948.
61. "No Man Is an Island," *NYHT*, October 19, 1948, 1948 Scrapbook, OPLOC; "Human ingenuity" quote from Paul Hoffman's advance notes in "Our Imperiled Resources: Discussion Guide on Our Renewable Natural Resources," *NYHT* booklet, 1948, found in "Organizational Material Mentioning Fairfield Osborn," Container 8, OPLOC; "Osborn Says We Must Guard Resources, or Face Socialism," *NYHT Forum Section*, October 24, 1948; "Speeches—Our Plundered Planet" folder, Box 1, RG 2, OPNYZS.
62. Harry S. Truman, "Inaugural Address," January 20, 1949, *Public Papers of the Presidents of the United States, Containing the Public Messages, Speeches, and Statements of the President, Harry S. Truman, January 1 to December 31, 1949* (Washington, D.C.: Government Printing Office, 1964), 112–116.
63. "Concern with resource development" quotation from Outline of Remarks of Secretary Krug for Briefing Session of Participants to the UN Scientific Conference on the Conservation and Utilization of Resources, n.d., U.N. Scientific Conference folder, Box 76, JKP; *Proceedings of the United Nations Scientific Conference on the Conservation and Utilization of Resources*, vol. 1 (Lake Success, N.Y.: United Nations Department of Economic Affairs, 1950), 7; Krug to Acheson, November 30, 1949, U.N. Scientific Conference folder, Box 76, JKP; *Proceedings of the United Nations Scientific Conference on the Conservation and Utilization of Resources*, vol. 1, 6.
64. William Vogt, "Point Four Propaganda and Reality," manuscript for *American Perspective*, 1950, Box 3, Folder 35, VP; William Vogt, "Let's Examine Our Santa Claus Complex," *Saturday Evening Post*, July 23, 1949, 17–19; "Address by Fairfield Osborn at Joint Invitation of Secretary of Interior and President of Carnegie Institution, February 9, 1949," Container 1, "Speeches, 1946–1949" folder, OPLOC. After Osborn again critiqued Point Four programs in his 1953 book, President Truman sent a response: "Wish I could discuss some Point IV programs with you that *will meet* the situation." Truman to Edwin Seaver, Little Brown and Company, December 16, 1953, "Correspondence, 1924–1969," Container 1, OPLOC.
65. President's Materials Policy Commission, *Resources for Freedom*: vol. 1, *Foundation for Growth and Security* (Washington, D.C.: Government Printing Office, 1952), 21, 169.
66. *Partners in Progress: A Report to President Truman by the International Development Advisory Board* (New York: Simon and Schuster, 1951), 1, 9, 12, 4, and 8. For another

Point Four report that acknowledges population growth but sees economic growth as more important, see Office of Public Affairs, *Point Four: Cooperative Program for Aid in the Development of Economically Underdeveloped Areas*, Booklet, Department of State, January, 1950. Reproduced in Dennis Merrill, ed., *Documentary History of the Truman Presidency*, vol. 27, *The Point Four Program: Reaching Out to Help the Less Developed Countries* (Frederick, Md.: University Publishers of America, 1999).

67. Harry S. Truman, press release, April 18, 1951, in Merrill, ed., *The Point Four Program*, 589; Harry S. Truman, *Off the Record: The Private Papers of Harry S. Truman* (New York: Harper & Row, 1980), 264; emphasis in original.

68. Richard H. Pough, "Book Notes," *Audubon* 50 (May–June, 1948); "Books You Should Read," *Outdoor America*, no date, 1948 Scrapbook, OPLOC; "Forgotten the Earth," *The Living Wilderness* (Autumn 1948), Container 10, OPLOC; Paul Sears, *NYHT Book Review*, 1948 Scrapbook, OPLOC; Harold Ickes, "Man to Man," *New York Post*, September 26, 1948; Bernard DeVoto, "Crisis of Man in Relation to His Environment," *NYHT Weekly Book Review*, August 8, 1948. For DeVoto, see Bernard De Voto to Fairfield Osborn, April 6, 1948, "Correspondence, 1924–1969," Container 1, OPLOC. For the professor's comments, see J. Russell Whitaker to William Vogt, July 27, 1948, Box 5, FF9, VP. For Leopold quote, see Thomas R. Dunlap, *Saving America's Wildlife* (Princeton: Princeton University Press, 1988), 99; Leopold to William Vogt, December 8, 1947, Box 3, "Vogt" folder, LP.

69. "Forty Thousand Frightened People," *New Republic*, October 4, 1948, 8–10; Sinnott quoted in Herbert Nichols, "Greed Held Check to Stretching Natural Resources," *Christian Science Monitor*, September 15, 1948.

CHAPTER 3   ABUNDANCE IN A SEA OF POVERTY

1. Paul R. Ehrlich, *The Population Bomb* (New York: Ballantine Books, 1968), 11. Emphasis added.
2. Ibid., 133.
3. *Partners in Progress: A Report to President Truman by the International Development Advisory Board* (New York: Simon and Schuster, 1951), 11; John Kenneth Galbraith, *The Nature of Mass Poverty* (Cambridge: Harvard University Press, 1979), 29, quoted in Nick Cullather, "Development: Its History," *DH* 24 (Fall 2000): 641–653; 641. For the third world, see H. W. Brands, *The Specter of Neutralism: The United States and the Emergence of the Third World, 1947–1960* (New York: Columbia University Press, 1989); Thomas McCormick, *America's Half-Century: United States Foreign Policy in the Cold War and After* (Baltimore: Johns Hopkins University Press, 1995), chapter 5 and pp. 135–139; John Lewis Gaddis, *Strategies of Containment: A Critical Appraisal of Postwar American National Security Policy* (New York: Oxford University Press, 1982), 176–180; Jeremi Suri, *Power and Protest: Global Revolution and the Rise of Détente* (Cambridge: Harvard University Press, 2003), 133, 245–246; and Odd Arne Westad, *The Global Cold War: Third World Interventions and the Making of Our Times* (Cambridge: Cambridge University Press, 2007).
4. Carl Sauer, "The Agency of Man on Earth," in William Thomas, *Man's Role in Changing the Face of the Earth* (Chicago: University of Chicago Press, 1956), 68.
5. Walter Prescott Webb, *The Great Frontier* (Boston: Houghton Mifflin, 1952), 13; Walter Prescott Webb, "Ended: Four Hundred Year Boom," *Harper's* 103, no. 1217 (October 1951): 25–35; David Potter, *People of Plenty: Economic Abundance and the American*

Character (Chicago: University of Chicago Press, 1954), 161–163, 89, 157. As will be discussed in chapter 4, Potter's views laid an intellectual foundation for the "modernization" programs in President Kennedy's "New Frontier" administration.

6. U.S. Census Bureau, International Database. Also see "Transitions in World Population," *PB*, March 2004, 3–40; 5. Also World Population Division, *World Population Prospects: The 2002 Revision* (New York: United Nations Population Division, 2003); and Lester Brown, *World without Borders* (New York: Random House, 1972), 5.

7. Potter, *People of Plenty*, 162, 164.

8. Most of the literature on the population movement in the 1950s focuses on fertility control and Cold War politics. Without overlooking those concerns, this chapter examines the population movement from the perspective of natural resources.

9. The Paley Commission issued its report in June 1952, but all the pieces were in place by January 1952. Quotation from John Harr and Peter J. Johnson, *The Rockefeller Century* (New York: Scribner, 1988), 459. For Rockefeller, see *ANB*, vol. 18 (1999): 700–701; and especially John Harr and Peter Johnson, *The Rockefeller Conscience: An American Family in Public and in Private* (New York: Scribner, 1991). Rockefeller gave $10,000 to the Conservation Foundation during its first two years. He gave IPPF $2,500 for its inaugural meeting in England in 1948. He was one of Planned Parenthood's largest donors. In 1947 and 1948, he gave a total of $22,000. For a summary of Rockefeller's gifts in the population area before 1952, see "Appendix A: Agencies and Services Currently Active in the Field of Population" 1952, Folder 667, Box 80, RG 5, Series 1, Subseries 5, JDR. The best secondary source on the Office of Population Research (OPR) is Simon Szreter, "The Idea of Demographic Transition and the Study of Fertility Change: A Critical Intellectual History," *PDR* 19, no. 4 (December 1993): 659–701; 676–678. Also see the Rockefeller Foundation's Annual Report from 1944, which describes the foundation's grant to the OPR; Frank Notestein, "The Office of Population Research," Rockefeller Foundation Confidential Monthly Report, no. 86, June 1, 1946, RFP; and "Princeton University Population Research, 1943–1945 "folder, A82, RFP; "The Population Council, 1952–1964, revised draft, March 15, 1965," Folder 2366, "Twelve Year Report," Box 128 Population Council Organization File, PCP.

10. "Appendix A: Agencies and Services Currently Active in the Field of Population" 1952, Folder 667, Box 80, RG 5, Series 1, Subseries 5, JDR. For interwar demography, see Derek Hoff, *The State and the Stork: The Population Debate and Policymaking in United States History* (Chicago: University of Chicago Press, forthcoming).

11. "National Academy of Sciences Conference on Population Problems: Morning Session, Friday, June 20, 1952," Folder 720, Box 85, Subseries 5, Series 1, RG 5, JDR, 2.

12. Matthew Connelly, *Fatal Misconception: The Struggle to Control World Population* (Cambridge: Harvard University Press, 2008), 157–158, 160.

13. Frank Notestein, "Population—the Long View," in *Food for the World*, ed. Theodore W. Schultz (Chicago: University of Chicago Press, 1945); Kingsley Davis, "The World Demographic Transition," *Annals of the American Academy of Political and Social Science* 237 (January 1945): 1–11. For a summary of transition theory, see Avery Guest and Gunnar Almgren, "Demographic Transition," in Edgar F. Borgatta and Rhonda J. V. Montgomery, *Encyclopedia of Sociology*, 2nd ed. (New York: Macmillan Reference USA, 2000). Also see Dennis Hodgson, "Demography as Social Science and Policy Science," *PDR* 9, no. 1 (March 1983): 1–34; Hodgson, "Orthodoxy and Revisionism in American Demography," *PDR* 14, no. 4 (December 1988): 541–569; and Simon Szreter, "The Idea

of Demographic Transition and the Study of Fertility Change: A Critical Intellectual History," *PDR* 19, no. 4 (December 1993): 659–701.

14. Szreter, "The Idea of Demographic Transition and the Study of Fertility Change," 665–666. Szreter draws from Dennis Hodgson, "The Ideological Origins of the Population Association of America," *PDR* 17, no. 1 (1991): 664.

15. Frank Notestein, "Summary of the Demographic Background of Problems of Underdeveloped Areas," in *International Approaches to Problems of Underdeveloped Areas* (New York: Milbank Memorial Fund, 1948), 13; and Notestein, "The Population of the World in the Year 2000," *American Statistical Association Journal* 45 (September 1950): 335–345; 343; Kingsley Davis, "Population and Change in Backward Areas," *Columbia Journal of International Affairs* 4, no. 2 (Spring 1950): 41–51; 49. All quoted in Dennis Hodgson, "Demography as Social Science and Policy Science," *PDR* 9, no. 1 (March 1983): 1–34; 24, 25.

16. Hodgson, "Demography as Social Science and Policy Science"; Frederick Osborn, "Population: An International Dilemma" (Princeton: Princeton University Press, 1958), reprinted in *Three Essays on Population: Thomas Malthus, Julian Huxley, Frederick Osborn* [1958] (New York: Mentor Books, 1960), 119.

17. Ansley J. Coale and Edgar Hoover, *Population Growth and Economic Development in Low-Income Countries: A Case Study of India's Prospects* (Princeton: Princeton University Press, 1958).

18. John Sharpless, "Population Science, Private Foundations, and Development Aid: The Transformation of Demographic Knowledge in the United States, 1945–1965," in Frederick Cooper and Randall M. Packard, *International Development and the Social Sciences: Essays on the History and Politics of Knowledge* (Berkeley: University of California Press, 1997), 183; Osborn, "Population: An International Dilemma," 122, 119.

19. "Documentation for Conference on Population Problems," Folder 718, Box 85, RG 5, Series 1, Subseries 5, JDR; "National Academy of Sciences Conference on Population Problems: Morning Session, Friday, June 20, 1952," Folder 720, Box 85, RG 5, Series 1, Subseries 5, JDR; "National Academy of Sciences Conference on Population Problems: Resolution Adopted by the Members of the Conference," June 22, 1952, Folder 667, Box 80, RG 5, Series 1, Subseries 5, JDR. Also see Connelly, *Fatal Misconception*, 157. The conference transcript reveals that the workbook was written by Dr. Hajnal. The biologist Marston Bates also pushed ecological thinking within the Rockefeller Foundation in the late 1940s.

20. During the 1920s, Sanger aligned with different eugenicists. Historians disagree about whether this was a reluctant, pragmatic alliance or a more sincere concern, but what is clear is that Sanger's Malthusian views were long-standing and deep, lasting from the 1920s into the 1960s. A good overview is Dennis Hodgson and Susan Watkins, "Feminists and NeoMalthusians: Past and Present Alliances," *PDR* 23 (September 3, 1997): 469–523. Also see James Reed, *From Private Vice to Public Virtue: The Birth Control Movement and American Society since 1830* (New York: Basic Books, 1978); Esther Katz, "Margaret Sanger," in *ANB*, 264–268; Ellen Chesler, "Margaret Sanger, 1879–1966," in *Research Guide to American Historical Biography*, ed. Robert Muccigrosso and Suzanne Niemeyer (Washington, D.C.: Beacham, 1988); Ellen Chesler, *Woman of Valor: Margaret Sanger and the Birth Control Movement in America* (New York: Simon & Schuster, 2007); Connelly, *Fatal Misconception*, 50–53.

21. Sanger quoted in Selma Robinson and Anne L. Goodman, "Man's Dilemma: Too Many Mouths, Too Little Food," *New York Star*, September 5, 1948, 1948 Scrapbook, OPLOC; Hodgson and Watkins, "Feminists and NeoMalthusians," 484.
22. For the IPPF, see correspondence between Hugh Moore and Tom Greissemer, Box 1, Folder 23, HMF; Beryl Suitters, *Be Brave and Angry: Chronicles of the International Planned Parenthood Federation* (London: International Planned Parenthood Federation, 1973); and Connelly, *Fatal Misconception*, 166–168.
23. For Planned Parenthood's public education campaigns, see its *Annual Report 1955*, PPFA, Inc, Box 101, PPFA. For the quotation on 85 million babies, see "By Choice instead of by Chance," 1953, Folder 40, Box 92, PPFA. For the Vox Population quotation, see *Planned Parenthood News*, June 1950, Planned Parenthood Association Chicago Area, MSP. For the "international population" quotation, see *Planned Parenthood News*, May 1960, Planned Parenthood Association Chicago Area, PPFA.
24. Eleanor B. Pillsbury, President, "Greetings!" *Planned Parenthood News*, Fall 1952, New York PPFA, PPFA II. For the IPPF and natural resources, see "Stockholm Conference," *Planned Parenthood News*, Fall 1953, 1; and Suitters, *Be Brave and Angry*, 131.
25. "By Choice Instead of by Chance," Folder 40, Box 92, PPFA II; "Stockholm Conference," *Planned Parenthood News* (Fall 1953), 1.
26. *Conservation Foundation 1956 Annual Report*, Annual Reports, 1949–1957, CFP. For the research by Davis and the others, see *New York Zoological Society Annual Report 1955* (New York, 1956), 55. Fairfield Osborn, *Our Crowded Planet: Essays on the Pressures of Population* (Garden City, N.Y.: Doubleday, 1962), 7. In 1961, Osborn gave the keynote address to the International Union for the Conservation of Nature, the world's leading international conservation organization. The papers of the Conservation Foundation are split between the New York Zoological Society, the Fairfield Osborn collection at the Library of Congress, and its successor organization, the World Wildlife Fund.
27. Eugene Odum, *Fundamentals of Ecology* (Philadelphia: Saunders, 1953), 3.
28. Ibid., 348–349. Human ecology was a growing field at the time. See Odum's citations and Frank Fraser Darling, "Ecological Approach to the Social Sciences," *American Scientist* 39 (1951): 244–254.
29. Harrison Brown, *The Challenge of Man's Future: An Inquiry Concerning the Condition of Man during the Years That Lie Ahead* (New York: Viking, 1954), xi. For Brown, see Roger Revelle, "Harrison Brown," *Biographical Memoirs* (Washington, D.C.: National Academy of Sciences, 1994), 41–55; and Harrison Brown et al., *Earth and the Human Future: Essays in Honor of Harrison Brown* (Boulder, Col.: Westview, 1986). Many biology textbooks at this time also spoke of human population growth. An important example is Garrett Hardin, *Biology: Its Human Implications* (San Francisco: W. H. Freeman, 1949).
30. Brown, *Challenge of Man's Future*, 263, 105, 103, 105, 263.
31. Robert Cushman Murphy and Charles Hitchcock, "Conference on Man's Role in Changing the Face of the Earth," *Geographical Review* 45, no. 4 (October 1955): 583–586; 585; Thomas, *Man's Role*, xxxvi, 49–50.
32. Ibid., 66, 68, 1030, 937, 937, 1118.
33. Ibid., 433, 1119, 1126.
34. Ibid., 1087.
35. Julian Huxley, "World Population," *Scientific American*, March 1956, reprinted in *Three Essays on Population: Thomas Malthus, Julian Huxley, Frederick Osborn* [1958] (New York: Mentor Books, 1960), 67, 79, 80, 81, 67, and 73.

36. Dasmann's *Destruction of California* is discussed in chapter 5 below.
37. Raymond Dasmann, *Environmental Conservation* (New York: Wiley, 1959), vii, 79, 270, 267.
38. Ibid., 276–277. Also see 216–219.
39. Ibid., 270, 292, 280, 282; Raymond Dasmann, *The Last Horizon* (New York: Macmillan, 1963), 222.
40. Dasmann, *Environmental Conservation*, 294, 280; Dasmann, *The Last Horizon*, 225.
41. Dasmann, *Environmental Conservation*, 293; Dasmann, *The Last Horizon*, 221.
42. Many books on birth control, even otherwise careful accounts, overlook or downplay the connection with concern about overpopulation. See, for example, Bernard Asbell, *The Pill: A Biography of the Drug that Changed the World* (New York: Random House, 1995). For the Malthusianism of the developers of the birth control pill, see Moira Davison Reynolds, "Katherine McCormick," in *Women Advocates of Reproductive Rights* (Jefferson, N.C.: McFarland, 1994), 107–116; Gregory Pincus, *The Control of Fertility* (New York: Academic Press, 1965), preface, viii; John Rock, *The Time Has Come: A Catholic Doctor's Proposals to End the Battle over Birth Control* (New York: Knopf, 1963); and James Reed, *From Private Vice to Public Virtue: The Birth Control Movement and American Society since 1830* (New York: Basic Books, 1978). Rock quotation from Planned Parenthood press release, May 6, 1954, Box 66, PPFA. *Estelle T. Griswold and C. Lee Buxton v. Connecticut*, 381 U.S. 479 (1965).
43. "National Academy of Sciences Conference on Population Problems: Morning Session, Friday, June 20, 1952," Folder 720, Box 85, Subseries 5, Series 1, RG 5, JDR; John D. Rockefeller III, "Proposed Establishment of Population Council," November 20, 1952, adopted at the first board meeting on the same date, quoted in Harr and Johnson, *The Rockefeller Conscience*, 38.
44. "The Population Council, 1952–1964, Revised Draft, March 15, 1965," Folder 2366 "Twelve Year Report," Box 128, Population Council Organization File, PCP, 1. Emphasis added.
45. Fairfield Osborn, *The Limits of the Earth* (Boston: Little, Brown, 1953), 76; Fairfield Osborn, "The Quality of Life," *Around the World: News of Population and Birth Control* (IPPF, May 1954); William Vogt, *Road to Survival* (New York: W. Sloane Associates, 1948), 80; Odum, *Fundamentals*, 350; Huxley, "World Population," 67; Joseph Wood Krutch, "A Naturalist Looks at Overpopulation," in Osborn, *Our Crowded Planet*, 207–214; 211.
46. Brown, *Challenge of Man's Future*, 259, 219; Osborn quotation in Thomas, *Man's Role*, 1086; Lynton Caldwell, "Politics and Population Control: A Report to the Conservation Foundation," February 20, 1966, Box 8, FF10, VP, 13; Dasmann, *The Last Horizon*, 246.
47. Samuel P. Hays, *Beauty, Health, and Permanence: Environmental Politics in the United States, 1955–1985* (New York: Cambridge University Press, 1987), 22.
48. Dasmann, *Environmental Conservation*, 7, 256. Emphasis in original.
49. Ibid., 7. Emphasis added.
50. Some historians have argued that Leopold's emphasis on land health carried eugenic overtones. See Gregg Mitman, "In Search of Health: Landscape and Disease in American Environmental History," *EH* 10, no. 2 (2005): 184–209.
51. Ibid., 256, 257, 262, 262, 256, 258.
52. "Wilderness Conference," *SCB* 44 (April 1959): 12, quoted in quoted in Michael P. Cohen, *The History of the Sierra Club, 1892–1970* (San Francisco: Sierra Club Books, 1988), 232.

53. John Kenneth Galbraith, *The Affluent Society* (Boston: Houghton Mifflin, 1958), 1; John Kenneth Galbraith, *A Life in Our Times: Memoirs* (Boston: Houghton Mifflin, 1981), chapter 21, "First View of India and 'The Affluent Society'"; Richard Parker, *John Kenneth Galbraith: His Life, His Politics, His Economics* (New York: Farrar, Straus, and Giroux, 2005), 276–280.
54. Galbraith, *Affluent Society*, 253, 355. See Adam Rome, "Give Earth a Chance: The Environmental Movement and the Sixties," *JAH* 90, no. 2 (2003): 525–554; and John Kenneth Galbraith, "How Much Should a Country Consume?" in Henry Jarrett, ed., *Perspectives on Conservation: Essays on America's Natural Resources* (Baltimore: Johns Hopkins University Press, 1958), 89–99.
55. Stuart Udall, *Quiet Crisis* (New York: Henry Holt, 1963), 186.
56. Richard N. Gardner, "Remarks by Richard Gardner," *PB* 21, no. 6 (December 1965): 141–143, 142.

CHAPTER 4   "FEED 'EM OR FIGHT 'EM"

1. For an overview of these changes, see Donald T. Critchlow, *Intended Consequences: Birth Control, Abortion, and the Federal Government in Modern America* (New York: Oxford University Press, 1999), chapter 2. For Johnson's policy shifts, see Phyllis Piotrow, *World Population Crisis: The U.S. Responds* (New York: Praeger, 1973), chapter 10. Also see note 5 below.
2. William Draper, "Parks—or More People?" *NPM* 40 (April 1966): 10–13. For more on parks and ghettoes, see the next chapter.
3. Ibid.
4. Kennedy's quotation is from Richard J. Barnet, *Intervention and Revolution: The United States in the Third World* (New York: World, 1968), 27, as quoted in Dennis Merrill, *Bread and the Ballot: The United States and India's Economic Development, 1947–1963* (Chapel Hill: University of North Carolina Press, 1990), 169.
5. To make this argument, I am building on the work of historians who have connected concern with population growth to the Cold War, such as James Reed, Eric Ross, Donald Critchlow, Ronald Greene, Björn-Ola Linnér, and Matthew Connelly. In addition to differing from them on some of the details, this chapter spells out the relationship between modernization programs and population programs, especially from the perspective of Malthusian environmentalists. James Reed, *From Private Vice to Public Virtue: The Birth Control Movement and American Society since 1830* (New York: Basic Books, 1978); Eric B. Ross, *The Malthus Factor: Population, Poverty, and Politics in Capitalist Development* (New York: Zed Books, 1998); Donald T. Critchlow, *Intended Consequences: Birth Control, Abortion, and the Federal Government in Modern America* (New York: Oxford University Press, 1999); Ronald Greene, *Malthusian Worlds: US Leadership and the Governing of the Population Crisis* (Boulder, Col.: Westview, 1999); Björn-Ola Linnér, *The Return of Malthus: Environmentalism and Post-War Population-Resource Crises* (Isle of Harris, UK: White Horse Press, 2003); Matthew Connelly, *Fatal Misconception: The Struggle to Control World Population* (Cambridge: Harvard University Press, 2008).
6. For Moore, see Critchlow, *Intended Consequences*, 18, 30–31, 47, 69–70, 151–153. Lawrence Lader offers an informative, if not very objective, biography of Moore in *Breeding Ourselves to Death* (New York: Ballantine Books, 1971). For Moore's thinking before World War II, see Clark Eichelberger to Hugh Moore, May 12, 1939, Box 24,

Folder 23, HMF. For Moore on Vogt and Burch, see Moore to Malcolm Davis, November 10, 1948, Box 14, Folder 17, HMF. Guy Irving Burch and Elmer Pendell, *Human Breeding and Survival* (New York: Penguin Books, 1947). "Really waking me up" quotation from Moore to Boudreau, November 9, 1948, Box 14, Folder 17, HMF.

7. Moore mailed 1,000 copies in the fall of 1954, 20,000 copies in 1956, 50,000 copies in 1957, 100,000 copies in 1958, and 200,000 copies in 1959. See Box 16, Folder 17, HMF. Also see Critchlow, *Intended Consequences*, 31. In 1959 Moore claimed he had sent out 700,000 copies of "The Population Bomb" during the 1950s. Moore to Notestein, August 20, 1959, Folder 1893, "Hugh Moore Fund," Box 101, Population Council Organization File, PCP.

8. By the early 1950s, Vogt himself was claiming that victory in the Cold War depended on controlling population growth. William Vogt, Address to 31st Annual Meeting Luncheon of the PPFA, October 24, 1951, Box 5, FF15, VP. Quotations are from Hugh Moore, "The Population Bomb," published by the Hugh Moore Fund, New York, in over ten editions between 1954 and 1968, 1957 edition, 13.

9. Hugh Moore, "The Population Bomb," 12; Critchlow, *Intended Consequences*, 32.

10. Hugh Moore, "The Population Bomb," 14.

11. *Partners in Progress: A Report to President Truman by the International Development Advisory Board* (New York: Simon and Schuster, 1951), 11, 31. Although ultimately not Malthusian, this booklet developed ideas that Moore and others added their own conclusions to. For the domino theory, see Frank Ninkovich, *Modernity and Power: A History of the Domino Theory in the Twentieth Century* (Chicago: University of Chicago Press, 1994); and Edwin E. Moise, "The Domino Theory," *Encyclopedia of American Foreign Policy*, 2nd ed., vol. 2, ed. Alexander Deconde, Richard Burns, and Frederick Logevall (New York: Scribner, 2002), 551–559.

12. For Draper, see Critchlow, *Intended Consequences*, 41–42.

13. I rely on evidence from 1958 and 1959 or as close as possible. Moore to Draper, April 25, 1959, Box 15, Folder 28, HMF. For Draper's request for more copies, see Moore to Draper, December 19, 1958, Box 15, Folder 28, HMF; Draper quotation from Draper to Moore, February 13, 1963, Box 1, Folder 11, HMF, quoted in Critchlow, *Intended Consequences*, 42. For Moore's role in Draper's decision, see Moore's brief but revealing account of his involvement with the Draper Committee in Moore, "Draper Committee," unpublished manuscript, August 20, 1959, Box 1, HMF. For Cook's role, see Robert Cook to Frederick Osborn, Population Council, "Draper Committee" Folder 1848, Box 99, Population Council Organization File, PCP; Critchlow, *Intended Consequences*, 43. A report to the committee by Draper's assistant Robert R. Adams in May 1959 shows the influences on the Committee. Robert R. Adams, "The Population Explosion," revised draft, May 27, 1959, "Draper Committee" folder 1848, Box 99, Population Council Organization File, PCP, 4 and 7.

14. For the Eisenhower administration, see Dennis Merrill, *Bread and the Ballot: The United States and India's Economic Development, 1947–1963* (Chapel Hill: University of North Carolina Press, 1990), 98–100 and chapter 5. For the Khrushchev quotation, see Richard Goodwin, *Remember America: A Voice from the Sixties* (Boston, 1988), 99, as quoted in Robert M. Collins, *More: The Politics of Economic Growth in Postwar America* (New York: Oxford University Press, 2000), 46. "New Look at Foreign Aid," *NYT*, November 26, 1958, 28.

15. Phyllis Piotrow, *World Population Crisis*, 39; President's Committee to Study the United States Military Assistance Program, *Composite Report of the President's Committee to*

Study the United States Military Assistance Program (Washington, D.C.: August 17, 1959), 94–95, as quoted in Critchlow, *Intended Consequences*, 43. "Communist domination" quotation from President's Committee to Study the United States Military Assistance Program, Letter to the President of the United States from the President's Committee to Study the United States Military Assistance Program and the Committee's Final Report; Conclusions Concerning the Mutual Security Program (Washington, D.C., August 17, 1959), 1, as quoted in Critchlow, *Intended Consequences*, 43. "Firecracker" quotation from Moore's notes, Box 1, Folder 1, HMF.

16. Quoted in John D. Morris, "Eisenhower Bars Birth-Curb Help," *NYT*, December 3, 1959, 1. Critchlow supports my interpretation of Eisenhower. Critchlow, *Intended Consequences*, 44–45.

17. Ernest Gruening, *Many Battles: The Autobiography of Ernest Gruening* (New York: Liveright, 1973), 28.

18. Ernest Gruening, *Mexico and Its Heritage* (New York: Century, 1928); Gruening, *Many Battles*, 200. Other historians have criticized U.S. family planning programs in Puerto Rico. See Laura Briggs, *Reproducing Empire: Race, Sex, Science, and U.S. Imperialism in Puerto Rico* (Berkeley: University of California Press, 2002).

19. Ernest Gruening, "To Foster Self-Help through Foreign Aid," *NYT*, February 2, 1968, 34; Gruening, *Many Battles*, 502.

20. Fairfield Osborn, *Our Plundered Planet* (Boston: Little, Brown, 1948), 34–35; Osborn quoted in Frank Rasky, "Vogt and Osborn: Our Fighting Conservationists," *Tomorrow* 9, no. 10 (June 1950), 5–9; 9; "The Real Enemy is Hunger," *Christian Leader*, October 1948, Container 10, OPLOC.

21. Aldous Huxley, "The Politics of Ecology: The Question of Survival," Center for the Study of Democratic Institutions, 1963, Folder 13, Box 143, NP, 6–7; Dr. George Wald, Speech, March 4, 1969, *New Yorker*, March 22, 1969, 28–29; and Series 6, Folder 46, Box 26, EP; John Fischer, "How I Got Radicalized: The Making of an Agitator for Zero," *Harper's*, April 1970, 18–22; 18; Ehrlich, Transcript/Notes of "The Environmental Crisis: Where We Stand," presented at the University of Dar es Salaam, Tanzania, September 17, 1971, Series 6, Folder 31, Box 26, EP. Russian physicist Andrei Sakharov made similar statements about population and environmental problems. See "Text of Essay by Russian Nuclear Physicist Urging Soviet-American Cooperation," *NYT*, July 22, 1968.

22. Stevenson quoted in "Building Stronger Families, Happier Children: Planned Parenthood World Population Fiftieth Anniversary" 1966, Box 21, Folder 14, HMF; Thant quoted in Donald King, *The US Role in the Greening of the World*, unpublished manuscript, June 1997, Box 1, DKP, 22.

23. For Kennedy's third-world policy, see John Lewis Gaddis, *Strategies of Containment: A Critical Appraisal of Postwar American National Security Policy* (New York: Oxford University Press, 1982), 198–203; and Jeremi Suri, *Power and Protest: Global Revolution and the Rise of Détente* (Cambridge: Harvard University Press, 2003), 17–25. For the White House task force, see Merrill, *Bread and the Ballot*, 172. For the 1962 NSC Report, see *Bread and the Ballot*, 184; John F. Kennedy, "Inaugural Address," *Public Papers of the Presidents, John F. Kennedy, 1961*, vol. 1 (Washington, D.C.: Government Printing Office, 1962), 1–3.

24. David Potter, *People of Plenty: Economic Abundance and the American Character* (Chicago: University of Chicago Press, 1954). For Rostow, see Mark Haefele, "Walt Rostow's Stages of Economic Growth: Ideas and Action," in David C. Engerman, *Staging Growth: Modernization, Development, and the Global Cold War* (Amherst: University of

Massachusetts Press, 2003), 81–103; Merrill, *Bread and the Ballot*, 154–156; and Gaddis, *Strategies of Containment*, 200–201, 203, 208–210. For modernization, see Engerman, *Staging Growth*; Nils Gilman, *Mandarins of the Future: Modernization Theory in Cold War America* (Baltimore: Johns Hopkins University Press, 2003); Michael E. Latham, *Modernization as Ideology: American Social Science and "Nation Building" in the Kennedy Era* (Chapel Hill: University of North Carolina Press, 2000). Also see J. Michael Armer and John Katsillis, "Modernization Theory," in *Encyclopedia of Social Science*, ed. Edgar F. Borgotta and Mauri Borgotta (New York: Macmillan, 1992), 1883–1887.

25. W. W. Rostow, *The Stages of Economic Growth: A Non-Communist Manifesto* (Cambridge: Cambridge University Press, 1960), 4. Michael E. Latham identifies four similar characteristics in *Modernization as Ideology*, 4.

26. Rostow quote from Gerald Meier, ed., *Leading Issues in Economic Development* (New York, 1964), 601, as quoted in Dennis Merrill, *Bread and the Ballot*, 154; Arthur Schlesinger, Jr., *A Thousand Days: John F. Kennedy in the White House* (Boston: Houghton Mifflin, 1965), 522, as quoted in Merrill, *Bread and the Ballot*, 170.

27. Rostow, *Stages*, 19, 22. For modernization and the demographic transition theory, see Gilman, *Mandarins of the Future*, 105.

28. Rostow, *Stages*, 19, 10.

29. See Phyllis Piotrow, *World Population Crisis*, chapter 4; Critchlow, *Intended Consequences*, 41–45; and Connelly, *Fatal Misconception*, 196–197.

30. For American Cold War strategy and India, see H. W. Brands, *The Specter of Neutralism: The United States and the Emergence of the Third World, 1947–1960* (New York: Columbia University Press, 1989), 13–140; Robert J. McMahon, *The Cold War on the Periphery: The United States, India, and Pakistan* (New York: Columbia University Press, 1994); Merrill, *Bread and the Ballot*; and Andrew J. Rotter, *Comrades at Odds: The United States and India, 1947–1964* (Ithaca: Cornell University Press, 2000). For Eisenhower's trip to India, see Merrill, *Bread and the Ballot*, 151. For the NSC policies, see Merrill, *Bread and the Ballot*, 139. For U.S. aid to India under Kennedy, see *Bread and the Ballot*, 205.

31. For India's importance to the United States, see Merrill, *Bread and the Ballot*, 1–3. Quotation from NSC 5909, "U.S. Policy toward South Asia," July 22, 1959, OSANSA, NSC Series, Box 68, WHO, as cited in *Bread and the Ballot*, 150.

32. The literature on specific applications of modernization programs is best developed for Vietnam. See Gaddis, *Strategies of Containment*, 224; and Suri, *Power and Protest*, 137–146. For Millikan on India, see Arthur Smithies to Draper Committee Members, February 25, 1959, enclosing Millikan, "India," February 20, 1959, Box 2, Records of U.S. President's Committee to Study the Military Assistance Program, Eisenhower Library, as quoted in Merrill, *Bread and the Ballot*, 155. For Nehru, see *Bread and the Ballot*, 161.

33. Merrill, *Bread and the Ballot*, 156 and 176.

34. John Kenneth Galbraith, *A Life in Our Times: Memoirs* (Boston: Houghton Mifflin, 1981), 328.

35. For the crisis, see B. M. Bhatia, *Famines in India: A Study in Some Aspects of the Economic History of India, 1860–1990*, 3rd rev. ed. (Delhi: Konark Publishers, 1990); and Nick Cullather, *The Hungry World: America's Encounter with Rural Asia* (Cambridge: Harvard University Press, 2010). Cullather argues that Lyndon Johnson exaggerated the problem for political leverage.

36. Director general and Ewell quoted in Lawrence Lader, *Breeding Ourselves to Death* (Santa Ana, Cal.: Seven Locks Press, 1971), 25. Revelle quotation from J. Anthony Lukas,

"45 Million in India Face Lean Fare," *NYT*, March 27, 1966, 1; "Johnson vs. Malthus," *NYT*, January 24, 1966, 29. For other journalists' take on how economic development in much of Asia was faltering, see Seymour Topping, "Asia's Rocky Road to Stability," *NYT*, January 18, 1965, 32; and James Reston, "Washington: Fight 'Em or Feed 'Em," *NYT*, February 11, 1966, 32. See also headlines from *NYT*, August 7, 1965; *NYT*, December 10, 1965, 46; *NYT* March 27, 1966; *NYT*, October 14, 1966, 42; *NYT*, April 1, 1966, 1; *NYT*, December 9, 1965, 3. Ronald Segal, *The Crisis of India* (London: J. Cape, 1965), 171–180.

37. Piotrow, *World Population Crisis*, 103. Also see Reed, *From Private Vice to Public Virtue*, 377–378.

38. Johnson U.N. quotation from Connelly, *Fatal Misconception*, 213; Lyndon B. Johnson, "Special Message to the Congress Proposing an Emergency Food Aid Program for India, March 30, 1966," *Public Papers of the Presidents, Lyndon Baines Johnson*, vol. 1 (Washington, D.C.: Government Printing Office, 1967), 366; 70 million quote from "Text of Johnson's Message to Congress Urging Action on Food Help for India," *NYT*, February 3, 1967, 12; Rostow, *Diffusion of Power*, 422, as quoted in Nick Cullather, "Miracles of Modernization," *DH* 28, no. 2 (2004): 227–254; 247; "America's job" quote from "Text of President's Omaha Address on Vietnam and World Food Shortages," *NYT*, July 1, 1966, 11. For the Indian famine and the U.S. relief effort, see Lester R. Brown, *Seeds of Change: The Green Revolution and Development in the 1970s* (New York: Praeger, 1970), 7.

Historians disagree why Johnson first decided to take up the population issue in 1965. Donald Critchlow emphasizes the importance of lobbying by John D. Rockefeller III and General Draper on Secretary of State Dean Rusk. Connelly downplays the 1965 statement about population and instead sees comments later that year as more important; he attributes Johnson's change to new economic cost/benefit arguments put forward by the economist Stephen Enke. My research suggests that arguments made about India and the larger third world in the context of the Cold War by Secretary of Agriculture Orville Freeman and USAID Director David Bell directly to Johnson in late 1964 were also important, if not the decisive factor. Johnson had tremendous faith in Freeman's judgment. See Critchlow, *Intended Consequences*, 71, and Connelly, *Fatal Misconception*, 209–213. See Bell to LBJ, December 9, 1964, Department of State, *Foreign Relations of the United States, 1964–1968*, vol. 9, *International Development and Economic Defense Policy* (Washington, D.C.: Government Printing Office, 1997), 77; Freeman to LBJ, November 23, 1964, USDA Subject Files, 1964, Reports, Box 1, LBJL; Orville Freeman to LBJ, April 27, 1964, National Security File, National Security Council History, Indian Famine, Box 25, vol. 1, Background Tab 3, LBJL; Freeman Oral History, LBJL, 5–6.

39. "Mrs. Gandhi Tries Again," *NYT*, March 13, 1967, 36, and "Fighting Famine in India," *NYT*, December 10, 1965, 46; James Reston, "Washington: Fight 'Em or Feed 'Em," *NYT*, February 11, 1966, 32. See "Text of President's Omaha Address on Vietnam and World Food Shortages," *NYT*, July 1, 1966, 11.

40. Gunnar Myrdal, *Asian Drama: An Inquiry into the Poverty of Nations* (New York: Pantheon, 1968); David C. Engerman and Corinna R. Unger, "Introduction: Towards a Global History of Modernization," *DH* 33, no. 3 (2009): 375–385; 376.

41. Edwin L. Dale, "McNamara, at World Bank, Decries Population Boom," *NYT*, October 1, 1968, 1; Willard Wirtz, "Optimum Population and Environment," Victor Bostrom Fund Report 13, Spring 1970, Series 1, Box 23, "Victor Bostrom" folder, 29, EP.

42. "Johnson vs. Malthus," *NYT*, January 24, 1966, 29.

43. First quotation from "Text of President's Omaha Address on Vietnam and World Food Shortages," *NYT*, July 1, 1966, 11. Subsequent quotations are from "Text of Johnson's Message to Congress Urging Action on Food Help for India," *NYT*, February 3, 1967, 12.
44. For budget numbers, see Critchlow, *Intended Consequences*, 79, 83; and Lara Marks, *Sexual Chemistry: A History of the Contraceptive Pill* (New Haven: Yale University Press, 2001), 29–31. For an overview of these changes, see *Intended Consequences*, chapter 2; Connelly, *Fatal Misconception*, 209–214, 227–228; and Phyllis Piotrow, *World Population Crisis*, chapter 10.
45. Cullather, *The Hungry World*; John H. Perkins, *Geopolitics and the Green Revolution: Wheat, Genes, and the Cold War* (New York: Oxford University Press, 1997).
46. "Text of Johnson's Message to Congress Urging Action on Food Help for India," *NYT*, February 3, 1967.
47. Jane Brody, "Overpopulation War Escalated," *NYT*, January 6, 1969; "The Green Revolution," *NYT*, May 21, 1968.
48. Bhatia, *Famines in India*, 341, 350–352; Paul Brass, *The Politics of India since Independence*, 2nd ed. (New York: Cambridge University Press, 1994), 270, 277–279, 291–299, quotations from 291. For the U.S. role, see Freeman to LBJ, November 23, 1964, USDA Subject Files, 1964, Reports, Box 1, LBJL.

CHAPTER 5   THE "CHINIFICATION" OF AMERICAN CITIES, SUBURBS, AND WILDERNESS

1. William Draper, "Parks—or More People?" *NPM* 40 (April 1966): 10–13; 12, 10.
2. "Fertility Splurge" quotation from Paul Light, *Baby Boomers*, as quoted in Edward D. Berkowitz, *Something Happened: A Political and Cultural Overview of the Seventies* (New York: Columbia University Press, 2006), 68; Herbert S. Klein, *A Population History of the United States* (New York: Cambridge University Press, 2004), 174; Michael French, "The U.S. Population since 1945," in his *U.S. Economic History since 1945* (New York: Manchester University Press, 1997), 4; *Economic Report of the President* (Washington, D.C.: Government Printing Office, 1995), 311, Table B-32.
3. Steven M. Gillon, *Boomer Nation: The Largest and Richest Generation Ever and How It Changed America* (New York: Free Press, 2004), 19.
4. Philip Hauser, "Our Population Crisis Is Here and Now," *Reader's Digest*, February 1962, 148. For towns of fewer than 2,500 people, see James T. Patterson, *Grand Expectations: The United States, 1945–1974* (New York: Oxford University Press, 1996), chapter 1. For metropolitan areas, see French, *U.S. Economic History since 1945*, 15. For central cities losing population, see Richard A. Easterlin, "Growth and Composition of the American Population in the Twentieth Century," in Michael R. Haines and Richard H. Steckel, *A Population History of North America* (New York: Cambridge University Press, 2000), 522.
5. Robert Rienow and Leona Train Rienow, *Moment in the Sun: A Report on the Deteriorating Quality of the American Environment* (New York: Dial Press, 1967), viii; William Paddock and Paul Paddock, "The Explosion in Humans," in *Topics in Animal Behavior, Ecology and Evolution*, ed. Vincent Dethier (New York: Harper & Row, 1971), 123. The Rienows also write of "Chinification"; see p. 7.
6. Council of Economic Advisors, "The Problem of Poverty in America," *Economic Report of the President* (Washington D.C.: Government Printing Office, 1964), 60, as quoted in James T. Patterson, *America's Struggle against Poverty, 1900–1980* (Cambridge: Harvard University Press, 1981), 137.

7. William Vogt, "Address to 31st Annual Meeting Luncheon of the Planned Parenthood Federation of America," October 24, 1951, Box 5, FF15, CONS76, VP; Frederick Osborn, "Population: An International Dilemma," reprinted in *Three Essays on Population: Thomas Malthus, Julian Huxley, Frederick Osborn* [1958] (New York: Mentor Books, 1960), 94.
8. Council of Economic Advisors, "The Problem of Poverty in America," 60, as quoted in Patterson, *Struggle against Poverty*, 137.
9. Donald T. Critchlow, *Intended Consequences: Birth Control, Abortion, and the Federal Government in Modern America* (New York: Oxford University Press, 1999), chapter 2.
10. Ibid., 56–60. For poll data on Catholics, see Phyllis Piotrow, *World Population Crisis: The United States Response* (New York: Praeger, 1973), 26.
11. Daniel Moynihan, "The Negro Family: The Case for National Action," in Lee Rainwater and William L. Yancey, *The Moynihan Report and the Politics of Controversy: A Transaction Social Science and Public Policy Report* (Cambridge: M.I.T. Press, 1967), 25–27. See Nicholas Lemann, *The Promised Land* (New York: Knopf, 1991), 202–218.
12. Michael Harrington, "Poverty, Family Planning, and the Great Society," address delivered at the annual meeting of Planned Parenthood-World Population, October 1965, Folder 43, Box 98, PPFA II.
13. For Watts, see Allen J. Matusow, *The Unraveling of America: A History of Liberalism in the 1960s* (New York: Harper & Row, 1984), 360; and Patterson, *Grand Expectations*, 588–589.
14. Thomas J. Sugrue, *The Origins of the Urban Crisis: Race and Inequality in Postwar Detroit* (Princeton: Princeton University Press, 1996); Kenneth T. Jackson, *Crabgrass Frontier: The Suburbanization of the United States* (New York: Oxford University Press, 1985), 244; Theodore Steinberg, *Down to Earth: Nature's Role in American History* (New York: Oxford University Press, 2002), 217; and Thomas J. Sugrue and John D. Skrentny, "The White Ethnic Strategy," in Bruce J. Schulman and Julian E. Zelizer, *Rightward Bound: Making America Conservative in the 1970s* (Cambridge: Harvard University Press, 2008), 176.
15. Matusow, *The Unraveling of America*, 360–367; Robert Lekachman, "Death of a Slogan—The Great Society of 1967," *Commentary*, as quoted in Matusow, *The Unraveling of America*, 379. For a narrative of the Detroit riot, see Sidney Fine, *Violence in the Model City: The Cavanagh Administration, Race Relations, and the Detroit Riot of 1967* (Ann Arbor: University of Michigan Press, 1989); Patterson, *Struggle against Poverty*, 126.
16. Clayborne Carson, "Race, Rights, and Reform," in *The 1960s and 1970s*, part 7 of *Encyclopedia of American Cultural and Intellectual History*, ed. Mary Kupiec Cayton and Peter W. Williams (New York: Scribner, 2001), 123. Other historians have commented on the overlap of U.S. domestic and international antipoverty programs. See Michael E. Latham, *Modernization as Ideology: American Social Science and "Nation Building" in the Kennedy Era* (Chapel Hill: University of North Carolina Press, 2000), 214; Nils Gilman, "Modernization Theory, the Highest Stage of American Intellectual History," in *Staging Growth: Modernization, Development, and the Global Cold War*, ed. David C. Engerman (Amherst: University of Massachusetts Press, 2003), 56. For poverty in the 1960s, see Nicholas Lemann, *The Promised Land* (New York: Knopf, 1991); and Patterson, *Struggle against Poverty*.
17. Critchlow, *Intended Consequences*, 84.
18. Linda Gordon, *Woman's Body, Woman's Right: A Social History of Birth Control in America* (New York: Grossman, 1976); Ronald Greene, *Malthusian Worlds: US Leadership and the*

Governing of the Population Crisis (Boulder, Col.: Westview, 1999); Matthew Connelly, *Fatal Misconception: The Struggle to Control World Population* (Cambridge: Harvard University Press, 2008); Derek Hoff, *The State and the Stork: The Population Debate and Policymaking in United States History* (Chicago: University of Chicago Press, forthcoming); Critchlow, *Intended Consequences*, 84.

19. Quotations from Critchlow, *Intended Consequences*, 61; Dr. Mary Smith, "Birth Control and the Negro Woman," *Ebony*, March 1968, 29–30. For an excellent overview of African-American attitudes toward birth control and abortion in the late 1960s, including arguments both for and against, see Jennifer Nelson, *Women of Color and the Reproductive Rights Movement* (New York: New York University Press, 2003), chapter 3. Also see Critchlow, *Intended Consequences*, 60–61 and 141–145; Bernard Asbell, *The Pill: A Biography of the Drug That Changed the World* (New York: Random House, 1995).

20. Harrington, "Poverty, Family Planning, and the Great Society."

21. Beth L. Bailey, *Sex in the Heartland* (Cambridge: Harvard University Press, 1999), 109, 112–114.

22. Ibid., 121, 135, 125.

23. Ibid., 125; Johanna Schoen, *Choice & Coercion: Birth Control, Sterilization, and Abortion in Public Health and Welfare* (Chapel Hill: University of North Carolina Press, 2005), 73.

24. "Women Rap Haden, Want Birth Control," *Pittsburgh Press*, August 7, 1968, Planned Parenthood of Central Ohio Papers (Mss 505), Box 1, Scrapbook 1968–1969, Ohio Historical Society (Columbus, Ohio). Also see "Negroes Fighting," *NYT*, August 11, 1968. For other groups, see Black Women's Liberation Group, Mount Vernon, New York, "Statement on Birth Control," printed in Robin Morgan, ed., *Sisterhood Is Powerful: An Anthology of Writings from the Women's Liberation Movement* (New York: Vintage Books, 1970), 360–361. Also see Schoen, *Choice & Coercion*.

25. Alan Gregg, "Population Problems," *Science* 121 (May 13, 1955): 681–682; 682.

26. William Thomas, *Man's Role in Changing the Face of the Earth* (Chicago: University of Chicago Press, 1956), 1142.

27. John Calhoun, "Population Density and Social Pathology," *Scientific American*, February 1962, 139–148.

28. Hudson Hoagland, "Cybernetics of Population Control," *Bulletin of the Atomic Scientists* 20, no. 2 (February 1964): 2–6; "A Self-Corrective for the Population Explosion?" *Time*, February 28, 1964, 56; Sally Carrighar, *Wild Heritage* (Boston: Houghton Mifflin, 1965); "A Scientist Looks at 'The Human Zoo,'" *US News and World Report*, March 2, 1970. For a summary, see John R. Wilmoth and Patrick Ball, "The Population Debate in American Popular Magazines, 1946–90," *Population and Development Review* 18, no. 4 (December 1992), 631–668; 649. Also see "2 Britons Discern a 'Sardine Syndrome,' Linking Recent Violence to the Pressure of Overpopulation," *NYT*, August 16, 1970; and Claire and W.M.S. Russell, "The Sardine Syndrome: Overcrowding and Social Tension," *Ecologist* 1, no. 2 (August 1970): 4–9.

29. Rienow and Rienow, *Moment in the Sun*, 97.

30. Gordon Rattray Taylor, "People Pollution," *Ladies' Home Journal*, October 1970, 74–80, quoted in Elaine Tyler May, *America and the Pill: A History of Promise, Peril, and Liberation* (New York: Basic Books, 2010), 46. Emphasis added.

31. Wilmoth and Ball, "The Population Debate in American Popular Magazines," 657.

32. Jennifer Kalish, "Suburbanization," in Paul Boyer, ed., *The Oxford Companion to US History* (New York: Oxford University Press, 2001), 753.

33. Patterson, *Grand Expectations*, 76. For suburbs, see Jackson, *Crabgrass Frontier*, chapter 13; David Halberstam, *The Fifties* (New York: Villard Books, 1993); G. Scott Thomas, *The United States of Suburbia* (New York: Prometheus Books, 1998); Adam Rome, *The Bulldozer in the Countryside: Suburban Sprawl and the Rise of American Environmentalism* (New York: Cambridge University Press, 2001).
34. Rome, *Bulldozer*, 16; Jackson, *Crabgrass Frontier*, 234.
35. Jackson, *Crabgrass Frontier*, 120–127. For lawns, see Theodore Steinberg, *American Green: The Obsessive Quest for the Perfect Lawn* (New York: Norton, 2006).
36. "Statement on Population Policy," *SCB* 50 (April 1965), 2; "The Poverty of Abundance: American Business and the World Population Crisis," Pamphlet, Planned Parenthood–World Population, January 1966, Folder 40, Box 98, PPFA, 25.
37. Raymond Dasmann, *The Destruction of California* (New York: Macmillan, 1965), 2, 206, 21, 19. Dasmann's autobiography, for which Paul Ehrlich wrote an introduction, is entitled *Called by the Wild: The Autobiography of a Conservationist* (Berkeley: University of California Press, 2002).
38. James Hudnut-Beumler, "The Culture and Critics of the Suburb and Corporation," in Cayton and Williams, eds., *Encyclopedia of American Cultural and Intellectual History*, 27–34; Dwight MacDonald, "A Theory of Mass Culture." *Diogenes* I (June 1953), 2. For a critique of highbrow response to suburbs, see Scott Donaldson, *The Suburban Myth* (New York: Columbia University Press, 1969).
39. Whyte, *The Last Landscape*, 199, as quoted in Rome, *Bulldozer*, 131.
40. Rome, *Bulldozer*, 128–135. For Brower, see Kevin Starr, *Golden Dreams: California in an Age of Abundance, 1950–1963* (New York: Oxford University Press, 2009), chapter 15, "Largest State in the Nation: A Rebellion against Growth and the Destruction of Environment," 413–435; 423. Rienow and Rienow, *Moment in the Sun*, 53, 13.
41. John Kenneth Galbraith, *A Life in Our Times: Memoirs* (Boston: Houghton Mifflin, 1981), 337; "The Economy: We are All Keynesians Now," *Time*, December 31, 1965, 64–67. For 1960s growth, see Bailey and Farber, *Columbia Guide to America in the 1960s* (New York: Columbia University Press, 2003), 55; Edward Berkowitz, *Something Happened*, 55; Lawrence Buell, *The Environmental Imagination: Thoreau, Nature Writing, and the Formation of American Culture* (Cambridge: Harvard University Press, 1995), 149.
42. Dasmann, *Destruction*, 214; Rienow and Rienow, *Moment in the Sun*, 6; Kenneth Boulding, "The Economics of the Coming Spaceship Earth," in Kenneth E. Boulding and Henry Jarrett, *Environmental Quality in a Growing Economy* (Baltimore: Johns Hopkins University Press, 1966), 3–14. For Boulding, see Anne Chisholm, *Philosophers of the Earth: Conversations with Ecologists* (New York: Dutton, 1972), 27–36; Debora Hammond, *The Science of Synthesis: Exploring the Social Implications of General Systems Theory* (Boulder: University Press of Colorado, 2003); Joseph De Steiguer, *The Origins of Modern Environmental Thought* (Tucson: University of Arizona Press, 2006), 88–98. For another critique of growth, see E. J. Mishan, *The Costs of Economic Growth* (London: Staples, 1967).
43. Rome, *Bulldozer*, 33, 42. For the importance of consumption to suburban culture, see Lizabeth Cohen, *A Consumers' Republic: The Politics of Mass Consumption in Postwar America* (New York: Knopf, 2003).
44. Reisman quotation from Daniel Seligman and Lawrence A. Mayer, "The Future Population 'Mix,'" *Fortune*, February 1959, 222–223. For Levittown, see Jackson, *Crabgrass Frontier*, 235. Gillon, *Boomer Nation*, 25; Starr, *Golden Dreams*, 17.

45. Gillon, *Boomer Nation*, 2; Wilmoth and Ball, "The Population Debate in American Popular Magazines," 656; "Babies Equal Boom," *Reader's Digest*, August 1951, 6; "Rocketing Birth, Business Bonanza," *Life*, June 16, 1958; Rienow and Rienow, *Moment in the Sun*, 9.

46. Dasmann, *Destruction*, 202; Robert Rienow, "Political Thickets Surrounding Wilderness: A Summary of the Conference," in Phillip S. Berry, ed., *Wilderness and the Quality of Life* (San Francisco, Sierra Club, 1967), 221–222; Rienow and Rienow, *Moment in the Sun*, 4.

47. Letter to the Editor, *SCB* 49 (June 1964): 15; Rienow, "Political Thickets," 222.

48. In 1972, the National Commission on Population Growth and the American Future identified ways in which the population argument was flawed, pointing out that the urban and suburban areas has spread faster than population and that poor planning had helped create sprawl. See Rome, *Bulldozer*, 143. For automobile numbers, see Jackson, *Crabgrass Frontier*, 246. "The Poverty of Abundance: American Business and the World Population Crisis," PPWP, January 1966, Folder 40, Box 98, PPFA II, 25.

49. William R. Catton, Jr., "Letting George Do It Won't Do It," *NPM* 38 (March 1964): 4–7.

50. Bruce Welch, "The Real Threat to Wilderness—Population," *NPM* 37 (January 1963): 10–11.

51. Ibid., 10–11; Robert Cook, "Outdoor Recreation Threatened by Excess Procreation," *PB* 20, no. 4 (June 1964): 89–94.

52. Harvey Broome, "Spring in a Mountain Notebook," *Living Wilderness* 17 (Spring 1952): 9–17. Echoing this concern in 1952, Joseph Wood Krutch warned of destroying wilderness for nothing more than "a more numerous breed." Joseph Wood Krutch, *The Desert Year* (New York: William Sloane, 1952), quoted in Olaus Murie, "Wholesome, Conditioning Environment," *Living Wilderness* 17 (Spring 1952): 18–19.

53. Brower quoted in Daniel Luten, "Numbers against Wilderness," *SCB* 49 (September 1964): 43; Lowell Sumner, "Are Beavers Too Busy?" *SCB* 42 (June 1957): 19. For Lowell, see Michael P. Cohen, *The History of the Sierra Club, 1892–1970* (San Francisco: Sierra Club Books, 1988), 215.

54. Raymond Cowles, "Population Pressures and Natural Resources," *The Meaning of Wilderness to Science: Proceedings, Sixth Biennial Wilderness Conference* (San Francisco: Sierra Club, 1960), 79–94, quotations from 81, 88, and 88.

55. Ibid., 81–83.

56. Ibid., 85.

57. Ibid., 86 and 85.

58. Ibid., 92–93.

59. Ibid., 93–94.

60. Ibid., 112; David Brower, *For Earth's Sake: The Life and Times of David Brower* (Salt Lake City: Peregrine Smith Books, 1990), 274. Cohen suggests that Brower's strong advocacy of population control was one of many issues that led to his dismissal as director. Cohen, *History of the Sierra Club*, 232.

61. Ansel Adams and Nancy Newhall, *This Is the American Earth* (San Francisco: Sierra Club Books, 1960), 44, 36. A book that David Brower edited in 1964, *Wildlands in Our Civilization*, included a similar juxtaposition. Although primarily about the United States, it contained an essay by Lee Merriman Talbot about the destruction of the wild lands of north India since independence called "The Wail of Kashmir." Talbot, "The

Wail of Kashmir" in David Brower and John Collier, eds., *Wildlands in Our Civilization* (San Francisco: Sierra Club, 1964).

62. Daniel B. Luten quoted in David Brower, ed. *Wilderness, America's Living Heritage: Proceedings of the Seventh Wilderness Meeting, Sierra Club, 1961* (San Francisco: Sierra Club, 1962), 180; Luten, "How Dense Can Man Be?" *SCB* 48 (September 1963), 80–93; 82; Letter to the Editor, *SCB* 49 (June 1964): 15. Luten provides a short history of Sierra Club thinking about population in "Numbers against Wilderness," *SCB* 49 (September 1964): 43–48.

63. Letter to the Editor, *SCB* 50 (June 1965): 20.

64. Gerald Piel, "Wilderness and the American Dream," in *Wilderness: America's Living Heritage*, 26–31; 27–28. Brower's comment is on page 31. For Osborn's speech, see *Tomorrow's Wilderness: Proceedings of the Eighth Wilderness Meeting* (San Francisco: Sierra Club, 1963), 149.

65. "Population Policy of the Sierra Club of San Francisco," *PB* 22, no. 2 (June 1966): 29–57; "The Poverty of Abundance: American Business and the World Population Crisis," Planned Parenthood-World Population, January 1966, Folder 40, Box 98, PPFA II, 27.

66. Rienow, "Political Thickets Surrounding Wilderness," 223, 224.

67. David Brower, "Foreword," in Paul R. Ehrlich, *The Population Bomb* (New York: Ballantine Books, 1968), 13–14.

CHAPTER 6   PAUL EHRLICH, THE 1960S, AND THE POPULATION BOMB

1. Paul R. Ehrlich, *The Population Bomb* (New York: Ballantine Books, 1968), 11.

2. Adam Rome argues that historians need to do more to connect the environmentalism of the late 1960s to the larger historical currents of the 1960s. Adam Rome, "'Give Earth a Chance': The Environmental Movement and the Sixties," *Journal of American History* 90, no. 2 (2003): 525–554.

3. David Darlington, "The Man Who Made Babies Disappear."

4. Gregg A. Mitman and Ronald L. Numbers, "Evolutionary Theory," *Encyclopedia of the United States in the Twentieth Century*, vol. 2, ed. Stanley I. Kutler (New York: Scribner's, 1996), 871, 859.

5. Seminal works of the Darwinian Synthesis—also known as the "modern" or "evolutionary" synthesis—include Theodosius Dobzhansky, *Genetics and the Origin of Species* (New York: Columbia University Press, 1937); Ernst Mayr, *Systematics and the Origin of Species from the Viewpoint of a Zoologist* (New York: Columbia University Press, 1942). Also see Ernst Mayr and William B. Provine, *The Evolutionary Synthesis: Perspectives on the Unification of Biology* (Cambridge: Harvard University Press, 1980); Stephen J. Gould, "The Hardening of the Modern Synthesis," in *Dimensions of Darwinism: Themes and Counterthemes in Twentieth-Century Evolutionary Biology*, ed. M Greene (Cambridge: Cambridge University Press, 1983), 71–96; Peter Bowler, *The Environmental Sciences* (New York: Norton, 1992); and Mitman and Numbers, "Evolutionary Theory."

6. Mitman and Numbers, "Evolutionary Theory"; Susan R. Schrepfer, *The Fight to Save the Redwoods: A History of Environmental Reform, 1917–1978* (Madison: University of Wisconsin Press, 1983), 96.

7. Paul R. Ehrlich and Ilkka Hanski, *On the Wings of Checkerspots: A Model System for Population Biology* (New York: Oxford University Press, 2004), vii. For the Lake Erie work, see J. H. Camin and P. R. Ehrlich, "Natural Selection in Water Snakes (Natrix

sipedon L.) on Islands in Lake Erie," *Evolution* 12 (1958): 504–511; Ehrlich and Camin, "Natural Selection in Middle Island Water Snakes (Natrix sipedon L.)," *Evolution* 14 (1960): 136; Paul R. Ehrlich and Richard W. Holm, *The Process of Evolution* (New York: McGraw-Hill 1963), 137–139. This study paralleled and slightly postdates Bernard Kettlewell's famous moth experiments in England.

8. Some biologists, including some pioneers of the Synthesis, maintained ideas of human specialness and progress. See Michael Ruse, *Monad to Man: The Concept of Progress in Evolutionary Biology* (Cambridge: Harvard University Press, 1996), 410–455.

9. Julian S. Huxley, *Evolution: The Modern Synthesis*, 2nd ed. (London: Allen & Unwin, 1962), 576. Also see William B. Provine, "Progress in Evolution and Meaning in Life," in *Evolutionary Progress*, ed. Matthew H. Nitecki (Chicago: University of Chicago Press, 1988), 65; John C. Greene, *Science, Ideology and World View: Essays in the History of Evolutionary Idea* (Berkeley: University of California Press, 1981); and James R. Moore, *History, Humanity, and Evolution: Essays for John C. Greene* (New York: Cambridge University Press, 1989).

10. Schrepfer, *Fight to Save the Redwoods*, 101, 96, 97, 96, 98, 80.

11. Raymond Dasmann, *The Destruction of California* (New York: Macmillan, 1965), 21. Emphasis added.

12. Ehrlich and Holm, *The Process of Evolution*, 295–296.

13. Paul R. Ehrlich and Anne H. Ehrlich, *Betrayal of Science and Reason: How Anti-Environmental Rhetoric Threatens Our Future* (Washington, D.C.: Island Press, 1996), 3.

14. Paul Ehrlich and L. C. Birch, "The 'Balance of Nature' and 'Population Control,'" *American Naturalist* 101, no. 918 (March–April 1967):97–107; 106.

15. Interview with Paul Ehrlich, Palo Alto, California, April 6, 2004. Ehrlich also says he was denied a job because of his religion. See "Paul Ehrlich, Famed Ecologist, Answers Questions," *Grist* (August 9, 2004), available at http://www.grist.org/article/ehrlich.

16. Paul Ehrlich, "Has the Biological Species Concept Outlived Its Usefulness?" *Systematic Zoology* 10 (1961), 167–176; Paul R. Ehrlich and Richard W. Holm, "Patterns and Populations," *Science* 137, no. 3531 (1962): 652–657.

17. Erlich and Holm, "Patterns and Populations," 657; Paul Ehrlich, "Career Summary," Series 1, Box 5, "Career Summary" folder, EP.

18. Paul Ehrlich, "Numerical Taxonomy," *Papua and New Guinea Scientific Society Annual Report and Proceedings* 17 (1965): 10–14, ACCN 2000–296, Folder 11, EP. For computers, see Ehrlich and Holm, *The Process of Evolution*, 303; also see 298.

19. Ehrlich to Ashley Montagu, January 11, 1963, Box 37, Folder 3, EP; Richard Holm and Paul Ehrlich, "A Biological View of Race" in Ashley Montagu, ed., *The Concept of Race* (New York: Free Press of Glencoe, 1964), 153–179, quotation from 173; Paul R. Ehrlich and S. Shirley Feldman, *The Race Bomb: Skin Color, Prejudice, and Intelligence* (New York: Quadrangle, 1977).

20. Theodosius Dobzhansky, *The Biological Basis of Human Freedom* (New York: Columbia University Press, 1956); and Dobzhansky and L. C. Dunn, *Heredity, Race, and Society* (New York: New American Library, 1952). Ehrlich considered Dobzhansky a mentor and convinced him to comment on the entire manuscript for Ehrlich and Holm, *The Process of Evolution*.

21. Ehrlich and Holm, "Patterns and Populations," 657; Ehrlich to Bryant Mather, March 13, 1962, Series 1, Box 28, "Bryant Mather" folder, EP.

22. Thomas S. Kuhn, *The Structure of Scientific Revolutions* (Chicago: University of Chicago Press, 1962); Ehrlich to Michener, September 11, 1963, Box 37, Folder 4, EP; Ehrlich, "Numerical Taxonomy."
23. Donald Critchlow noticed this pattern, too. Critchlow, *Intended Consequences: Birth Control, Abortion, and the Federal Government in Modern America* (New York: Oxford University Press, 1999), 156–157.
24. William Cronon, "Landscapes of Abundance and Scarcity," in Clyde A. Milner, Carol A. O'Connor, and Martha A. Sandweiss, *Oxford History of the American West* (New York: Oxford University Press, 1994), 603–638. For urbanization and its discontents in the West, see William Graves, "California: The Golden Magnet," *National Geographic*, May 1966, 595–639; Amy Scott, "Cities and Suburbs," in David R. Farber and Beth L. Bailey, *The Columbia Guide to America in the 1960s* (New York: Columbia University Press, 2001), 265–266; John M. Findlay, *Magic Lands: Western Cityscapes and American Culture after 1940* (Berkeley: University of California Press, 1992); Kevin Starr, *Golden Dreams: California in an Age of Abundance, 1950–1963* (New York: Oxford University Press, 2009), 15–16, and chapter 15, "Largest State in the Nation: A Rebellion against Growth and the Destruction of Environment," 413–435; Richard Walker, *The Country in the City: The Greening of the San Francisco Bay Area* (Seattle: University of Washington Press, 2007), 83, 144; and Herbert S. Klein, *A Population History of the United States* (New York: Cambridge University Press, 2004), 199.
25. Rebecca S. Lowen, *Creating the Cold War University: The Transformation of Stanford* (Berkeley: University of California Press, 1997), 2, 6. Also see Roger W. Lotchin, *Fortress California, 1910–1961: From Warfare to Welfare* (Urbana: University of Illinois Press, 2002); and Starr, *Golden Dreams*, 226–227.
26. Paul Ehrlich, "The Food from the Sea Myth: The Natural History of a Red Herring," manuscript of speech delivered to the Commonwealth Club of California, April 21, 1967, Series 6, Box 26, "Articles 1965–1967" folder, EP, 10; Ehrlich to Robert E. Beer, September 28, 1961, Series 1, Box 30, "Robert Beer" folder, EP; Ehrlich to Paul Anderson, February 2, 1965, Series 1, Box 30, "Paul Anderson" folder, EP.
27. *San Jose News*, March 4, 1963. For butterfly extinctions, see Ehrlich and Ehrlich, *Betrayal of Science*, 7. Ehrlich quotation from Heather Newbold, ed., *Life Stories* (Berkeley: University of California Press, 2000), 22. Development also infringed upon Ehrlich's Rocky Mountain research station.
28. Quotation from "Ehrlich Career Summary, 1977," Series 7, Box 5, "Paul R. Ehrlich—Bio Sketches and C.V.s" folder, EP. For coevolution, see Paul Ehrlich and Peter Raven, "Butterflies and Plants: A Study in Coevolution," *Evolution* 18 (1964): 586–608; and Paul Ehrlich, *The Machinery of Nature* (New York: Simon & Schuster, 1986), 145–146.
29. For Ehrlich's youth, see Ehrlich and Ehrlich, *Betrayal of Science*, 3. Also see "Following Their Footsteps: Teens on Their Way Talk to Five Who Have Made It," *Seventeen*, January 1971, 94–95. For fire ants, see Paul R. Ehrlich and Anne H. Ehrlich, *Population, Resources, Environment: Issues in Human Ecology* (San Francisco: W. H. Freeman, 1970), 646–647, 979–981. For contemporaneous references, see Ehrlich to Klots, March 5, 1958, Series 1, Box 28, "Alexander Klots" folder, EP. Ehrlich got to know the biologist E. O. Wilson as part of this campaign. For Ehrlich and Wilson, see Christopher Lewis, "Progress and Apocalypse: Science and the End of the Modern World," Ph.D. dissertation, University of Minnesota, 1991, 254.
30. Rachel Carson, *Silent Spring* (Boston: Houghton Mifflin, 1962).

31. Paul Ehrlich, "The Biological Revolution," conference address at "A New Look at 1984—Forces and Ideas Shaping the Future," Riverside, California, January 31, 1965, ACCN 2000–296, Box 1, Folder 11, EP, quotation from 10 and 9. Also printed in the *Stanford Review* (September–October 1965), Series 8, Box 1, "Press Clippings 1965–1968, Folder 3 of 6," EP.

32. Paul Ehrlich, "The Biological Revolution," 13.

33. Ehrlich, *The Population Bomb*, 36.

34. Ehrlich to Susan K. Ingraham, November 12, 1968, Box 22, "Correspondence, Outgoing," November 1968, EP.

35. Brown watched India closely as a USDA analyst of international agricultural trends. Lester Brown, *Man, Land & Food: Looking ahead at World Food Needs* (Washington, D.C.: U.S. Dept. of Agriculture, Economic Research Service, Regional Analysis Division, 1963); Lester Brown, *World without Borders* (New York: Random House, 1972); Lester Russell Brown and Erik P. Eckholm, *By Bread Alone* (New York: Praeger, 1974). *U.S. News and World Report* even featured Brown in a cover story on the India's agricultural problems in early 1964. Carl Pope was a Peace Corps volunteer in north India in the late 1960s; Carl Pope, *Sahib: An American Misadventure in India* (New York: Liveright, 1972). Garrett Hardin, "Tragedy of the Commons," *Science* 162 (December 1968): 1243–1248; and "Living on a Lifeboat," *Bioscience* 24, no. 10 (1974): 561–568. Hardin visited South Asia in 1970. For "no more Indias," see Donald Fleming, "Roots of the New Conservation Movement," *Perspectives in American History* 6 (1972): 7–91; 52.

36. Ehrlich and Ehrlich, *Betrayal of Science*, 5; Ehrlich to Friends, from Srinagar, June 29, 1966, and Ehrlich to Friends, from Delhi, July 8, 1966, both in SC 223 2001–2, Box 1, "Fieldnotes, 65–66" folder, EP; Ehrlich to L. C. Birch, September 12, 1966, Box 5, "Charles Birch—Before 1972" folder, EP. "*Every* educated Indian we talked to thought the situation damn near hopeless, or worse," Ehrlich noted. Ehrlich to Friends, from Kenya, July 1966, EP, SC 223 2001–2, Box 1, "Fieldnotes, 65–66" folder, EP. Also see Paul Ehrlich to L.C. Birch, September 12, 1966, and Box 5, "Expenses June 23–July 14, 1966 Thailand, Cambodia, and India" folder, EP.

37. Kingsley Davis, "Population Policy: Will Current Programs Succeed?" *Science*, November 10, 1967, 730–739; Davis, *The Population of India and Pakistan* (Princeton: Princeton University Press, 1951). For Davis, see William Peterson, "Kingsley Davis," *International Encyclopedia of Social Science*, vol. 18 (1979), 139; and Dennis Hodgson, "Demography as Social Science and Policy Science," *Population and Development Review* 9, no. 1 (March 1983): 1–34; 17.

38. William and Paul Paddock, *Famine 1975!* (Boston: Little, Brown 1967), 64–91.

39. Paul Ehrlich, "Paying the Piper," *New Scientist* 36 (December 14, 1967): 652; Paul Ehrlich, "Population, Food, and Environment: Is the Battle Lost?" manuscript of speech at the University of Texas at Austin, November 16, 1967, Series 6, Box 26, "Articles 1965–1967" folder, EP, 2.

40. William Whyte, *The Last Landscape* (Garden City, N.Y.: Doubleday, 1968); Marshall Herskovits, quoted in Steven M. Gillon, *Boomer Nation: The Largest and Richest Generation Ever and How It Changed America* (New York: Free Press, 2004), 85.

41. Paul Ehrlich, "Population and Environment," manuscript of a speech given on the Stanford campus, October 18, 1967, Series 6, Box 26, "Articles 1965–1967" folder, EP, 2.

42. Ibid.; Paul Ehrlich, "The Population Crisis: Is Our End of the Boat Sinking?" manuscript of speech given at the 72nd Annual Scientific Meeting of the Alumni Association, University of California School of Dentistry, San Francisco, January 14, 1968, Series 6, Box 26, Folder 22, EP, 7; Paul Ehrlich, "Birth Control: Too Little and Much Too Late," manuscript of speech presented to Planned Parenthood of the Bay Area, February 7, 1968, Series 6, Box 26, Folder 16, EP; National Advisory Commission on Civil Disorders, *Report of the National Advisory Commission on Civil Disorders* (Washington, D.C.: Government Printing Office, 1968).
43. Paul Ehrlich, "Population, Food, and the Environment: The Problem and the Consequences," University of Utah, April 4, 1968, Series 6, Box 26, EP.
44. Paul Ehrlich to Charles Birch, May 20, 1968; and Ehrlich to Charles Birch, July 24, 1968, Series 1, Box 5, "Charles Birch" folder, EP.
45. Ehrlich later noted that he and his wife co-wrote the book, but that Ballantine insisted on a single author. For the campaign quotation, see Ehrlich, "The Population Crisis: Is Our End of the Boat Sinking?"
46. Ehrlich, *The Population Bomb*, 23, 28.
47. Ibid., 25.
48. Ibid., 29, 34, 31; for titles, see Ehrlich to Ian Ballantine, March 14, 1968, Series 6, Box 2, Folder 52, EP.
49. Ibid., 33.
50. Ibid., 37, 44.
51. Ibid., 41, 43.
52. Ibid., 46–47, 49, 49, 50, 77, 53.
53. "Intimate relationship" quotation from Ehrlich, "The Food from the Sea Myth," 8; also in Ehrlich, *The Population Bomb*, 49. Other quotations from Ehrlich, *The Population Bomb*, 46–47.
54. Daniel McKinley, "The Population Bomb," *Defenders of Wildlife News*, October–December, 1968, SC 223, Series 6, Box 2, Folder 40, EP.
55. Ehrlich called climate change an "even more important" problem than the others. He wrote, "All of the junk we dump into the atmosphere, all of the dust, all of the carbon dioxide, have effects on the temperature balance of the Earth." This change might bring cooling or heating, but the results could be "catastrophic." With just a few degrees of heating, he said the polar ice caps would melt, perhaps raising ocean levels 250 feet. "Gondola to the Empire State Building, Anyone?" Ehrlich, *The Population Bomb*, 60–61.
56. Ibid., 67. Emphasis in original.
57. Ehrlich, *The Population Bomb*, 64, 168; Paul Ehrlich, "Overcrowding and Us," *National Parks Magazine*, April, 1969, 10, 11. In early 1968, Ehrlich and Jonathan Freedman, a professor of psychology at Stanford, began planning five years of research on the effects of crowding on human beings. In the *Bomb*, Ehrlich warns that extrapolating from the behavior of rats to the behavior of man is risky.
58. Ehrlich, *The Population Bomb*, 65–66.
59. Ibid., 72–78.
60. Both quotes in Ehrlich, *The Population Bomb*, 195, 196.
61. Paul Ehrlich to John Weaver, April 25, 1968 letter, Box 40, "John Weaver" folder, EP.
62. Ehrlich, *The Population Bomb*, 83.

63. Historian Timothy May describes the 1950s and 1960s as the heyday of "gee whiz" science. May, "The End of Enthusiasm: Science and Technology," in Farber and Bailey, *The Columbia Guide to America in the 1960s*, 305. Ehrlich, *The Population Bomb*, 108.
64. Paul Ehrlich, "One-Dimensional Ecology," *Ecologist* 2, no. 8 (August 1972): 11–21; 13.
65. Paul Ehrlich, "Overcrowding and Us," *NPM*, April 1969, 10–12.
66. Ehrlich, *The Population Bomb*, 135, 140.
67. Ibid., 169, 172.
68. Ibid., 126, 142, 127.
69. Ibid., 170. Lynn White, Jr., "The Historical Roots of Our Ecologic Crisis," *Science* 155, no. 3767 (1967): 1203–1207.
70. Ehrlich, *The Population Bomb*, 149.
71. Ibid., 151, 151.
72. Ibid., 72, 82, 90, 89, 93, 123.
73. Ibid., 161, 161, 166.
74. Ibid., 166.
75. Ibid., 170.
76. Ehrlich, "The Food from the Sea Myth"; Ehrlich, *The Population Bomb*, 132; Ehrlich to Ian Ballantine, March 14, 1968, Series 6, Box 2, Folder 52, EP. Ehrlich used the Spaceship Earth metaphor in his October 1967 Stanford University speech, and his November 1967 Texas speech. See Kenneth Boulding, "The Economics of the Coming Spaceship Earth," in Kenneth E. Boulding and Henry Jarrett, *Environmental Quality in a Growing Economy* (Baltimore: Johns Hopkins University Press, 1966), 3–14; Barbara Ward, *Spaceship Earth* (New York: Columbia University Press, 1966), vii and 15; and Buckminster Fuller, *An Operation Manual for Spaceship Earth* (Carbondale: Southern Illinois University Press, 1969).
77. Ehrlich, *The Population Bomb*, 15.

CHAPTER 7   STRANGE BEDFELLOWS

1. Donald Fleming, "Roots of the New Conservation Movement," *Perspectives in American History* 6 (1972): 7–91; 7.
2. Harold Bostrom to Hugh Moore, December 15, 1969, Box 1, Folder 13, HMF.
3. Wade Green uses the terms "voluntarist," "compulsionist," and "population controller." Wade Green, "The Militant Malthusians," *Saturday Review*, March 11, 1972, 40–44. For a useful catalogue of measures beyond voluntary contraception, see Bernard Berelson, "Beyond Family Planning," *Science* 163, no. 3867 (1969): 533–543. Paul R. Ehrlich and Anne H. Ehrlich, *Population Resources Environment: Issues in Human Ecology* (San Francisco: W. H. Freeman, 1970), 274; also see 254–256. The Ehrlichs displayed a sense of reluctance: "One would hope that population size can be controlled in the United States without resort to such distasteful and socially disruptive measures."
4. Charles McCabe, "The Fearless Spectator," *San Francisco Chronicle*, September 2, 1970; Allan Chase, *The Legacy of Malthus: The Social Costs of the New Scientific Racism* (New York: Knopf, 1977), 339. Garrett Hardin, "The Tragedy of the Commons," reprinted in Wes Jackson, *Man and the Environment*, 2d ed. (Dubuque, Iowa: W. C. Brown, 1973); originally in *Science* 162 (December 13, 1968): 1243–1248.
5. Garrett Hardin to Perce S. Barrows, January 28, 1969, Box 6, GHP.

6. Hardin, "The Tragedy of the Commons," in Jackson, *Man and the Environment*, 246. Hardin had little experience in the developing world, although his writings, including the "The Tragedy of the Commons," had global implications. In early 1968, however, not long before he wrote this famous essay about pastures, Hardin traveled to Africa for a safari. He went, he wrote at the time, to go "before all the wild animals are gone." Garrett Hardin letter, December 18, 1967, Box 3, "Correspondence, Outgoing December 1967" folder, GHP.
7. Hardin, "The Tragedy of the Commons," in Jackson, *Man and the Environment*, 247, 245. For early analysis of Hardin's framework, see "The Tragedy of the Commons Revisited," *Science* 166 (November 28, 1969): 1103–1107.
8. Hardin, "The Tragedy of the Commons," in Jackson, *Man and the Environment*, 245, 253, 249.
9. Ibid., 249, 252, 252, 252.
10. Pope Paul VI, *Humanae Vitae: Encyclical of Pope Paul VI on the Regulation of Birth*, July 25, 1968, *Acta Apostolicae Sedis* 60 (1968): 481–503, English translation from *The Pope Speaks* 13 (Fall 1969): 329–346; "The Pope and Malthus," *NYT*, August 4, 1968, E 14.
11. Ballantine press release, "Interest in 'The Population Bomb' Growing!" August 8, 1968, March 14, 1968, Series 6, Box 2, Folder 54, EP. Pope Paul VI *Humanae Vitae* in *The Pope Speaks*, 334, 336. After noting that "the rapid increase in population" has "made many fear that world population is going to grow faster than available resources," Pope Paul warned that this fear "can easily induce public authorities to be tempted to take even harsher measures to avert this danger;" Pope Paul VI, *Humanae Vitae* in *The Pope Speaks*, 329.
12. Paul R. Ehrlich and Richard W. Holm, *The Process of Evolution* (New York: McGraw-Hill, 1963), 292; Paul Ehrlich to Karl Sax, August 8, 1968, Series 1, Box 40, "Karl Sax" folder, EP. "Night baseball" quotation from David Darlington, "The Man Who Made Babies Disappear," *Cal Today: San Jose Mercury News*, October 24, 1982, ACCN 1994–007, Box 7, "1982 Darlington" folder, EP. Paul R. Ehrlich, *The Population Bomb* (New York: Ballantine Books, 1968), 140–142.
13. Pope Paul VI, *Humanae Vitae* in *The Pope Speaks*, 337.
14. Ibid., 329–335.
15. Paul Ehrlich to Charles Michener, November 26, 1968, Box 37, Folder 4, EP; Ernst Mayr to Ehrlich, April 22, 1968, Series 1, Box 28, "Ernst Mayr" folder, EP; "Scientist's Statement on Birth Control Encyclical," ACCN 2000–296, Box 2, Folder 8, EP.
16. Shulamith Firestone, *The Dialectic of Sex: The Case for Feminist Revolution* (New York: Morrow, 1970), 192. For an overview of relations between feminists and Malthusians, although one that does not pay much attention to environmental Malthusians, see Dennis Hodgson and Susan Watkins, "Feminists and NeoMalthusians: Past and Present Alliances," *PDR* 23, no. 3 (September 1997): 469–523.
17. Lucinda Cisler, "Unfinished Business: Birth Control and Women's Liberation," in Robin Morgan, ed., *Sisterhood Is Powerful: An Anthology of Writings from the Women's Liberation Movement* (New York: Random House, 1970), 246, 288.
18. Ibid., 248, 285, 286.
19. Although there are several good biographies of Douglas, no one has examined the interplay between his passionate environmentalism and his civil liberties thinking.
20. For Hardin, see David Garrow, *Liberty and Sexuality: The Right to Privacy and the Making of Roe v. Wade* (New York: Macmillan, 1994), 293, 295. Garrow, however, does not

investigate the connections between Hardin's concern about overpopulation and his abortion rights advocacy. For NARAL, see Donald T. Critchlow, *Intended Consequences: Birth Control, Abortion, and the Federal Government in Modern America* (New York: Oxford University Press, 1999), 134–135. Derek Hoff also discusses Hardin in Hoff, *The State and the Stork: The Population Debate and Policymaking in United States History* (Chicago: University of Chicago Press, forthcoming). Ehrlich, *The Population Bomb*, 147–148; George Wald, "A Better World for Fewer Children," in *The Crisis of Survival* (a compilation of the Spring 1970 issue of *The Progressive*) (Glenview, Ill.: Scott, Foresman, 1970), 114–119; 117.

21. Ehrlich, *The Population Bomb*, 140–141. For the "should be thrown open" quote, see Paul Ehrlich, "Are There Too Many of Us?" *McCall's*, July 1970, 104. For the "give women something to do" quote, see BBC interview, January 14, 1972, Series 5, Box 1, Folder 28, EP. Demographer Judith Blake articulated this position best. See Judith Blake, "Population Policy for Americans: Is the Government Being Misled?" *Science* 164, no. 3879 (1969): 522–529.

22. Cisler, "Unfinished Business," 286–287.

23. Richard Bowers, "Excerpts from Zero Population Growth, Inc. Prospectus," no date, ACCN 2000–296, Box 2, Folder 4, EP. A full history of ZPG has yet to be written. Critchlow provides some background, and Kluchin points to ZPG's many efforts to expand birth control and abortion rights, but emphasizes its coercive aspects. Hoff includes an excellent analysis of ZPG's early years. Critchlow, *Intended Consequences*; Rebecca M. Kluchin, *Fit to Be Tied: Sterilization and Reproductive Rights in America, 1950–1980* (New Brunswick: Rutgers University Press, 2009), 116, 123; Hoff, *State and the Stork*. For ZPG and immigration, see chapter 8 of this book.

24. Statement adopted by ZPG board of directors, September 30, 1969, reprinted in *ZPG National Reporter*, August 1970, Abortion Collection, Folder 27, Box 3, PPFA, 8; "Statement of Purposes and Goals," February 10, 1969, "ZPG" folder, Box 130, PPFA.

25. Green, "The Militant Malthusians," 42. For Bowers, see Critchlow, *Intended Consequences*, 156. *ZPG National Reporter*, August 1970, Abortion Collection, Folder 27, Box 3, PPFA, 23.

26. *US News and World Report*, March 2, 1970; "A New Movement Challenges the U.S. to Stop Growing," *Life*, April 17, 1970, 32–37; Hugh More to Robert G. Wehle, January 2, 1970, HMF, Box 1, Folder 13; Pennefield Jensen, "A Student Manifesto on the Environment," *Natural History* 79 (April 1970): 22.

27. The Associated Press, for instance, declared 1964 "the year of the kid." Steven M. Gillon, *Boomer Nation: The Largest and Richest Generation Ever and How It Changed America* (New York: Free Press, 2004), 7, 73.

28. Quoted in Adam Rome, "'Give Earth a Chance': The Environmental Movement and the Sixties," *JAH* 90, no. 2 (September 2003): 525–554; "Earth Day 1970—Miscellaneous Participation" folder, Box 157, NP.

29. Anne Chisholm, *Philosophers of the Earth: Conversations with Ecologists* (New York: Dutton, 1972), 142, 143.

30. Stephanie Mills, "Mills College Valedictory Address," in Bill McKibben, *American Earth: Environmental Writing since Thoreau* (New York: Literary Classics of the United States, 2008), 469–471. Also see Mills, *Whatever Happened to Ecology?* (San Francisco: Sierra Club Books, 1989); Mills, "Why I Chose to Be Childless," *More* Magazine Web site, http://www.more.com/2050/7097-why-i-chose-to-be.

31. Mills, "Valedictory Address," 470; Mills, "O and All the Little Babies in the Alameda Gardens, Yes," in *Ecotactics: The Sierra Club Handbook for Environment Activists*, ed. John G. Mitchell (New York: Pocket Books, 1970), 81.
32. Mills, "O and All the Little Babies in the Alameda Gardens, Yes," 80, 81.
33. Phyllis Piotrow, *World Population Crisis: The United States Response* (New York: Praeger, 1973), viii.
34. "Prologue," from *The Crisis of Survival*, 2.
35. William Buckley, "The Birth Rate," *National Review*, March 23, 1965, 231. Also see the *National Review*'s special issue, "The Population Explosion," July 27, 1965. For the Republican platform, see Critchlow, *Intended Consequences*, 88. A Korean War veteran and lawyer, McCloskey was a Congressman from Paul Ehrlich's district in California and a future candidate for president within the Republican Party. For his population views, see Paul N. McCloskey, *Stanford Alumni Almanac*, 1969, Series 6, Box 2, Folder 58, EP.
36. For Bush, see Piotrow, *World Population Crisis*.
37. Republican Task Force on Earth Resources and Population, "Earth Resources and Population—Problems and Directions," *CR* 116, no. 114 (July 8, 1970); George Bush, "Remarks Introducing Republican Task Force on Earth Resources and Population's Report 'Earth Resources and Population—Problems and Directions,'" *CR* 116, no. 114 (July 8, 1970). Later in his career, Bush claimed that he was a supporter of Planned Parenthood until it took up abortion. In the 1970 report, Bush showed no great support for legalization, but not much animosity toward it, either. Instead he argued that abortion was a state, not a federal, issue and strongly emphasized the disparity of services between rich and poor. See George Bush, *All the Best, George Bush: My Life in Letters and Other Writings* (New York: Scribner, 1999), 116.
38. J. Brooks Flippen, *Nixon and the Environment* (Albuquerque: University of New Mexico Press, 2000), 34. Republicans, of course, had a strong claim to being leaders on conservation matters. Theodore Roosevelt and Gifford Pinchot had both been Republicans. In 1969, no one could know whether the environment would become a Democratic or Republican issue.
39. On March 26, John D. Rockefeller wrote the head of the president's Science office, Lee DuBridge, about the population problem. DuBridge's assistant in charge of environmental issues, Don King, wrote a memo on population, which he delivered on April 7. King concluded, "Excessive population growth has been characterized as a major problem of our time. . . . The problem is of sufficient importance that a positive position should be taken." In later memos, King mixed international and domestic concerns. On April 17, apparently prompted by a letter from Rockefeller and a memo from Lee DuBridge, Moynihan convened a meeting to discuss population. Russell Train, Nixon's advisor on environmental issues and a past president of the Conservation Foundation, also supported a strong population policy. See Donald King memo to DuBridge, April 7, 1969, Box 1, "Reader File, Jan–June, 1969," DKP. Also see Box 1, "Reader File, Jan-June, 1969," DKP. Also see Critchlow, *Intended Consequences*, 89–90; Piotrow, *World Population Crisis*, 163.
40. Rogers and Hannah memo cited in Peter J. Donaldson, *Nature against Us: The United States and the World Population Crisis, 1965–1980* (Chapel Hill: University of North Carolina Press, 1990), 26; Memorandum of Conversation, from National Security Meeting, September 2, 1969, San Clemente, California, National Archives, Nixon Presidential Materials, NSC Files, Presidential/HAK Memcons, Box 1026, June to December 1969, reprinted in David S.

Patterson and Bruce F. Duncombe, eds., *Foreign Relations of the United States 1969–1976*, vol. 4, *Foreign Assistance, International Development, Trade Policies, 1969–1972* (Washington, D.C.: Government Printing Office, 2002), 282.

41. Donald King, *The US Role in the Greening of the World*, unpublished manuscript, June 1997, Box 1, DKP, 20.
42. Richard Nixon, "Special Message to the Congress on Problems of Population Growth, July 18, 1969," in *Public Papers of the Presidents of the United States* (Washington, D.C.: Government Printing Office, 1969), 529, 525.
43. Ibid., 522.
44. Ibid., 524.
45. Ibid., 526.
46. Phyllis Piotrow emphasizes the shift to domestic concerns in Piotrow, *World Population Crisis*, 163. Donald Critchlow seconds this analysis; Critchlow, *Intended Consequences*, 91. The Nixon administration also put together several conferences and reports, including the Task Force on World Food and Environment (1969), and "Protecting the World Environment in the Light of Population Increases," a 1970 report by the Office of Science and Technology.
47. Nixon, "Special Message to the Congress on Problems of Population Growth, July 18, 1969," 525.
48. Public Law 190, 91st Congress 1st session (1 January 1970), sec. 101.
49. Lynton Caldwell, "Environment: A New Focus for Public Policy," *Public Administration Review* 23, no. 3 (September 1963): 132–139; 132. For Caldwell, see Karen Erdman, "Preventing Catastrophe," *Progressive* (August 1985); and Lynton Caldwell, *Environment: A Challenge for Modern Society* (Garden City, N.Y.: Doubleday, 1970).
50. Caldwell, "Environment: A New Focus for Public Policy," 138; Caldwell, "Politics and Population Control: A Report to the Conservation Foundation," February 20, 1966, Box 8, FF 10, VP.
51. Robert Bartlett and James Gladden, "Lynton K. Caldwell and Environmental Policy: What Have We Learned?" in Lynton Caldwell, Robert V. Bartlett, and James N. Gladden, *Environment as a Focus for Public Policy* (College Station: Texas A&M University Press, 1995), 4. For Caldwell's quotation, see 329–330.
52. F. Fraser Darling, "A Wider Environment of Ecology and Conservation," *Daedalus* 96, no. 4 (Fall 1967): 1003–1019, reprinted in *The Crisis of Survival* (New York: Morrow, 1970), 34. For NEPA's legislative and political history, see Flippen, *Nixon and the Environment*, 48–49; Matthew J. Lindstrom and Zachary A. Smith, *The National Environmental Policy Act: Judicial Misconstruction, Legislative Indifference & Executive Neglect* (College Station: Texas A&M University Press, 2001), 21.
53. Gaylord Nelson, Susan Campbell, and Paul R. Wozniak, *Beyond Earth Day: Fulfilling the Promise* (Madison: University of Wisconsin Press, 2002), chapter 1. For Nelson, see "Gaylord Nelson," *Current Biography*, 1960; Bill Christofferson, *The Man from Clear Lake: Earth Day Founder Gaylord Nelson* (Madison: University of Wisconsin Press, 2004), 302–312.
54. Nelson, Campbell, and Wozniak, *Beyond Earth Day*, chapter 1. For Nelson's early thinking on population, see Christofferson, *The Man from Clear Lake*, 141; and Nelson to John Fitzgerald Kennedy, August 29, 1963, Name File, White House Central Files, LBJ Library, quoted in Christofferson, *The Man from Clear Lake*, 177. For the 1970 letter, see Gaylord Nelson to Mrs. Marylu Raushenbush, February 25, 1970, Box 136, "Population,

1969–1970" folder, NP. The "breeding" quote is from Gaylord Nelson, "Our Polluted Planet," in *The Crisis of Survival*, 193. The "frontier" quote is from Gaylord Nelson, "Remarks accompanying S. 3151, The Introduction of Environmental Quality Education Act," *CR* 115, no. 191 (November 19, 1969). The Ehrlich quotation was from Paul Ehrlich, "Eco-Catastrophe!," *Ramparts*, September 1969, 24–28; see *CR*, September 12, 1969.

55. A complete list is impossible, but a sample would include urban planner Ian L. McHarg, Environmental Defense Fund lawyer Charles F. Wurster, geographer Wes Jackson, ecologist Paul Sears, former governor of Delaware and future director of the Audubon Society Russell Peterson, and photographer Phil Hyde. Ian L. McHarg, "Man and Environment," in Leonard J. Duhl, ed., *The Urban Condition: People and Policy in the Metropolis* (New York: Basic Books, 1963), 45; Charles F. Wurster to Ehrlich, January 24, 1969, Box 44, "Environmental Defense Fund" folder, EP; Wes Jackson to Ehrlich, March 1970, Series 1, Box 28, "Wes Jackson" folder, EP. For Sears, see his review of Ehrlich's *The Population Bomb* in the March 1969 *Audubon*, 94. "The Environmental Books Hall of Fame," *Not Man Apart*, November 1973, ACCN 1994–007, Box 7, "Not Man Apart" folder, EP.

56. Michael P. Cohen, *The History of the Sierra Club, 1892–1970* (San Francisco: Sierra Club Books, 1988), 440; Michael Egan, *Barry Commoner and the Science of Survival: The Remaking of American Environmentalism* (Cambridge: MIT Press, 2007), 121; Jeremy Main, "Conservationists at the Barricades," *Fortune* 81 (February 1970): 144–147.

57. Hayes warns of a population that is "expanding geometrically" in "Earth Day: A Beginning," in *The Crisis of Survival*, 211. For Hayes, see Flippen, *Nixon and the Environment*, 9. For Sydney Howe, see Nelson, Campbell, and Wozniak, *Beyond Earth Day*, chapter 1. News release from the Office of Environmental Teach-In, Inc., December 4, 1969, Box 44, "Earth Day/Environmental Action" folder, EP.

58. "Drive to Stop Population Growth," *US News and World Report* 68 (March 2, 1970); M. Mead, "Crisis of Our Overcrowded World," *Redbook*, October 1969, 40; R. Gordon, "In My Opinion: We Must Stop Multiplying!" *Seventeen*, May 1970, 234; A. T. Day, "Population Increase: A Grave Threat to Every American Family," *Parents Magazine*, October 1969, 56–57; *Look*, April 21, 1970; Lord Ritchie-Calder, "Mortgaging the Old Homestead," *Foreign Affairs* 48 (January 1970): 207–220; Charles Ogburn, "Catastrophe by the Numbers: American Population," *American Heritage* 21 (December 1969): 114–117; Morris Udall, "Standing Room Only on Spaceship Earth," *Reader's Digest*, December 1969, 131–135; "Squeezing into the '70s," *Life*, January 9, 1970, 9.

59. "A Day Devoted to a Better Earth," *Life*, April 24, 1970, 41; Flippen, *Nixon and the Environment*, 8; "Dawning of Earth Day," *Time*, April 27, 1970, 46; "Derailing Demographers of Doom," *WSJ*, August 27, 1970.

60. In general, see Environmental Crisis Week Calendar of Activities, "Earth Day 1970—Miscellaneous Participation" folder, Box 157, NP; and "Earth Day Teach-In," *Daily Universe*, Tuesday April 21, 1970, "Earth Day Clips, 1970" folder, Box 157, NP. For San Jose State, see Andrew Sansom, "Students Mobilize for Eco-Action, Parks and Recreation," April 1970, Box 156, Folder 38, NP. For Louisville, see "Save-the-Earth Week," *Louisville Courier-Journal and Times*, Sunday April 19, 1970, "Earth Day Clips, 1970" folder, Box 157, NP. For high school events, see Environmental Crisis Week Calendar of Activities, "Earth Day 1970—Miscellaneous Participation," Box 157, NP. For Concerned Demographers, see "Earth Week Lifestyle on Trial," UW Organizations and Madison Civic Groups, Box 157, Folder 13, NP.

61. Ehrlich quoted in "The Unquiet Scientist: A Profile," *Observer* (London), September 9, 1971, Series 6, Box 26, Folder 32, EP; "ZPG: New Movement Challenges the U.S. to Stop

Growing," *Life*, April 17, 1970, 32–37. According to a 1972 article, ZPG had 350 local chapters totaling 35,000 members; 65 percent were Protestant, 15 percent were Catholic. Almost all were white and very educated: 98 percent had education beyond high school; 29 percent had a master's or Ph.D. Many had science backgrounds. Green, "The Militant Malthusians."

62. *Population Growth and Ecology*, E-Day Issue, Zero Population Growth—Madison, April 1970, Box 156, Folder 38, NP. For the popularity of the Kaibab incident, see C. John Burk, "The Kaibab Deer Incident: A Long-Persisting Myth," *BioScience* 23, no. 2 (February 1973): 113–114; Christian C. Young, *In the Absence of Predators: Conservation and Controversy on the Kaibab Plateau* (Lincoln: University of Nebraska Press, 2002), 3, 207.

63. Sierra Club publicity letter, July 30, 1968, Series 6, Box 2, Folder 54, EP; Morris K. Udall, "Preparing for Peace—VI: Our Spaceship Earth—Standing Room Only," *Congressman's Report*, July 30, 1969; originally in the *Arizona Republic*, July 27, 1969, Box 16, "Opinions" folder, EP; "Nelson Sees Population as No. 1 Problem," *Milwaukee Sentinel*, April 17, 1970, quoted in Christofferson, *The Man from Clear Lake*, 308; Environmental Teach in, April 19–23, 1970, Handout, Box 157, Folder 13, NP.

64. Gaylord Nelson, "Our Polluted Planet," *Progressive*, November 1969, 13–17; Nelson, Earth Day Speech, University of Wisconsin, Madison, April 21, 1970, videotape, visual materials archive, Wisconsin Historical Society, quoted in *Progressive*, November 1969, 6. For other Nelson views on the Cold War, see Christofferson, *The Man from Clear Lake*, 309.

65. For early 1970 sales numbers, see Lois Brown, Ballantine Books, to Paul Ehrlich, January 7, 1970, Series 6, Box 2, Folder 53, EP. For later numbers, see "Ballantine Books Press Release," ACCN 2000–296, Box 1, Folder 20, EP. For Ehrlich's activities, see D. M. Rorvick, "Ecology's Angry Lobbyist," *Look*, April 21, 1970, 42–44; and "Some are Paying Heed," *New Scientist and Science Journal*, September 16, 1971, 634–636. In 1970, Ehrlich contributed several environmental articles each to *Saturday Review* and the Hearst Papers, and had individual articles in *TWA Ambassador, McCalls, The Progressive, National Wildlife, The Plain Truth,* the *NYT, National New Democratic Coalition Newsletter, Bulletin of the Field Museum of Natural History, Field and Stream, Biological Conservation, Natural History,* and *Center Magazine* (Center for the Study of Democratic Institutions, Santa Barbara, Cal.). With his wife Anne, he also published a popular environmental studies textbook: Ehrlich and Ehrlich, *Population, Resources, Environment: Issues in Human Ecology* (San Francisco: Freeman, 1970).

66. Dick Gregory, "Rebellious Controlled," *Palo Alto Times*, January 18, 1969, and Paul Ehrlich to Dick Gregory, January 20, 1969, both in ACCN-1994–007, Box 8, "Dick Gregory" folder, EP; Paul Ehrlich to Walter Thompson, November 26, 1969, Box 9, "PRE Statement re Genocide" folder, EP. For Thompson, see Critchlow, *Intended Consequences*, 61, 152–153.

67. Elaine Wheatley, "One You Can't Afford to Miss," *Wichita Falls Times*, Texas, April 12, 1970, Series 6, Box 2, Folder 37, EP. Letter from Gordon and May Slifer, Black River Falls, Wisconsin, to *Olathe Daily News*, March 18, 1970, Series 6, Box 2, Folder 38, EP.

68. Melvin Laird, "Remarks," in *Proceedings of Symposium on Human Ecology, Consumer Protection and Environmental Health Service* (Washington, D.C.: U.S. Department of Health, Education, and Welfare, 1968), 98; Udall, "Preparing for Peace VI: Our Spaceship Earth"; Walter E. Howard, "The Population Crisis Is Here Now," *BioScience*, September 1969, reprinted in Wes Jackson, *Man and the Environment* [1971] 2nd ed. (Dubuque, Iowa: Wm. C. Brown, 1973), 185. Some Malthusian biologists, such as

Marston Bates, rejected Calhoun's arguments. Marston Bates, *A Jungle in the House: Essays in Natural and Unnatural History* (New York: Walker, 1970), 130.

69. Paul Ehrlich, "Morality and Politics: A Hastily Prepared Personal Position Paper," circa 1969, Series 6, Box 26, Folder 40, EP; Paul Ehrlich, "Population Control" manuscript, June 5, 1972, Series 1, Box 23, "Saturday Evening Post" folder, EP, 15.

70. Paul Ehrlich to Editor, *Los Angeles Times*, April 17, 1969, ACCN 1994–007, Box 7, "Los Angeles Times" folder, EP; "Fighting to Save the Earth from Man," *Time*, February 2, 1970, 56–57; *US News and World Report*, March 2, 1970, 37.

71. Richard A. Watson, Letter to the Editor, no date, *FOCUS/Midwest* 8, no. 52, Box 16, "Opinions" folder, EP, 4; Udall, "Preparing for Peace." When two American officials visited South Vietnamese Vice-Prime Minister Vien in 1972, he pulled a copy of Ehrlich's *Population, Resources, Environment* (1970) from his desk and quoted from it. One of these American officials was pleased because "this top Vietnamese leader had such a good understanding of the fundamental development problem of this country." See Charles B. Green to A. E. Farwell, August 21, 1972, Series 1, Box 26, Folder 33, EP.

72. Paul Ehrlich address, Madison, Wisconsin, March 13, 1970, University of Wisconsin–Madison Archives; "MAN is the Endangered Species," *National Wildlife* 8 (April–May 1970): 39; Peter Collier, "An Interview with Ecologist Paul Ehrlich," *Mademoiselle*, April 1970, 188–189.

73. Howard, "The Population Crisis Is Here Now," in Jackson, *Man and the Environment*, 190; George Wald, "A Better World for Fewer Children"; Brower quoted in National Staff of Environmental Action, *Earth Day: The Beginning* (New York: Arno Press/NYT Books, 1970), 218; *Newsweek*, January 26, 1970, 32.

74. Ben Wattenberg, "The Nonsense Explosion: Overpopulation as a Crisis Issue," *New Republic*, April 4 and 11, 1970, 18–23. Urban planner Jane Jacobs made a similar argument. See Reports on Speech Earth Day, Milwaukee, WISN-TV 12, May 6, 1970, Box 136, "Population, 1969–1970" folder, NP. For a new left critique, see James Ridgeway, *The Politics of Ecology* (New York: Dutton, 1970).

75. Steve Weissman, "Why the Population Bomb Is a Rockefeller Baby," in *Eco-Catastrophe* (San Francisco: Canfield Press, 1970), 27–28.

76. "The Plowboy Interview: Paul Ehrlich," *Mother Earth News*, July 1974, ACCN 1994–007, Box 7, "Mother Earth News" folder, EP, 7.

CHAPTER 8   WE'RE ALL IN THE SAME BOAT?!

1. Don Mitchell, M.D., to President Richard M. Nixon, April 14, 1970, Box 40, "Nixon" folder, EP.

2. Tri-Cities, Washington ZPG Chapter, "Earth Day Newsletter," April 21, 1970, Box 157, "Earth Day 1970 Nelson Clips" folder, NP; "Ecology: A Cause Becomes a Mass Movement," *Life*, January 30, 1970, 22–30.

3. Matthew Connelly, *Fatal Misconception: The Struggle to Control World Population* (Cambridge: Harvard University Press, 2008), 50.

4. Björn-Ola Linnér, *The Return of Malthus: Environmentalism and Post-War Population-Resource Crises* (Isle of Harris, UK: White Horse Press, 2003), 196.

5. Beck and Kolankiewicz also focus on the political and cultural changes of the 1970s but tend to dismiss them as political correctness, not sound arguments. Roy Beck and Leon Kolankiewicz, "The Environmental Movement's Retreat from Advocating U.S.

Population Stabilization (1970–1988): A First Draft of History," *Journal of Policy History* 12, no.1 (2000): 123–156; 127, 128.

6. First National Congress on Optimum Population and Environment (COPE) Program, 1970, Box 16, EP, 5; Carl Rowan, "'Genocide' Fear Deserves Study," *Washington Evening Star*, June 24, 1970; "Position Statement—Black Caucus," First National Congress on Population and Environment, Booklet, Fall 1970, p. 20, SCP 71–103, Box 106, Folder 3, "Population Clippings," PPFA; "Black Caucus Statement on Withdrawing from First National Congress on Optimum Population and Environment," June 1970, "Black Attitudes" folder, Box 107, PPFA. For a conference summary, see "Population Activities in the U.S.," *PB* 6, no. 6 (December 1970). For more, see Joseph Reilly, "Canon Shaw's Quiet Mission: Defuse the Population Bomb," *Chicago Sun-Times*, May 12, 1970; "Scientific and Welfare Groups Open a 4-Day Study of Population Growth," *NYT*, June 9, 1970; Sierra Club Papers 71–103, Box 136, Folder 1, SCP.

7. Dick Gregory, "Answer to Genocide," *Ebony* 26, no. 12 (October 1971): 66–72; Robert G. Weisbord, *Genocide? Birth Control and the Black American* (Westport, Conn.: Greenwood, 1975), 129.

8. Commission on Population Growth and the American Future, *Population Growth and the American Future* (Washington, D.C.: Government Printing Office, 1972), 72.

9. Ronald Walters, "Population Control and the Black Community: Part II," *The Black Scholar*, June 1974, 29; Roy Innis, "Speaking Out: The Zero Population Growth Game," *Ebony*, November 1974, 110.

10. Carl T. Rowan, "Birth Control Issue," March 28, 1971, Field Enterprises, Inc., "Black Attitudes" folder, Box 107, PPFA.

11. Paul Ehrlich, transcript of First National Congress on Optimum Population and Environment (COPE) speech, June 9, 1970, Series 1, Box 44, "MS-First National Congress on Optimum Population and Environment" folder, EP. Ehrlich published a version of this speech in Paul R. and Anne H. Ehrlich, "Population Control and Genocide," *New Democratic Statesman*, July 1970, and in Paul R. Ehrlich and John P. Holdren, *Global Ecology: Readings toward A Rational Strategy for Man* (New York: Harcourt Brace Jovanovich, 1971).

12. Ibid. Carl Rowan reacted favorably to Ehrlich's COPE speech; Rowan, "'Genocide' Fear Deserves Study." For the revised edition with ODC, see Paul Ehrlich, *The Population Bomb* (New York: Ballantine, 1971), 7. Ehrlich made similar arguments in Paul R. Ehrlich and Richard L. Harriman, *How To Be a Survivor: A Plan to Save Spaceship Earth* (New York: Ballantine, 1971); Paul R. Ehrlich and Anne H. Ehrlich, *The End of Affluence: A Blueprint for Your Future* (New York: Ballantine, 1974).

13. James T. Patterson, *Restless Giant: The United States from Watergate to Bush v. Gore* (New York: Oxford University Press, 2005), 133; Robert M. Collins, *More: The Politics of Economic Growth in Postwar America* (New York: Oxford University Press, 2000), 98, 143; E. F. Schumacher, *Small Is Beautiful; Economics as if People Mattered* (New York: Harper & Row, 1973); Donella H. Meadows et al., *The Limits to Growth: A Report for the Club of Rome's Project on the Predicament of Mankind* (Washington, D.C.: Potomac Associates, 1972). Hoff explores the rise of the ecological economists in his chapter 6; Derek Hoff, *The State and the Stork: The Population Debate and Policymaking in United States History* (Chicago: University of Chicago Press, forthcoming).

14. Donald Fleming, "Roots of the New Conservation Movement," *Perspectives in American History* 6 (1972): 40; *Time*, February 2, 1970. The best source on Commoner is Michael

Egan, *Barry Commoner and the Science of Survival: The Remaking of American Environmentalism* (Cambridge: MIT Press, 2007).

15. Barry Commoner, *The Closing Circle: Nature, Man, and Technology* (New York: Knopf, 1971), 183, 214. Commoner also criticized Ehrlich's position in testimony at the National Commission on Population Growth and the American Future in November 1970, and at the annual meeting of the American Association for the Advancement of Science in December 1970.
16. Commoner, *The Closing Circle*, chapter 11, "The Question of Survival."
17. *The Closing Circle*, 131, 129, and 144.
18. Paul Ehrlich and Richard Holden, "One-Dimensional Ecology," *Ecologist* 2, no. 8 (August 1972): 11; this first appeared in the *Bulletin of the Atomic Scientists*; Commoner, *Closing Circle*, 177. As this debate was ongoing, Ehrlich devised his IPAT formula: environmental impact equals population multiplied by affluence multiplied by technology.
19. "Dr. Ehrlich's Backlash," *Nature* 237 (June 16, 1972): 359. Also see John Maddox, *The Doomsday Syndrome* (New York: McGraw-Hill, 1972).
20. David Harvey, "Population, Resources, and the Ideology of Science," *Economic Geography* 50, no. 3 (1974): 256–257.
21. Pete Seeger, Letter to the Editor, *Environment* 4, no. 5 (June 1972): 40.
22. Donald T. Critchlow, *Intended Consequences: Birth Control, Abortion, and the Federal Government in Modern America* (New York: Oxford University Press, 1999), 161. I discuss both abortion and immigration in greater detail later in this chapter.
23. For the ecological faction, see ibid., 163. Paul Ehrlich to John D. Rockefeller III, August 6, 1971, Box 44, "Presidential Commission of Population Growth and the American Future" folder, EP.
24. Commission on Population Growth and the American Future, *Population Growth and the American Future* (Washington, D.C.: Government Printing Office, 1972); Ehrlich to Rockefeller, April 13, 1972, Box 44, "Presidential Commission on Population Growth and the American Future" folder, EP; Ehrlich, "Is the Tide Turning?" *Bioscience* 22, no. 5 (May 1972): 277.
25. Richard M. Nixon, "Statement by the President on the Report of the Commission on Population Growth and the American Future," May 5, 1972, in *Public Papers of the Presidents of the United States: Richard M. Nixon* (Washington, D.C.: Government Printing Office, 1974), 576–577; Critchlow, *Intended Consequences*, 169, 171. For Buchanan, see Derek Hoff, "'Kick that Population Commission in the Ass': The Nixon Administration, the Commission on Population Growth and the American Future, and the Defusing of the Population Bomb," *Journal of Policy History* 22, no. 1 (2010): 43–44.
26. In the fullest treatment of the NCPGAF, Hoff convincingly argues that the conservative "market-knows-best" approach to population emerged in the 1960s, not the 1970s, as many scholars have assumed. See Hoff, "Kick that Population Commission in the Ass," 36–38; Harold Barnett and Chandler Morse, *Scarcity and Growth: The Economics of Natural Resource Availability* (Baltimore: Johns Hopkins University Press for Resources for the Future, 1963); Esther Boserup, *The Conditions of Agricultural Growth: The Economics of Agrarian Change under Population Pressure* (New York: Aldine, 1965); "Derailing Demographers of Doom," *WSJ*, August 27, 1970.
27. Wade Green, "The Militant Malthusians," *Saturday Review*, March 11, 1972, 41; Paul Ehrlich to John D. Rockefeller III, August 6, 1971; J. Brooks Flippen, *Nixon and the*

Environment (Albuquerque: University of New Mexico Press, 2000), 102; Hoff, "Kick That Population Commission in the Ass," 31.

28. Critchlow, *Intended Consequences*, 169–170; Mrs. T. W. Hodges to John D. Rockefeller III, May 22, 1972, Commission on Population, Box 8, RG 220, National Archives, quoted in Critchlow, *Intended Consequences*, 163.

29. Critchlow, *Intended Consequences*, 171.

30. Hoff, "Kick That Population Commission in the Ass," 45. For more on Republicans and population, see the next chapter.

31. Larry Prior, "Environmentalists Surge Onward, Strengthened by Oil Cutoff," *Denver Post*, 1974; Linnér, *The Return of Malthus*, 185.

32. NSSM 200, 10, 4, 24. For more on NSSM 200, see John Sharpless, "Population Science, Private Foundations, and Development AID: The Transformation of Demographic Knowledge in the United States, 1945–1965," in Frederick Cooper and Randall M. Packard, *International Development and the Social Sciences: Essays on the History and Politics of Knowledge* (Berkeley: University of California Press, 1997), 176–200. Kissinger had a long history with the population issue, having worked with the Rockefeller Foundation on international development problems in the 1950s. See John Rockefeller III to Nelson Rockefeller, July 11, 1956, Folder 669, Box 80, RG 5, Series 1, Subseries 5, JDR.

33. Garrett Hardin, "Living on a Lifeboat," *BioScience* 24, no. 10 (October 1974): 561–568; quotations from 565, 566.

34. Commoner, *Closing Circle*, 235–246.

35. Ehrlich, "One-Dimensional Ecology," 11, 19, 19–20.

36. Connelly, *Fatal Misconception*, 310–314, 317. Also see Critchlow, *Intended Consequences*, 149–150, 181–183.

37. Frances Moore Lappé and Joseph Collins, *Food First: Beyond the Myth of Scarcity* (Boston: Houghton-Mifflin, 1977), 7.

38. Warren Belasco, *Appetite for Change: How the Counterculture Took on the Food Industry, 1966–1988* (New York: Pantheon Books, 1989), 46; "Frances Moore Lappé," in *Contemporary Authors* (Detroit: Gale Research, 2003); Frances Moore Lappé, *Diet for a Small Planet* (New York: Ballantine Books, 1975).

39. Lappé and Collins, *Food First*, 65, 112.

40. Ibid., 17–19, 7.

41. Ibid., 62–63.

42. Ibid., 19, 8. In making this argument, Lappé and Collins echoed and expanded a similar argument made by Susan George in *How the Other Half Dies*; both books anticipated similar arguments from Nobel prize–winning Amartya Sen. Susan George, *How the Other Half Dies: The Real Reasons for World Hunger* (New York: Penguin, 1976); Amartya Sen, *Poverty and Famines: An Essay on Entitlement and Deprivation* (New York: Oxford University Press, 1981).

43. Lappé and Collins, *Food First*, 46.

44. Ibid., 14.

45. Ibid., 71.

46. Walter E. Howard, "The Population Crisis Is Here Now," *BioScience*, September 1969, reprinted in Wes Jackson, *Man and the Environment* [1971], 2nd ed. (Dubuque,

Iowa: Wm. C. Brown, 1973), 189 and 191. Paul N. McCloskey, *Stanford Alumni Almanac*, 1969, Series 6, Box 2, Folder 58, EP. Howard, "The Population Crisis Is Here Now," 189.

47. Garrett Hardin, "The Tragedy of the Commons," *Science*, n.s. 162, no. 3859 (December 13, 1968), 1243–1248, 1252–1253; Garrett Hardin, "Parenthood: Right or Privilege?" *Science* 169 (1970): 427.
48. Bruce J. Schulman, *The Seventies: The Great Shift in American Culture, Society, and Politics* (New York: Free Press, 2001), 172, 174.
49. John D'Emilio and Estelle B. Freedman, *Intimate Matters: A History of Sexuality in America* (New York: Harper & Row, 1988), 315. Also see Critchlow, *Intended Consequences*, 142–144.
50. Barbara Seaman, *The Doctors' Case against the Pill* (New York: P. H. Wyden, 1969), 45–46.
51. Jennifer Nelson, *Women of Color and the Reproductive Rights Movement* (New York: New York University Press, 2003), 1, 4, and 18; and Philip Reilly, *The Surgical Solution: A History of Involuntary Sterilization in the United States* (Baltimore: Johns Hopkins University Press, 1991), 41–55.
52. "Bomb Ehrlich," Women against Genocide, circa 1970, Handout, Series 6, Box 26, Folder 30, EP.
53. Connelly, *Fatal Misconception*, 315; and Critchlow, *Intended Consequences*, 149–150.
54. Linda Gordon, "The Politics of Population: Birth Control and the Eugenics Movement," *Radical America* 8, no. 4 (July–August 1974): 61–97; Gordon, "The Politics of Birth Control, 1920–1940: The Impact of Professionals," *International Journal of Health Services* 5, no. 2 (Fall 1975): 253–277. Linda Gordon, *Woman's Body, Woman's Right: A Social History of Birth Control in America* (New York: Grossman, 1976), 393, 398, 401. Also see Dennis Hodgson and Susan Watkins, "Feminists and NeoMalthusians: Past and Present Alliances," *PDR* 23, no. 3 (September 1997): 469–523.
55. For Gandhi's quotation, see William Borders, "India Reports Gains in Population Drive," *NYT*, September 16, 1976. For more, see Connelly, *Fatal Misconception*, 318–334; Ian Robert Dowbiggin, *The Sterilization Movement and Global Fertility in the Twentieth Century* (Oxford: Oxford University Press, 2008), 188–192; and Davidson R. Gwatkin, "Political Will and Family Planning: The Implications of India's Emergency Experience," *PDR* 5 (March 1979): 29–59.
56. Kai Bird, "Indira Gandhi Uses Force," *Nation*, June 19, 1976, 747–749; Jeremiah Novak, "Fear, Force and Sterilization in India," *America*, November 27, 1976, 362–365. My research has added two bits of information. The first is a letter from Garrett Hardin to Chandrasekhar in 1975 about his upcoming visit to Santa Barbara. Hardin included a copy of "Living on a Lifeboat" in the letter. Garrett Hardin to S. Chandrasekhar, April 11, 1975; Box 13, HP. Second, a U.S. National Security Council report on population from July 1976, just as India was launching its program, argued for improving the status of women and integrating family planning with broader health programs. It also called for "strong direction from the top": "Population programs have been particularly successful where leaders have made their positions clear, unequivocal, and public, while maintaining discipline down the line from national to village levels, marshalling government workers (including police and military), doctors, and motivators to see that population policies are well administered and executed." Strong direction also included incentives for sterilization. About India's involuntary sterilization program, the report laid out pros and cons. Mentioning "reservations," it concluded that U.S. officials should avoid public criticism. "Population policies are a matter for each country to decide for itself." NSC Under Secretaries Committee, "Memo for the

President: First Annual Report on U.S. International Population Policy," July 29, 1976, accessed through CREST, National Archives II, 1, 5, 47–48.

57. Paul R. Ehrlich, Anne H. Ehrlich, and John P. Holdren, *Ecoscience: Population, Resources, Environment* (San Francisco: W. H. Freeman, 1977), xiii, xiv.

58. Ibid., 745, 376, 793, 793.

59. For population hawks, see Ehrlich, Ehrlich, and Holdren, *Ecoscience*, 793–794. For coercion, see 738, 785–789. The authors mentioned the 1976 sterilization campaign in India under Indira Gandhi without strong judgment either way. It's not known what they knew about it by the time of the book's publication.

60. Ehrlich, Ehrlich, and Holdren, *Ecoscience*, 922. Also see Stephanie Mills, "The Ehrlichs Talk: Equity and Apocalypse," *Not Man Apart*, October 1977, ACCN 1994–007, Box 7, "Not Man Apart" folder, EP.

61. P. R. Ehrlich and A. H. Ehrlich, "ZPG: Where to Now?" *Zero Population Growth National Reporter* 10, no. 1 (1978): 4–5.

62. *US News and World Report* interview with Leonard Chapman, July 22, 1974, 27–30; Elena R. Gutiérrez, *Fertile Matters: The Politics of Mexican-Origin Women's Reproduction* (Austin: University of Texas Press, 2008), 78; David M. Reimers, *Still the Golden Door: The Third World Comes to America* (New York: Columbia University Press, 1985), ix. Because immigration had been severely restricted since the 1920s, the percentage of foreign-born Americans in 1970 was the lowest it had been in the twentieth century and maybe longer.

63. Elaine Stansfield, president of the Los Angeles ZPG, "Testimony at Senate Commission on Immigration and Refugee Policy Hearings," quoted in Gutiérrez, *Fertile Matters*, 79; Anthony W. Smith, "The Gates of the Republic," *NPM*, October 1975, 2; general ZPG information from Gutiérrez, *Fertile Matters*, 85. For more on the emergence of immigration among environmentalists, see Kingsley Davis, "The Migration of Human Populations," *Scientific American* 231, no. 3 (September 1974): 93–105; Gerda Bikales, "Immigration Policy: The New Environmental Battlefield," *NPM* 51 (December 1977): 13–16; Bikales, "Illegal Immigration: An Environmental Issue," *NPM* 52 (June 1978): 2. Anyone interested in the deep history should see Edward Murray East, *Mankind at the Crossroads* (New York: Scribner, 1923); William Vogt, *Road to Survival* (New York: W. Sloane Associates, 1948).

64. *ZPG California Newsletter*, November 1978, quoted in Gutiérrez, *Fertile Matters*, 82.

65. Hardin, "Living on a Lifeboat," *BioScience* 24, no. 10 (October 1974): 561–568; quotations from 561, 562; Garrett Hardin, "The Survival of Nations and Civilization," *Science* 172, no. 3990 (June 25, 1971): 1297. An argument similar to "Living on a Lifeboat" appeared in "Lifeboat Ethics: The Case against Helping the Poor," *Psychology Today* 8 (September 1974): 38–43.

66. John H. Tanton, "International Migration as an Obstacle to Achieving World Stability," *Ecologist* 6, no. 6 (July 1976): 221–227; quotations from 227, 225, 226. For Tanton, see Gutiérrez, *Fertile Matters*, 75–79; Reimers, *Still the Golden Door*, 48.

67. Tanton, "International Migration as an Obstacle to Achieving World Stability." James Crawford, *Hold Your Tongue: Bilingualism and the Politics of English Only* (Reading, Mass.: Addison-Wesley, 1992), 151.

68. Gutiérrez, *Fertile Matters*, 81, 91.

69. Paul R. Ehrlich, Loy Bilderback, and Anne H. Ehrlich, *The Golden Door: International Migration, Mexico, and the United States* (New York: Ballantine Books, 1979), 360. For a

summary, see Anne Ehrlich and Paul Ehrlich, "Ecoscience: Immigration in the Future," *Mother Earth News*, January–February 1980, 150–151.

70. Ehrlich, Bilderback, and Ehrlich, *The Golden Door*, 355; Herman Daly, "Review of The Golden Door," *International Migration Review* 14, no. 1 (Spring 1980): 124–126.

71. Graham, *Unguarded Gates*, 104. On FAIR, compare Reimers, *Still the Golden Door*, 43, 48; Gutiérrez, *Fertile Matters*, 92; Crawford, *Hold Your Tongue*, 152; and Mark Dowie, *Losing Ground: American Environmentalism at the Close of the Twentieth Century* (Cambridge: MIT Press, 1995), 163.

CHAPTER 9   RONALD REAGAN, THE NEW RIGHT, AND POPULATION GROWTH

1. "Most Americans have failed to recognize the importance of the 1970s to today's confusing political universe," write Bruce Schulman and Julian Zelizer in a recent book on the 1970s. Bruce J. Schulman and Julian E. Zelizer, *Rightward Bound: Making America Conservative in the 1970s* (Cambridge: Harvard University Press, 2008), 4. For Carter, see Robert M. Collins, *More: The Politics of Economic Growth in Postwar America* (New York: Oxford University Press, 2000), 158.

2. Bruce J. Schulman, *The Seventies: The Great Shift in American Culture, Society, and Politics* (New York: Free Press, 2001), 193.

3. Historians have described many of the factors behind the New Right, but have mostly overlooked their rejection of Malthusian-inspired limits-to-growth thinking. For a sample of this literature, see Donald T. Critchlow, "The Conservative Ascendancy: How the GOP Right Made Political History" (Cambridge: Harvard University Press, 2007); Lisa McGirr, *Suburban Warriors: The Origins of the New American Right* (Princeton: Princeton University Press, 2001); Schulman and Zelizer, *Rightward Bound*. In a recent article, James Turner argues that debates about public lands, especially in the U.S. west, played a major role in the rise of the New Right. James Morton Turner, "'The Specter of Environmentalism': Wilderness, Environmental Politics, and the Evolution of the New Right," *JAH* 96, no. 1 (2009): 123–148.

4. Larry Prior, "Environmentalists Surge Onward, Strengthened by Oil Cutoff," *Denver Post* (1974).

5. William Shannon, "Quiet Issue," *NYT*, September 19, 1976. Jeffrey Stine, "Environmental Policy during the Carter Presidency," in Gary M. Fink and Hugh Davis Graham, *The Carter Presidency: Policy Choices in the Post-New Deal Era* (Lawrence: University Press of Kansas, 1998), 179–201.

6. *Public Papers of the Presidents of the United States: Jimmy Carter, Containing the Public Messages, Speeches, and Statements of the President, Volume 1: January 20, 1977 to June 24, 1977* (Washington, D.C.: Government Printing Office, 1977).

7. 1979 speech quoted in Collins, *More*, 158; "Transcript of President's Farewell Remarks on Major Issues Facing the Nation," *NYT*, January 15, 1981.

8. Collins, *More*, 157.

9. Schulman, *The Seventies*, 126. Historian Leo Ribuffo quoted in Collins, *More*, 157.

10. Oliphant and Rockefeller quoted in Collins, *More*, 149. For the conference, see Collins, *More*, 148–149.

11. Collins, *More*, 151–152.

12. Council on Environmental Quality (U.S.), *The Global 2000 Report to the President: Entering the Twenty-First Century* (Washington, D.C.: Government Printing Office, 1980).

13. "The Population Bomb, Reconsidered," *NYT*, November 30, 1978; "The Chicken Little Syndrome," *NYT*, July 28, 1980.
14. Schulman, *The Seventies*, 195.
15. James T. Patterson, *Restless Giant: The United States from Watergate to Bush v. Gore* (New York: Oxford University Press, 2005), 130.
16. Nelson Rockefeller, *Our Environment Can Be Saved* (New York: Doubleday, 1970), vi, 75. For John Rockefeller, see Donald T. Critchlow, *Intended Consequences: Birth Control, Abortion, and the Federal Government in Modern America* (New York: Oxford University Press, 1999).
17. For more conservatives' opposition to Malthusianism in the Nixon administration, see chapter 8 and Derek Hoff, *The State and the Stork: The Population Debate and Policymaking in United States History* (Chicago: University of Chicago Press, forthcoming), chapter 7.
18. Ronald Reagan, "Our Environment Crisis," *Nation's Business*, February 1970, 24–28. See Lou Cannon, *Governor Reagan: His Rise to Power* (New York: PublicAffairs, 2003), 177–178, 297–321.
19. For the conference, see "Reagan Will Budget Smog Fight," *Fresno Bee*, November 17, 1969, in "RR Conference on California's Changing Environment—November 15–17, 1969" folder, Box 66, Research Unit, RRL; Harold Gilliam, "Our Air Is Too Dirty, Our Water Too Polluted," *San Francisco Examiner*, November 30, 1969, in "RR Conference on California's Changing Environment" folder, ibid.
20. Reagan Press Conference, January 8, 1970, "Press Conference Transcripts 1/8/70–1/27/70," Reagan Gubernatorial Papers, Box P2, RRL.
21. Ronald Reagan, Remarks to the YMCA Model Legislature, January 30, 1970, "Speeches, 1970" folder, Box P18, RRL. Also see "Birth Control Information OK—Reagan," *San Francisco Examiner*, January 30, 1970.
22. Ronald Reagan, Remarks to American Petroleum Institute, November 16, 1971, "Speeches, 1970" folder, Box P18, RRL, all quotes from 2.
23. Ibid., 4, 5, 5.
24. Ibid., 5.
25. Ibid., 9.
26. Ronald Reagan, Speech to AFL CIO, March 7, 1973, Statements [Environment], Box RS 130, RRL.
27. For other conservatives, see Meg Jacobs, "The Conservative Struggle and the Energy Crisis," in Schulman and Zelizer, *Rightward Bound*, 193–209.
28. Ronald Reagan, Speech to Cattleman's Association, December 15, 1973, "RR Statements—Energy Crisis" folder, Box 130, RRL.
29. Paul Boyer, "The Evangelical Resurgence in 1970s American Protestantism," in Schulman and Zelizer, *Rightward Bound*, 29–51. Also see Beth Bailey, "Religion," in David R. Farber and Beth L. Bailey, *The Columbia Guide to America in the 1960s* (New York: Columbia University Press, 2001), 296–303. For more on the emergence of the cultural right, see John D'Emilio and Estelle B. Freedman, *Intimate Matters: A History of Sexuality in America* (New York: Harper & Row, 1988); Lisa McGirr, *Suburban Warriors: The Origins of the New Right* (Princeton: Princeton University Press, 2001); Janice M. Irvine, *Talk about Sex: The Battles over Sex Education in the United States* (Berkeley: University of California Press, 2002).

30. One evangelical leader pushing environmental issues has been Richard Cizik, leader of the National Association of Evangelicals in the first decade of the twenty-first century. See Deborah Solomon, "Earth Evangelist," *NYT*, April 3, 2005.
31. The association of population control and abortion rights persisted for years. Roy Beck and Leon Kolankiewicz, "The Environmental Movement's Retreat from Advocating U.S. Population Stabilization: A First Draft of History," *Journal of Policy History* 12, no. 1 (2000): 129.
32. John A. Steinbacher, *The Child Seducers* (Fullerton, Cal.: Educator Publications, 1971). For 1990s conservatives, see Matthew Connelly, *Fatal Misconception: The Struggle to Control World Population* (Cambridge: Harvard University Press, 2008), chapter 9.
33. D'Emilio and Freedman, *Intimate Matters*.
34. For conservatives, see Patterson, *Restless Giant*, 136; and Schulman, *The Seventies*, 201; Gerald N. Grob et al., *Interpretations of American History: Patterns and Perspectives*, 7th ed., vol. 2 (New York: Free Press, 2000), 405; Irvine, *Talk about Sex*, 35. Also see William C. Martin, *With God on Our Side: The Rise of the Religious Right in America* (New York: Broadway Books, 1996), chapters 4 and 5.
35. Steinbacher, *The Child Seducers*, 252; Constance Horner, "Is the New Sex Education Going Too Far?" *NYT*, December 7, 1980. Steinbacher also attacked secular humanism and bumper stickers such as "More Deviation, Less Population."
36. For SEICUS and the disputes in 1968 and 1969, see Irvine, *Talk about Sex*, 9. Ehrlich, *The Population Bomb*, 180 and 182.
37. Francis A. Schaeffer and C. Everett Koop, *Whatever Happened to the Human Race?* (Old Tappan, N.J.: Revell, 1979), 28, 124; Tim F. LaHaye, *The Battle for the Mind* (Old Tappan, N.J.: Revell, 1980), 63. For other attacks on secular humanism, see Edward Dobson quoted in Godfrey Hodgson, *The World Turned Right Side Up: A History of the Conservative Ascendancy in America* (Boston: Houghton Mifflin, 1996), 178. For conservative Protestants and Catholics, see Patterson, *Restless Giant*, 136; and Paul Boyer, "The Evangelical Resurgence in 1970s American Protestantism," in Schulman and Zelizer, *Rightward Bound*, 36–37.
38. Frances Frech, untitled, *Kansas City Star*, November 19, 1972.
39. Marjorie J. Spruill, "Gender and America's Right Turn," in Schulman and Zelizer, *Rightward Bound*, 71. Also see Patterson, *Restless Giant*, 135–136; Schulman, 201; Bailey, "'She Can Bring Home the Bacon?': Negotiating Gender in Seventies America," in Beth L. Bailey and David R. Farber, *America in the Seventies* (Lawrence: University Press of Kansas, 2004), 109. Editorial in *Kenosha News*, March 3, 1972.
40. Paul Weyrich, "Senate Hands Catholics a Bitter Pill," *Wanderer*, August 6, 1970.
41. Jerry Falwell, *Listen, America!* (Garden City, N.Y.: Doubleday, 1980), 56, 104–108.
42. For more on the changes in conservative economic thought in the 1960s and early 1970s, see chapter 8 above and Hoff, *State and the Stork*.
43. Julian Simon and Rita Simon, *A Life against the Grain: The Autobiography of an Unconventional Economist* (New Brunswick, N.J.: Transaction, 2002), chapter 20. In 1983, Simon became a professor of management at the University of Maryland.
44. Julian Simon, *The Ultimate Resource* (Princeton: Princeton University Press, 1981), 9 and 242.
45. On Simon, see A. C. Kelley, *International Encyclopedia of Social Science and Behavioral Science*, vol. 17, ed. Neil J. Smelser and Paul B. Baltes (Oxford: Elsevier Science, 2001), 11786. For Coale and Hoover, see chapter 3 above.

46. Allen C. Kelley, "The Population Debate in Historical Perspective: Revisionism Revised," in Nancy Birdsall, Allen Kelley, and Steven Sinding, eds., *Population Matters: Demographic Change, Economic Growth, and Poverty in the Developing World* (Oxford: Oxford University Press, 2001), 24–54.
47. Simon, *Ultimate Resource*, 4.
48. Ibid., 11. Ironically, Simon had long suffered from depression. See Julian Simon, *Good Mood: The New Psychology of Overcoming Depression* (LaSalle, Ill.: Open Court, 1993).
49. Patterson, *Restless Giant*, 112; Carter quoted in Collins, *More*, 161, 162.
50. "Transcript of Ronald Reagan's Remarks at New Conference," *NYT*, October 15, 1980.
51. Douglas E. Kneeland, "Teamsters Back Republican," *NYT*, October 10, 1980; Timothy Stanley, *Kennedy vs. Carter: The 1980 Battle for the Democratic Party's Soul* (Lawrence: University Press of Kansas, 2010), 202. Collins emphasizes the importance of Representative Jack Kemp of New York in the Republican Party's discovery of the growth agenda; Collins, *More*, 191.
52. Collins, *More*, 196; Republican Party Platform of 1980, adopted by the Republican National Convention July 15, 1980, Detroit, Michigan; http://www.presidency.ucsb.edu/ws/index.php?pid=25844#axzz1X13Wn5Wx. Also see Andrew Busch, *Reagan's Victory: The Presidential Election of 1980 and the Rise of the Right* (Lawrence: University Press of Kansas, 2005), 10–11.
53. "Review and Outlook: Earth Choice," *WSJ*, April 22, 1980. Also see Irving Kristol's attacks on environmentalism in Kristol, *Two Cheers for Capitalism* (New York: Basic Books, 1978), 46, 48–49.
54. Patterson, *Restless Giant*, 147; Collins, *More*, 197.

CONCLUSION

1. Samuel Hays, *Beauty, Health, and Permanence: Environmental Politics in the United States, 1955–1985* (Cambridge: Cambridge University Press, 1987); Adam Rome, "'Give Earth a Chance': The Environmental Movement and the Sixties," *JAH* 90, no. 2 (September 2003): 525–554.
2. Edmund Muskie, *Journeys* (Garden City, N.Y.: Doubleday, 1972), 103.
3. Walter E. Howard, "The Population Crisis Is Here Now," September 1969, reprinted in Wes Jackson, *Man and the Environment*, 2nd ed. (Dubuque, Iowa: Wm. C. Brown, 1973), 181.
4. Steven M. Gillon, *Boomer Nation: The Largest and Richest Generation Ever and How It Changed America* (New York: Free Press, 2004), 21.

EPILOGUE

1. Joel E. Cohen, "Human Population Grows Up," in Laurie Ann Mazur, ed., *A Pivotal Moment: Population, Justice, and the Environmental Challenge* (Washington, D.C.: Island Press, 2010), 27.
2. Laurie Mazur, "Introduction," ibid., 1.
3. Paul Ehrlich and Anne Ehrlich, "The Population Bomb Revisited," *Electronic Journal of Sustainable Development* 1, no. 3 (2009): 67, 68, 63, 68.
4. Fred Pearce, *The Coming Population Crash and Our Planet's Surprising Future* (Boston: Beacon, 2010), 210–213, 194.

INDEX

abortion: access to, 147, 158–159, 160, 177; conservatives on, 185, 186, 212; cultural right *vs.* population planners on, 211–212; "on demand," 10, 159, 191; eugenicists on, 74, 154; forced, 73; illegal, 74; and population control in Japan, 69, 78; and women's movement, 191
abundance, 2, 5, 61, 62–63, 65, 133
Affluent Society, The (Galbraith), 82–83
African Americans: criticism of population movement, 109, 111, 172, 173, 178–180; criticism of reproduction rate of, 107; and environmental risks, 179–180; forced sterilization of, 192–193; and racial riots, 108; and urbanization, 105, 113–114
aggregate demand, 31–32, 34, 68
agriculture: and foreign policy, 21, 138; green revolution, 66, 102–103, 127, 133, 154–155, 177, 226, 231; history of, 136; hybrid corn, 20, 21; hybrid seeds, 145–146, 231; permanent, 21; pesticides, 135, 142, 209; soil erosion, 21, 26, 45, 48, 57–58, 142, 199; subsistence, 54; war, effect on, 38, 45
American Association for the Advancement of Science (AAAS), 60
American Eugenics Society, 39, 67
American Museum of Natural History, 39
Anderson, Carol, xiii
Animal Ecology (Elton), 25, 49
animal/human divide, 15, 22, 23, 25, 27–28, 41, 49, 73, 81, 119, 128, 151, 217, 228
Anker, Peder, xiii, 25–26
apocalypticism, 20, 72, 76, 88–89, 136, 145, 209, 223
Australia, resource degradation in, 48–49
Avila, Enrique, 54

baby boom, 7, 79, 80, 84, 105, 117–118, 160–162, 221–222, 223, 227
Bailey, Beth, 110–111
Bangladesh, 187, 189, 196
Barnett, Harold, 185, 216
biology. *See* ecology; eugenics; evolution; population biology
biotic resistance, 48
"Biotic View of the Land, A" (Leopold), 26–27
birth control, 4, 9–10; access to, 147, 158, 169, 177; coercive, 153, 175; conservatives on, 211–212; dissemination to poor women, 123; domestic programs, 101–102, 110–111; and health risks, 192; international programs, 101–102; in Japan, 69, 78, 90; legislation, 109; for lower classes, 22; and racism, 121–122; for single persons, 158; technology for, 78, 121–122; Vogt on, 52, 158. *See also* abortion; sterilization
birth rate: declining, in U.S., 32, 209, 226; declining, worldwide, 189, 226; eugenic interest in, 19; increasing, in U.S., 105; modern medicine, effect on, 88–89. *See also* baby boom
Borlaug, Norman, 102
Borstelmann, Thomas, xiii
Boserup, Esther, 185, 216
Bostrom, Harold, 152–153
Boulding, Kenneth, 75–76, 116–117, 150
Bowers, Richard, 159, 160
Broome, Harvey, 120
Brower, David, 116, 120, 122, 124–125, 140, 174, 208
Brown, Harrison, 73–75, 80
Brown, Lester, 137
Buchanan, Pat, 185
Burch, Guy I., 45, 88
Bush, George H. W., 163, 164, 207; on population, 163–164

Caldwell, Lynton, 73, 80, 167–168
Calhoun, John, 112, 113, 143, 172
California, 114–115, 132–135, 136, 160, 207–208
cancer, as metaphor for population growth, 112, 115, 149
Carrighar, Sally, 112
Carr-Saunders, Alexander, 25–26, 49
carrying capacity, 15, 16, 20, 23, 38, 70, 72; defining, 18, 47; degradation of, 48–49; difference from population composition, 18–19; predicting, problems with, 73; of wildlife populations, 23–25, 27, 28–29, 49, 146, 170–171
Carson, Rachel, xiii, 2, 6, 25, 135, 144
Carter, Jimmy, 201, 203–206, 218
Catholics, 59, 92, 97, 107, 148, 162, 175, 212, 214
Chapman, Royal, 19, 24, 48

285

children: as consumers, 117–118, 148; as focus of society, 105, 133; and financial disincentives, 147; and financial incentives, 121–122
China, 48, 73, 98, 144, 188, 212
Chinification, 9, 105, 122–123
Cisler, Lucinda, 157, 158, 159
civil rights movement, xiii, 55, 92, 106, 109, 130–132, 192
Clark, Joseph, 92–93, 109
climate change, xv, 143, 205, 231
Clinton, Dale, 110–111
Coale, Ansley, 69, 216
coercion, in population programs: African American critique of, 178; Brown on, 73–74; Commoner critique of, 182; Ehrlich and, 126, 148–149, 153, 172, 175, 185, 195–196; environmentalists and, 10, 11, 75, 78, 94, 226, 227; feminist critique of, 190–194; Hardin on, 153–155; Johnson on, 101, Reagan on, 208; ZPG on, 160
Cold War: and ecology, xi–xii; as economic struggle, 90–91; and environmental movement, xiii; foreign assistance, effect on environment, 37–38; and Hugh Moore's "The Population Bomb," 88–91; and India, 97–103; and modernization theory, 94–97, 98, 103, 146; national security during, 88–91; and population programs, 9, 10, 85–88, 99–101, 103, 144–146, 151, 221–222, 224, 226–227
Collins, Joseph, 188, 189–190, 196
Collins, Robert, 32, 61, 180, 181, 204, 205
colonialism, 10, 54, 95, 190
Commoner, Barry, 181–184, 187–188
communism, 86, 88, 89, 90–91
Congress on Optimum Population and Environment (COPE), 178
Connelly, Matthew, 46, 177, 194
conservation movement, xiv, 2, 5, 6, 7, 34, 37; move toward environmentalism, 21, 23, 26, 28, 47, 50, 59, 80, 124, 129, 142, 143; Paley Commission, 58; post–World War II divisions within, 46, 57, 96; shifts in during World War II, 42, 60
Conservation Foundation, 72–73
conservatives. *See* Republican Party
consumption, xiv, 8, 11–12; Carter on, 204; cultural right and, 211, 215; disparities in, 76, 146, 199; Ehrlich on, 61, 148–150, 162, 172–173, 181; Ehrlich-Commoner debates and, 182; environmentalism and, 223, 227; interwar biologists and, 15, 18–19, 22–26, 28; Keynesian approach to, 29–34, 117; Lappe and Collins on, 190; Osborn on, 38, 50, 51; overconsumption, 2, 11–12, 37, 65, 72, 181, 203, 227; post–World War II, xii, 32–33; and suburbs, 116–118; underconsumption, 31; Vogt on, 38, 50–51
contraception. *See* birth control
Cook, Robert, 90
Council of Economic Advisors (CEA), 32
counterculture, 127, 147–149, 162

Cowles, Raymond, 120–122
Critchlow, Donald, 107, 109, 184, 185

Darling, Frank F., 74, 121, 122, 168
Darling, Jay N. "Ding," 14, 44, 45, 58
Darwin, Charles, 5, 25, 127–128
Darwinian Synthesis, 128–129, 141, 146, 223
Dasmann, Raymond, 118, 133; on change as not progress, 129; on environmental conservation, 80–82; global approach of, 76–77; on roles of women, 77, 159; on suburbanization, 114–115, 116; on urbanization, 134; on voluntary forms of population control, 78
Davis, Kingsley, 68, 70, 72, 133, 137, 144
DDT, xi, 2, 52, 135, 141, 142, 181, 209
decolonization, 62, 98
deer population irruptions. *See* Kaibab Plateau
demographic transition theory, 68–69
Destruction of California, The (Dasmann), 76, 114, 129
détente, 94, 165–166, 224
developing countries. *See* third world
Development Decade, 94, 97, 100
diminishing returns, 22
Dobzhansky, Theodosius, 128, 132, 133
domino theory, xi, 89
Draper, William, 85–87, 89–90, 91, 104
Draper Committee, 86, 91
DuBridge, Lee, 165, 270n39
Dudziak, Mary, xiii

Earth Day, xi, xiv, 2–3, 152, 168–171
East, Edward M., 13, 15, 19–23, 222
ecology, xii, 2, 11, 23, 27, 135, 223; defining, 37; human ecology, 73, 93, 96; Odum on, 79–80; postwar shift toward, 47, 50, 120, 136; wildlife ecology, 25, 28–29, 38, 39–40, 41, 43, 49
Economic Consequences of the Peace, The (Keynes), 33
economic growth, xiv–xv, 11–12, 15; baby as engine for, 116–118, 148; Carter on, 218; Ehrlich on, 133, 148; human capital theory on, 185; Keynes on, 29, 31–32, 69; post–World War II, 37–38, 46, 96, 115, 124; Reagan on, 210, 218–219; relation to agricultural growth, 102; suburban, 116–117; and wilderness, 121–122, 123
Ehrlich, Anne, 130, 132–133, 140, 153, 194–196, 231
Ehrlich, Paul, xiv, 2, 3, 4, 8, 60; on abortion, 126, 159, 212; on African Americans and population control, 179–180; on agricultural technology, 103; on biological categories, 130–132; on carrying capacity, 150; and civil rights movement, 130; and coercive population control, 147, 153, 190–191; criticized for coercion stance, 172; debate with Catholic Church, 157; on directed evolution, 129; early life/education, 127–128; and Earth Day, 169, 171–172; and economic growth, 133,

148; on environmental deterioration, 142; on food aid, 126, 149; on green revolution, 145–146; on human sexual drive, 141, 147; on immigration, 199; on India, 136, 137–138; on interconnectedness, 93–94; on lagging food supply, 141–142; mellowing of stance of, 194–196; on modern agriculture, 142–143; on modern health care, 163; on modernization, 103, 146, 151; on overconsumption, 11, 61, 148–150, 162, 173; popularity of, 162; on population growth, 141; on poverty as environmental problem, 151, 172; present-day views on *The Population Bomb*, 231; on racial unrest, 143; on roles of women, 159; and Spaceship Earth metaphor, 150, 176–177, 217; at Stanford University, 132–136; on sterilization, 126; and taxi ride metaphor, 150–151; on traditional religion, 147–148; on urban troubles, 139–140; on Vietnam, 139, 173–174; and war, 129–130

Ehrlich-Commoner debate, 181–184

Eisenhower, Dwight D., 59, 86, 91, 98, 164

elderly persons, increase in numbers, 230

elitism, 67, 69, 88, 173, 179, 190; of eugenics movement, 20, 22, 81; and immigration, 228; and universalism, 55–56

Elton, Charles, 25–26, 49

Ely, Richard, 51

energy crisis, 195, 210–211

Environmental Conservation (Dasmann), 76–77, 80

environmentalism: Brower and, 124; Caldwell and, 167–168; Earth Day, 168–171; origins of, xii–xiv, 1–4, 6, 8, 21, 23, 26–27, 37, 47, 66, 80, 105, 166, 224–225, 234n4; and suburbs, 114–119; and urban poverty, 111–113; and wilderness, 119–124. *See also* conservation movement

environmental Malthusians. *See* Malthusians, environmental

environmental resistance, 19, 47–48

eugenics, 10, 17–18; and elitism, 20, 22; emphasis on genetic quality, 81; and move to quality of life, 79

evolution: coevolution, 135, 142; cultural conservatives on, 213; Darwinian Synthesis, 128–129, 141, 146; directed, 129; evolutionary biology, 129, 130, 141, 156; and Malthus, 2; and natural selection, 128–129

Falwell, Jerry, 214–215

family, 9, 15, 71, 150, 219, 225; African American, 109; cultural conservatives on, 207, 214–215; and population, 107; single-parent, 107

family planning. *See* birth control

famine: in Africa, 195; as cause for war, 144; as check on overpopulation, 187; in India, 99, 100, 138, 195

Famine 1975! (Paddock and Paddock), 138

feminism, 157–158, 193–194

fertility rate, 96, 165, 197; differential fertility, 6, 19, 21, 22, 23

Firestone, Shulamith, 157–158

Fleming, Donald, 137, 152

food aid: as cause for overpopulation, 85; cutting off/scaling back, 11, 126, 149, 187; to India, 99–100, 141, 149; under Johnson, 102, 103; to third world, 121

Food and Agriculture Organization (FAO), 53, 99, 177

food chains, xi, 25, 26, 27, 73, 217, 223

food shortages, 85, 126, 127, 136, 189, 231

Ford, Gerald, 203

foreign aid: and Bretton Woods conference, 33, 46; conservatives on, 227; Caldwell and, 167–168; Draper on, 86–67, 90–91; Ehrlich on, 136–137, 145–146, 149; modernization programs, 95–97; 98–99, 100, 102–103, 146; Point Four programs, 33, 37, 46, 57–58, 59; and population, 67–72, 92–93, 212; Potter on, 64; Vogt on, 52–53, 57. *See also* food aid

Foreign Assistance Act, Title X, 101

foreign policy: agricultural, 21; and consumerism, 33–34; isolationism, 47, 206; New Right on, 207; and resource development, 57–59, 62, 88. *See also* immigration

forest conservation, 23–24

French, Frances, 213–214

frontier, xiii, 5, 63–65, 74, 77, 86, 94, 97, 98, 99, 133, 146, 152, 210

Fundamentals of Ecology (Odum), 73, 79–80

Game Management (Leopold), 26, 34

Galbraith, John Kenneth, 62, 82–83, 97, 98, 99, 116

Gandhi, Indira, 194

General Theory of Employment, Interest, and Money, The (Keynes), 30–31

genetic resistance, 142

genotype/phenotype, 16

Gillon, Steve, 117, 227

Global 2000 Report to the President, 205, 215

Goldwater, Barry, 164, 206

Gordon, Linda, 193

Great Depression, 23, 30–32, 37, 206

Great Society, 66, 83–84, 100, 108, 168, 204

green revolution, 66, 88, 102, 154–155, 177, 226, 231

Gregg, Alan, 111–112, 128

Gregory, Dick, 172, 178–179

Griswold v. Connecticut, 158

Grove, Richard, xiii

growth. *See* economic growth

Gruening, Ernest, 91–93, 99, 107, 109

Guano Islands, 42–43

Hannah, John A., 165

Hardin, Garrett, 2, 10, 129, 133; on abortion, 158–159, 212; on India, 137; and lifeboat metaphor, 187, 197–198; on population control, 153–155, 160, 190–191

Harrington, Michael, 107–108, 110

Hays, Samuel, 6, 9, 66, 80, 225

Hoff, Derek, 109, 185
Hoffman, Paul, 57, 64
Holdren, Richard, 188, 194–195
Hoover, Edgar, 69, 216
Hoover, Herbert, 17, 46
Hornaday, William, 24
Howard, Walter, 173, 174, 191
Howe, Sydney, 169
Humanae Vitae (Paul VI), 155–158
human capital theory, 185
humanism, 213
hunger, social causes of, 189. *See also* food aid
Huxley, Aldous, 93
Huxley, Julian, 76, 80, 129, 132
hybrid seeds, 145–146, 222, 231

immigration, 19, 196–200, 226; restrictions on, 13, 21, 23, 159, 228
India: and Ehrlich's taxi ride metaphor, 150–151; family planning in, 137, 144, 194; food aid to, 99–100, 141, 149; food crisis in, 136, 137; forced sterilization in, 10, 149; foreign aid to, 67; hunger/malnutrition in, 86, 101; industrialization in, 98; U.S. economic aid to, 97–98
industrialization: effect on population growth, 26, 77; effect on warfare, 38, 46, 47, 49; environmental consequences of, xii, 34–35, 38, 74–75, 113, 188, 207; in India, 98; in Soviet Union, 90
inner city, 4, 9, 26, 105, 106–113, 127, 133, 139–140, 143, 172, 177, 179–180, 200, 223, 226
Innis, Roy, 179
interconnectedness of world, xii, 13, 23, 25, 34, 47, 48, 50, 52, 89, 93–94, 196, 222–223
International Monetary Fund (IMF), 33, 46
International Planned Parenthood Federation (IPPF), 71
interwar period, and population growth: ethnic and racial consequences, 15, 22; political-economic aspects, 14–15; women and family consequences, 15

Jackson, Jesse, 179
Japan, birth control in, 69, 78, 90
Jasper Ridge Biological Preserve, 134–135
Johnson, Lyndon B., 1, 66, 83–84, 99–103, 106–107, 168, 204

Kaibab Plateau, 25, 27, 28–29, 49, 146, 170–171
Kennedy, John F., 86, 94–95, 97, 98–99
Keynes, John Maynard, xiv, 14–15, 29–32, 33, 34, 116, 123
Keynesian models, 69, 117
Khrushchev, Nikita, 90
King, Martin Luther Jr., 55, 98, 109, 138, 140
Kissinger, Henry, 187
Koop, C. Everett, 213
Korean War, 58–59
Krug, Julius, 46, 57, 96

Krutch, Joseph W., 7, 80
Kuhn, Thomas, 132

LaHaye, Beverly, 214
LaHaye, Tim, 213
Lappé, Frances Moore, 188–190, 196
Leopold, Aldo, xii, 15–16, 23–29, 34–35, 54, 60, 223
Leopold, Starker, 76, 82
lifeboat metaphor, 187, 196, 197–198
limits, 2, 5, 15, 47, 96, 164; Carter on, 203–206; divide among Malthusians about, 65, 88, 96, 103; Ehrlich on, 146; interwar biologists on, 18, 20, 27, 28, 32, 34, 60; Keynesian growthmanship and, 32, 58; limits-to-growth movement, 180–181, 201, 207, 210, 218–219; postwar conservationists and, 72–78, 223; Reagan and, 203; and wilderness, 119
Limits to Growth, The (Club of Rome), 180–181, 203
Lowen, Rebecca, 134

"malaise" speech on energy policy (Carter), 204
malaria, 40, 52–53
Malcolm X, 109
Malthus, Thomas, xiii, 4–5, 20, 25, 31, 128, 163
Malthusians, environmental, 11–12, 117–118, 133; appeal of, 151; Caldwell, 167; Commoner critique of, 181–184; critique of consumption, 148–150, 181; critique of Kennedy and Johnson administrations, 88, 96, 103, 126, 137, 142, 144; definition of, 4–6, 15, 20; on domestic population growth, 105; and Earth Day, 169; economic conservative critique of, 215–218; feminist critique of, 190–194; flawed diagnosis of, 10–11; Hardin, 153–155; Lappe and Collins critique of, 188–190; loss of appeal of, 226–228; New Right critique of, 211–215; origins of, 221–224; Reagan critique of, 202, 207–211. *See also* Ehrlich, Paul
"Man's Role in Changing the Face of the Earth" (1955 conference), 74–75, 77, 112
Marsh, George Perkins, xii, 50
Marshall Plan, 33, 37–38, 46, 57
mass behavior, 27–28
mass culture, 115, 223
Maynard, Joyce, 161–162
McCloskey, Paul ("Pete"), 164, 169, 191
McNamara, Robert, 100–101
medicine, modern, as cause of overpopulation, 22, 52–53, 64–65, 85, 141, 163
Mendel, Gregor, 20
Mexico, 92, 102, 144, 198–199
Milbank Memorial Fund, 67
military-industrial complex, 38, 46, 148
military-industrial-university complex, 134
Millikan, Max, 98
Mills, Stephanie, 1–2, 162–163
Mitman, Gregg, 128

modernization programs, 95–97; agricultural, 102–103, 133; in American cities, 106, 108; environmental Malthusian critique of, 103, 146, 151; in India, 98–99, 103, 146
Mondale, Walter, 218
Moore, Hugh, 88–89, 90, 91, 107, 160
Morris, Desmond, 112–113
Morse, Chandler, 185, 216
motherhood, 8, 77, 157–160, 162, 191, 201
Moynihan, Daniel P., 107, 109–110, 165
Muir, John, xi
Mumford, Lewis, 74, 112
Muskie, Edmund, 225
Myrdal, Gunnar, 100

National Association for the Repeal of Abortion Laws (NARAL), 159
National Commission on Population Growth and the American Future (NCPGAF), 184–185, 186
National Environmental Policy Act (NEPA), 166–168, 226
national parks, xii, 9, 85, 104, 113, 115, 119, 121, 133, 154
national security: and population, xiii, 87–91; relation to poverty, 224
National Security Council, 97, 98
natural selection, 128–129, 155
Nehru, Jawaharlal, 98, 103
Nelson, Gaylord, 2–3, 168–169, 171
New Deal, 23, 28, 43, 106
New Frontier, 94, 95, 97, 98–99
New Right, 201, 203, 206–211, 212, 214–215, 224, 228
New York Zoological Society, 39
Nixon, Richard, 1, 164–166, 177; ideological shift by, 184–186; on population, 164–166, 184–186
North American Treaty Organization (NATO), 165
Notestein, Frank, 68, 216
not in my backyard (nimbyism), 81
nuclear weapons, 73, 136, 154, 181, 183, 195, 204
Numbers, Ronald, 128

Odum, Eugene, 73, 79–80
Office of Population Research, 67
oil crisis, 186, 187, 210
oil spill, Santa Barbara, 164–165, 168
open space, shrinking, 83, 104, 112–133
open space movement, 114–116
Osborn, Fairfield, 6–7, 36–38; at Bronx Zoo, 39–40, 41; on carrying capacity, 47–48; Conservation Foundation of, 72–73; on consumption, 50, 51; early life of, 39; elitist views of, 69; on foreign aid, 57; on interconnectedness of world, 47, 48, 50, 52, 93; on overpopulation as cause for war, 49–40; on Point Four programs, 58; on population control, 67, 69; on population growth, 64; on population growth in U.S., 75; and quality of life, 79, 80; on racial/social difference/similarities, 54–55; on resource depletion/environmental degradation, 69–70; on resource protection, 41–42; on soil erosion, 48–49; universalism of, 55–56; on urban alienation, 50; on war, 222
Our Plundered Planet (Osborn), 29, 36–37, 41, 42, 47, 49, 50, 52, 54–55, 56, 60, 64, 93

Paley Commission, 58–59, 62, 96
"Partners in Progress," 59
Patterson, James, 108, 113–114, 180, 206, 218, 219
Paul VI (pope), 155–158, 162
Pearce, Fred, 231–232
Pearl, Raymond, 13, 15, 16–19, 222
Pearson, Karl, 16
Pendell, Elmer, 45, 88
People of Plenty (Potter), 5
permanent agriculture, 21
pesticides/DDT, 2, 52, 135, 141, 142, 181, 209
phenotype/genotype, 16
Phillips, Howard, 212
Piel, Gerald, 123–124
Pillsbury, Eleanor, 71–72
Pinchot, Gifford, 2, 5–6, 23, 51, 59, 244n25
Planned Parenthood, 70–72, 106; feminist criticism of, 158, 159; and geographical space, 124
Pogo (comic strip), xi
Point Four programs, 33, 38, 46, 57–58, 59
pollution, 83, 119, 134, 143, 179–180
Pope, Carl, 137
population, xiii–xiv, decrease in global, 226; decrease in global rate of growth, 230; increase in global, 7, 221; increase in underdeveloped countries, 64–65, 140–141; increase in U.S., 221–222; law of population growth, 17
population biology, 16–19, 133–135
Population Bomb, The (Ehrlich), xv, 1–2, 4, 8, 61, 66, 124, 126–127, 140–151, 158; background to, 136–140. *See also* Ehrlich, Paul
"Population Bomb, The" (Moore), 88–89, 90
Population Council, 66–68, 106
Population Crisis Committee, 107
Population Problem, The (Carr-Saunders), 49
Potter, David, 5, 63–64, 65, 70, 95
poverty: and access to birth control, 70, 110; cultural causes of, 95; Ehrlich on, 142, 150, 151; and environmentalism, 8, 10, 225; international, 61–62, 86; international war and, 94–95; as low national priority, 226–227; Malthus on, 4; national security and, 89, 91, 94, 100, 223–224; population growth as cause of, 104, 106–107, 123; urban, 111–113, 172–173
predator control, 27, 28, 49, 53
presidential election (1980), 218–219
President's Materials Policy Commission (Paley Commission), 58–59, 62, 96

Puerto Rico, 52–53, 54, 92
pyramid of numbers, 25, 26

quality of life, 7, 9, 65–66, 78–82, 154, 177; and 1960s liberalism, 82–84; population growth as threat to, 105, 225–226

race, xiv, 4, 6, 8, 9, 11, 15, 16; and birth control, 111, 165, 172–173, 175; East on, 22; Ehrlich on, 130–132; and environmentalism, 225, 227; and immigration, 198–199; and inner cities, 106, 143; Osborn and, 54; role in suburbanization, 113–114; Vogt and, 54; and wilderness, 121–122
racial unrest, 108, 143
racism, 20, 21, 67, 179; African American criticism of population movement as, 178–180
racism, scientific. *See* eugenics
Radl, Shirley, 160
Raven, Peter, 135
Reagan, Ronald, 8, 202–203, 207–211, 218–219
reductionism, biological, 10, 82, 116, 120, 122, 127, 139, 143, 146, 163, 171, 173, 223, 226
religious right, 211. *See also* Republican Party
Republican Party: and cultural conservatives, 211–215; and economic conservatives, 215–218; and New Right, 201, 203, 206–211, 212, 214–215, 224, 228; Rockefeller Republicans, 191, 206–207; and population growth, 163–166. *See also* Bush, George H. W.; Nixon, Richard; Reagan, Ronald
resources, natural, xiv, 8, 228; degradation of, 48–49, 69–70, 189, 204; and foreign policy, 57–59, 62, 88; protection of, 41–42; scarcity of, 1, 5, 6, 36, 58–59, 85, 87, 186; substitutes for, 185, 187, 216–217; during World War II, 44
Revelle, Roger, 99
Rienow, Leona, 113, 116–118
Rienow, Robert, 113, 116–118, 124
riots, urban, 106, 108, 139–140
Road to Survival (Vogt), 11, 29, 37, 45, 47, 50, 51, 52, 54, 55, 56, 60, 79
Rock, John, 78
Rockefeller, John D. III, 66–67, 79, 107, 165, 184, 186, 188, 193
Rockefeller, Lawrence, 169
Rockefeller, Nelson, 44, 186, 206, 207
Rockefeller Foundation, 102
Rogers, William P., 165
Rome, Adam, 2, 82–83, 117, 161–162, 225, 234n4
Roosevelt, Franklin D., 33, 51
Roosevelt, Theodore, xii, 2, 5–6, 51, 57
Rostow, Walt W., 95, 96, 98, 100
Rowan, Carl, 179

Sand County Almanac, A (Leopold), 23, 28–29, 35
Sanger, Margaret, 15, 70–71, 92

Sauer, Carl, 74, 77
scarcity. *See under* resources, natural
Schafly, Phyllis, 214
Schoen, Johanna, 10, 111
Schrepfer, Susan, 129
Schulman, Bruce, 191, 201, 204
science and scientists, 21, 22, 27, 34, 60, 111, 143, 156–157, 165, 175, 183–184
Scripps Foundation, 67
"S" curve, 17, 73
Seaman, Barbara, 192
Sears, Paul, 60, 75
Seeger, Pete, 170, 184
sex, 4, 8, 9, 21, 107, 128, 147, 173; New Right on, 212; postwar attitudes about, 222; sex education, 147, 212–213
sexual drive, human, xiv, 4, 8, 9, 21–22, 85, 107, 128, 141
Sexuality Information and Education Council of the United States (SIECUS), 212
Sierra Club, and population issues, 7, 120, 123–124, 169
Simon, Julian, 215–217
Singh, Karan, 188
Social Security Act of 1967, 101, 109
soil erosion, 21, 22, 26, 45, 48, 57–58, 142, 190, 199
South Africa, 120–121
Soviet Union, 93, 94, 98, 130, 144, 187, 204, 210, 226–227
Spaceship Earth metaphor, 150, 176–177, 187, 190, 196, 198, 217, 227. *See also* universalism
statistics: availability of, 13–14; Ehrlich on, 131; usefulness of, 17
Steinbacher, John, 212–213
sterilization: access to, 158, 160; incentivized, 54, 194; involuntary, 10, 126, 153, 192–193
Stevenson, Adlai, 44
Stoddard, Lothrop, 15, 22
suburbs, 4, 37, 113–119; growth of, 113–114, 105
Sumner, Lowell, 120
sustainability, xv, 21, 47, 228. *See also* carrying capacity
Szreter, Simon, 68

Taeuber, Conrad F., 186
Taft, Robert Jr., 164, 206
Tanton, John, 198–199
taxonomic classification, 175
technology, xi, 6, 8, 23, 26, 124, 227, 228; agricultural, 138, 231 (*see also* green revolution); birth control as, 78, 102, 109, 121–122, 222; Commoner on, 182; Ehrlich on, 136, 145–146, 150; in foreign aid, 68, 70, 88, 95–96, 103; Hardin on, 154; Malthus on, 4; Potter on, 64; Simon on, 217; Reagan on, 209, 218
Thant, U, 94
"Thinking Like a Mountain" (Leopold), 15–16, 25, 28–29

third world, xiv, 4, 8–9, 62, 86, 88–89, 91, 94, 97, 108, 177, 187–188, 189, 200; and American suburbs, 115; Ehrlich on, 136–137, 140–141, 144–146, 148–151; population growth in, 64, 68, 73, 165
Thompson, Warren, 68, 74, 172
Thoreau, Henry David, xi
total war, xii, xiv, 38, 46, 222, 226
trade, 4, 6, 62
"Tragedy of the Commons, The" (Hardin), 153
Train, Russell, 73
Truman, Harry S., 32, 46, 57, 58–59, 62, 91
Tucker, Richard, xiii, 38
Turner, Frederick Jackson, 5, 6
Tydings, Joseph, 109, 163

Udall, Morris, 169, 171, 172, 174
Udall, Stuart, 8, 83
universalism, xiv, 55–56, 132, 173, 175, 176, 178, 191, 198, 200, 215, 225
urban alienation, 50
urban growth, post–World War II, 105
urbanization: of California, 133, 134; increase in, 230; post–World War II, 223–224
urban poverty, 111–113, 172–173. *See also* inner city
U.S. Agency for International Development (USAID), 101–102, 212

Vandenberg, Arthur, 46
violence, 27, 41, 108, 113, 125, 143, 151, 223
Vietnam War, xi, 78, 86, 97, 108, 139–140, 144, 168, 173–174
Vogt, William, 6–7, 37–38; birding activities of, 40–41; on birth control, 52, 158; on capitalism, 51–52; on carrying capacity, 47–48; on consumption, 11, 50–51; diplomatic work, 44; on domestic poverty, 106; elitism of, 55–56; on foreign aid programs, 52–53; Guano Island studies, 42–43; on interconnectedness, 50; as national director of Planned Parenthood, 72; on Point Four program, 57–58; on population limitation, 52–54; on quality of life, 79; on resource degradation, 49, 69–70; on urban alienation, 50; on war, 222

Wald, George, 93, 159, 174
Wallace, Alfred R., 5
Walters, Ron, 179
war, xi, 8; as check on overpopulation, 187; and environmental thought, xi–xiv, 19, 26, 27, 29, 50, 124, 129, 222; environmental effects of, 41–42, 44, 114; famine as cause of, 20, 144; population growth as cause of, 5, 17, 38, 45, 49, 70–71, 78, 92, 144, 174; and wilderness, 120
War on Poverty, 106, 109, 204
Wattenberg, Ben, 175
Webb, Walter P., 62–63
Weinberger, Caspar, 188
Welch, Bruce, 119
welfare programs, in U.S., 106
Weyrich, Paul, 212, 214
Whyte, William, 114, 115–116, 139
wilderness: disappearance of, 83; overpopulation as threat to, 81, 119–124, 223; Wilderness Act of 1964, 119
Williamsburg Conference, 67, 69–70, 79
Wirtz, Willard, 100, 101
wolf extirpation programs, 28
women: and birth control, 110, 111, 123; and feminism, 193–194; Malthusians and, 10, 15, 77, 121, 123, 159; women's health movement, 191; women's liberation movement, 8, 152, 157–158, 162
World Bank, 33, 46, 100–101
World's Hunger, The (Pearson and Harper), 48
World War I, overpopulation as cause of, 17, 20, 49, 92
World War II: environmental effects of, 41–42, 44; environmental impact of technologies developed during, 114; and Leopold's ideas, 26, 27; overpopulation as cause of, 27, 70–71
Worster, Donald, 1

X, Malcolm, 109

Zero Population Growth (ZPG), 159–160, 170–171, 174, 176–177, 181, 184; and immigration, 197, 198–199

ABOUT THE AUTHOR

An environmental and foreign relations historian, Thomas Robertson attended Williams College and received his Ph.D. in history from the University of Wisconsin–Madison in 2005, where he studied with William Cronon, Paul Boyer, and Jeremi Suri. He currently teaches at Worcester Polytechnic Institute in Worcester, Massachusetts. His article "'This Is the American Earth': American Empire, the Cold War, and American Environmentalism" appeared in the September 2008 issue of *Diplomatic History*. His next project will examine the environmental, social, and political consequences of U.S. development projects in Nepal from the 1950s to the 1980s.

CPSIA information can be obtained at www.ICGtesting.com
Printed in the USA
BVOW030109280613

324555BV00002B/5/P